Transport, Persistence and Toxicity of Pollutants in the Sea

Transport, Persistence and Toxicity of Pollutants in the Sea

Editors

Mauro Marini
Anna Annibaldi

MDPI

Basel • Beijing • Wuhan • Barcelona • Belgrade • Novi Sad • Cluj • Manchester

Editors

Mauro Marini
National Research Council
Ancona, Italy

Anna Annibaldi
Università Politecnica delle Marche
Ancona, Italy

Editorial Office
MDPI
St. Alban-Anlage 66
4052 Basel, Switzerland

This is a reprint of articles from the Special Issue published online in the open access journal *Applied Sciences* (ISSN 2076-3417) (available at: https://www.mdpi.com/journal/applsci/special_issues/Transport_Pollutants_Sea).

For citation purposes, cite each article independently as indicated on the article page online and as indicated below:

Lastname, A.A.; Lastname, B.B. Article Title. *Journal Name* **Year**, *Volume Number*, Page Range.

ISBN 978-3-0365-8816-2 (Hbk)
ISBN 978-3-0365-8817-9 (PDF)
doi.org/10.3390/books978-3-0365-8817-9

Cover image courtesy of Mauro Marini

Contents

About the Editors

Mauro Marini

Following the completion of his degree in Agriculture at the University of Perugia, Italy, in 1990, Dr. Mauro Marini has been with the CNR for more than 30 years, studying the chemical processes of soil, groundwater and seawater organic and inorganic compounds. He is especially interested in the biogeochemical processes and in pollutant and xenobiotic interactions in marine sediments and organisms. He was Section Head of CNR-ISMAR (Ancona, Italy) from September 2009 until it became CNR-IRBIM in September 2018.

Anna Annibaldi

Anna Annibaldi is an Associate Professor in Analytical Chemistry at the Marche Polytechnic University, Ancona (Italy), where she teaches Analysis of Pollutants and Legislation and Environmental Monitoring. She graduated in Chemistry at the University of Bologna in 2001 and obtained a PhD degree in Biology and Marine Ecology in 2005 at Marche Polytechnic University. Her scientific interests have focused principally on the field of environmental analytical chemistry, particularly on the determination of organic and inorganic pollutants in different environmental matrices (seawater, sediments, marine organisms). Annibaldi takes part in different national and international research projects on the effect of pollutants in environmental ecosystems.

applied sciences

MDPI

Editorial

Transport, Persistence, and Toxicity of Pollutants in the Sea

Mauro Marini [1,2,*] and Anna Annibaldi [2,3,*]

1 Institute for Biological Resources and Marine Biotechnologies (IRBIM), National Research Council of Italy (CNR), 60125 Ancona, Italy
2 Fano Marine Center, The Inter-Institute Center for Research on Marine Biodiversity, Resources and Biotechnologies, 61032 Fano, Italy
3 Dipartimento di Scienze della Vita e dell'Ambiente (DISVA), Università Politecnica delle Marche, 60131 Ancona, Italy
* Correspondence: mauro.marini@cnr.it (M.M.); a.annibaldi@staff.univpm.it (A.A.)

1. Introduction

Eight research articles and two reviews are included in this Special Issue focused on the transport, persistence, and toxicity of pollutants in different seas: the Western and Central Mediterranean Sea, the Adriatic Sea, the Tyrrhenian Sea, the Baltic Sea, and the South China Sea.

The results, specifically described below, highlight different aspects of the main topic of the Special Issue, including:

- The implications of anthropogenic effects on the coastal environment and seabed;
- How climate change modifies surface water regimes and whether it affects the runoff of contaminants in the sea;
- Pollutants in sediments and marine organisms;
- Seasonal variations in pollutants in the sea;
- Estuarine pollution and the influence on seawater contamination.

The results provided in this Special Issue are grouped into three different topics that thoroughly cover the themes reported in all the papers.

2. Effects of Contamination on Marine Organisms

Coastal marine areas receive significant anthropogenic input, mainly derived from metropolitan areas, industries, and activities related to tourism. Many studies have been conducted on heavy metal accumulation and on its possible effects on different edible marine species; in particular, the most-studied sessile organisms are bivalves. Roveta et al. [1] conducted a review of heavy metals, focusing on other sessile taxa (sponges, cnidarians, bryozoans, polychaetes, barnacles, and tunicates), proposed to be bioindicators in shallow coastal waters. In a second paper, Roveta et al. [2] showed the concentrations of mercury in the tissues of different sponges collected in Montecristo and Giglio, two islands of the Tuscany Archipelago National Park in the Tirrenian Sea (Mediterranean Sea). These results underline the species-specificity of metal concentrations for Porifera and provide additional data to address the main input of the Marine Strategy guidelines in order to protect coasts, seas, and oceans.

De Giovanni et al. [3] summarized the results of epidemiological investigations on the genetic component of individual susceptibility in exposure to methylmercury and polycyclic aromatic hydrocarbons in the coastal population, and on the effects of these two pollutants on human epigenetic profiles (DNA methylation). The potential impact of methylmercury and polycyclic aromatic hydrocarbons on the human genome and Mediterranean epigenome was investigated, as this population is characterized by a traditionally high consumption of local fish; this influences the characteristics that, consequently, render the Mediterranean Sea particularly polluted [4,5].

Citation: Marini, M.; Annibaldi, A. Transport, Persistence, and Toxicity of Pollutants in the Sea. *Appl. Sci.* **2022**, *12*, 7017. https://doi.org/10.3390/app12147017

Received: 1 July 2022
Accepted: 11 July 2022
Published: 12 July 2022

In a second paper, De Giovanni et al. [6] collected and harmonized the results of several studies published over the years (1981–2019) in order to obtain a database [7] of georeferenced observations on polycyclic aromatic hydrocarbons (PAHs) in Western and Central Mediterranean seafood. For each observation, some information on the taxonomy and the ecology of the sampled species was reported, as well as details on the investigated hydrocarbon, and spatial and temporal information on the sampling. Moreover, two health risk indexes were calculated for each record and included in the database. The analysis of the data showed that, at a consumption rate typical of the Italian population, seafood caught from the area considered in the present work seems to pose minimal risk to health.

Rivoira et al. [8] studied the presence of 16 polycyclic aromatic hydrocarbons (PAHs) and 14 dioxin- and non-dioxin-like polychlorinated biphenyls (PCBs) in the neritic protected marine area of the Southern Ligurian Sea (Mediterranean Sea), affected by the impact of human activities. The study was focused on the possible partition of micropollutants within seawater, sediment, and zooplankton. The results showed that both seasonal and anthropogenic causes strongly affect contaminants' transfer behaviors, with summer being most impacted by PAH and PCB contamination.

To better understand the occurrence of preferential bioaccumulation pathways in zooplankton, partition studies were also performed in several taxa (hyperbenthic Isopoda, holoplanktonic crustacean copepods, and ichthyoplankton), through the calculation of bioaccumulation factor values. They observed that both living and feeding habits could influence the bioaccumulation process.

3. River Runoff into the Sea

Barra et al. [9] focused on the geochemical and sedimentological characterization of recent sediments from two marine sites located in the North Adriatic Sea, between the Po River prodelta and the Rimini coast. Major and trace metal concentrations reflect the drainage area of the Po River and its tributaries, considered one of the most polluted areas in Europe [4].

High values of Cr, Ni, Pb, and Zn were detected in sediments collected from the Po River prodelta. This suggests a Po River supply. Meanwhile, lower levels of these elements in sediments collected in front of the Rimini coast were measured, indicating the Apennines provenance of these.

The historical trends of Pb and Zn reconstructed from the sedimentary record around the Rimini coast document several changes that can be correlated with the industrialization prior to World War II.

Giglio et al. [10] examined the transport of sediments and their surficial pathways from the mouth of Neretva River toward the Eastern Adriatic Sea. Sediment dynamics were evaluated using several proxies such as organic matter, radiochemical isotopes, selected metal concentrations, and physical parameters. The data analysis showed that the influence of the river on particle distribution decreases northward, with an estimated sediment accumulation rate ranging from 1.9 to 8.5 mm/yr. We speculate that either the dispersion or accumulation of sediments is driven by an eddy in the waters of the Neretva River, triggered or intensified seasonally by the interaction of karstic springs, river input, and Eastern Adriatic Sea waters.

The Huangmao Island dumping area is adjacent to the Pearl River Estuary in the South China Sea. From its first dumping activity in 1986 to 2017, 6750×104 m^3 dredged materials were dumped in this dumping area. The evaluation results reported by Tao et al. [11] for the heavy metals revealed that the dumping area with a large dumping amount was more severely polluted. The dumping of dredged materials seemed to have a negative impact on the benthic community in the dumping area.

4. Dynamics of the Sediments

Karlonienė et al. [12] assessed trace heavy metals' temporal and spatial distribution in the sandy beach sediments along the south-eastern Baltic Sea coast (Lithuania). They studied how geochemical analysis can provide valuable information about the local and regional patterns of sediment transport, distribution, provenance, and coasts condition. The spatial analysis of trace elements indicated that the trace metal content depends on the coastal processes, showing a difference in the mainland and spit sea-coast.

Sediments represent a critical component of coastal marine ecosystems due to the toxic and long-lasting effects of the contaminants buried therein. Spagnoli et al. [13] studied the properties of the superficial sediments off the coast of the Southern Marche Region (Western Adriatic Sea); it is characterized by numerous minor rivers [14] and a bottom current that flows southwards, carrying out suspended and dissolved contaminants [15–17]. Some pollution indicators, such as the enrichment factor, the geoaccumulation index, and the pollution load index were calculated to assess the deviation from the natural background levels. The results showed pollution by As and Ba due to human activity in the 20th century.

5. Perspectives on Future of Research

The work presented in this Special Issue offers concrete advances and new insights for the study of anthropogenic chemical contaminants in marine ecosystems yet in-depth research will still be required for a complete understanding of the traces of these in the sea.

Over the next few years, water scarcity is expected to increase in light of global climate change and increased urbanization, particularly near coastal zones; moreover, extreme weather events, which have occurred more frequently in recent decades, could lead to increasing surface runoff via the collection of pollutants, including emerging ones, discharging them into the sea. Effective means of risk-based prioritization of contaminants, particularly in seafood and bathing water, are needed within this context. Ultimately, much additional work is needed to identify affordable and effective means of evaluating and determining these contaminants and sensitizing people to rational and conscious use, according to green deal priorities.

Author Contributions: M.M. and A.A. both conceived the idea and wrote the manuscript. All authors have read and agreed to the published version of the manuscript.

Funding: This research received no external funding.

Institutional Review Board Statement: Not applicable.

Informed Consent Statement: Not applicable.

Data Availability Statement: Not applicable.

Conflicts of Interest: The authors declare no conflict of interest.

References

1. Roveta, C.; Annibaldi, A.; Afghan, A.; Calcinai, B.; Di Camillo, C.G.; Gregorin, C.; Illuminati, S.; Mantas, T.P.; Truzzi, C.; Puce, S. Biomonitoring of Heavy Metals: The Unexplored Role of Marine Sessile Taxa. *Appl. Sci.* **2021**, *11*, 580. [CrossRef]
2. Roveta, C.; Pica, D.; Calcinai, B.; Girolametti, F.; Truzzi, C.; Illuminati, S.; Annibaldi, A.; Puce, S. Hg Levels in Marine Porifera of Montecristo and Giglio Islands (Tuscan Archipelago, Italy). *Appl. Sci.* **2020**, *10*, 4342. [CrossRef]
3. De Giovanni, A.; Giuliani, C.; Marini, M.; Luiselli, D. Methylmercury and Polycyclic Aromatic Hydrocarbons in Mediterranean Seafood: A Molecular Anthropological Perspective. *Appl. Sci.* **2021**, *11*, 11179. [CrossRef]
4. Frapiccini, E.; Annibaldi, A.; Betti, M.; Polidori, P.; Truzzi, C.; Marini, M. Polycyclic aromatic hydrocarbon (PAH) accumulation in different common sole (*Solea solea*) tissues from the North Adriatic Sea peculiar impacted area. *Mar. Pollut. Bull.* **2018**, *137*, 61–68. [CrossRef] [PubMed]
5. Girolametti, F.; Annibaldi, A.; Carnevali, O.; Pignalosa, P.; Illuminati, S.; Truzzi, C. Potential toxic elements (PTEs) in wild and farmed Atlantic bluefin tuna (*Thunnus thynnus*) from Mediterranean Sea: Risks and benefits for human consumption. *Food Control* **2021**, *125*, 108012. [CrossRef]
6. De Giovanni, A.; Abondio, P.; Frapiccini, E.; Luiselli, D.; Marini, M. Meta-Analysis of a New Georeferenced Database on Polycyclic Aromatic Hydrocarbons in Western and Central Mediterranean Seafood. *Appl. Sci.* **2022**, *12*, 2776. [CrossRef]
7. De Giovanni, A.; Abondio, P.; Frapiccini, E.; Luiselli, D.; Marini, M. *Polycyclic Aromatic Hydrocarbons in Seafood Caught in Western and Central Mediterranean from 1981 to 2019*; PANGAEA: San Francisco, CA, USA, 2022. [CrossRef]

8. Rivoira, L.; Castiglioni, M.; Nurra, N.; Battuello, M.; Mussat Sartor, R.; Favaro, L.; Concetta Bruzzoniti, M. Polycyclic Aromatic Hydrocarbons and Polychlorinated Biphenyls in Seawater, Sediment and Biota of Neritic Ecosystems: Occurrence and Partition Study in Southern Ligurian Sea. *Appl. Sci.* **2022**, *12*, 2564. [CrossRef]
9. Barra, E.; Riminucci, F.; Dinelli, E.; Albertazzi, S.; Giordano, P.; Ravaioli, M.; Capotondi, L. Natural Versus Anthropic Influence on North Adriatic Coast Detected by Geochemical Analyses. *Appl. Sci.* **2020**, *10*, 6595. [CrossRef]
10. Giglio, f.; Romano, S.; Albertazzi, S.; Chiarini, F.; Ravaioli, M.; Ligi, M.; Capotondi, L. Sediment Dynamics of the Neretva Channel (Croatia Coast) Inferred by Chemical and Physical Proxies. *Appl. Sci.* **2020**, *10*, 807. [CrossRef]
11. Tao, W.; Jiang, Z.; Peng, X.; Yang, Z.; Cai, W.; Yu, H.; Ye, J. Impact of Dredged Material Disposal on Heavy Metal Concentrations and Benthic Communities in Huangmao Island Marine Dumping Area near Pearl River Estuary. *Appl. Sci.* **2021**, *11*, 9412. [CrossRef]
12. Karlonienė, D.; Pupienis, D.; Jarmalavičius, D.; Dubikaltinienė, A.; Žilinskas, G. The Impact of Coastal Geodynamic Processes on the Distribution of Trace Metal Content in Sandy Beach Sediments, South-Eastern Baltic Sea Coast (Lithuania). *Appl. Sci.* **2021**, *11*, 1106. [CrossRef]
13. Spagnoli, F.; De Marco, R.; Dinelli, E.; Frapiccini, E.; Frontalini, F.; Giordano, P. Sources and Metal Pollution of Sediments from a Coastal Area of the Central Western Adriatic Sea (Southern Marche Region, Italy). *Appl. Sci.* **2021**, *11*, 1118. [CrossRef]
14. Campanelli, A.; Fornasiero, P.; Marini, M. Physical and chemical characterization of the water column in the Piceno Costal area (Adriatic Sea). *Fresenius Environ. Bull.* **2004**, *13*, 430–435.
15. Marini, M.; Fornasiero, P.; Artegiani, A. Variations of hydrochemical features in the coastal waters of Monte Conero: 1982–1990. *Mar. Ecol.* **2002**, *23*, 258–271. [CrossRef]
16. Marini, M.; Russo, A.; Paschini, E.; Grilli, F.; Campanelli, A. Short-term physical and chemical variations in the bottom water of middle Adriatic depressions. *Clim. Res.* **2006**, *31*, 227–237. [CrossRef]
17. Droghini, E.; Annibaldi, A.; Prezioso, E.; Tramontana, M.; Frapiccini, E.; De Marco, R.; Illuminati, S.; Truzzi, C.; Spagnoli, F. Mercury content in Central and Southern Adriatic Sea sediments in relation to seafloor geochemistry and sedimentology. *Molecules* **2019**, *24*, 4467. [CrossRef] [PubMed]

applied sciences

MDPI

Review

Biomonitoring of Heavy Metals: The Unexplored Role of Marine Sessile Taxa

Camilla Roveta [1], Anna Annibaldi [1,*], Afghan Afghan [1], Barbara Calcinai [1], Cristina Gioia Di Camillo [1], Chiara Gregorin [1,2], Silvia Illuminati [1,*], Torcuato Pulido Mantas [1], Cristina Truzzi [1] and Stefania Puce [1]

1 Department of Life and Environment Sciences, Università Politecnica delle Marche, via Brecce Bianche, 60131 Ancona, Italy; c.roveta@pm.univpm.it (C.R.); a.afghan@pm.univpm.it (A.A.); b.calcinai@staff.univpm.it (B.C.); c.dicamillo@staff.univpm.it (C.G.D.C.); c.gregorin@pm.univpm.it (C.G.); t.pulido@pm.univpm.it (T.P.M.); c.truzzi@staff.univpm.it (C.T.); s.puce@staff.univpm.it (S.P.)

2 Integrative Marine Ecology Department, Stazione Zoologica Anton Dohrn, Villa Comunale, 80121 Naples, Italy

* Correspondence: a.annibaldi@staff.univpm.it (A.A.); s.illuminati@staff.univpm.it (S.I.)

Abstract: Coastal areas are known to receive significant anthropogenic inputs, mainly deriving from metropolitan areas, industries, and activities related to tourism. Among these inputs, some trace elements are listed as priority pollutants in the European Water Framework Directive, due to their ability to bioaccumulate in organisms. Many studies have been conducted on heavy metals (HMs) accumulation and on their possible effects on different edible marine species. While the most studied sessile organisms are bivalves, in the current review, we focus our attention on other sessile taxa (sponges, cnidarians, bryozoans, polychaetes, cirripeds, and tunicates), proposed as bioindicators in coastal shallow waters. Although their potential as bioindicator tools has been repeatedly highlighted in the literature, these organisms are still poorly investigated and considered for monitoring. In this context, we analyze the available literature about this topic, in order to summarize the current knowledge and identify possible applications of these organisms in a bioremediation scenario.

Keywords: zoobenthos; sentinel species; suspension feeders; water pollutants; bioaccumulation; heavy metals

Citation: Roveta, C.; Annibaldi, A.; Afghan, A.; Calcinai, B.; Di Camillo, C.G.; Gregorin, C.; Illuminati, S.; Pulido Mantas, T.; Truzzi, C.; Puce, S.et al. Biomonitoring of Heavy Metals: The Unexplored Role of Marine Sessile Taxa. *Appl. Sci.* **2021**, *11*, 580. https://doi.org/10.3390/app11020580

Received: 17 December 2020
Accepted: 5 January 2021
Published: 8 January 2021

1. Introduction

Marine coastal areas are among the most exploited and vulnerable ecosystems, constantly subjected to multiple anthropogenic pressures [1,2], due to the intensification of metropolitan areas, tourism, and industrial and agricultural activities [3,4]. These human-induced perturbations cause in many cases the release of various contaminants, which can be transported in coastal waters through different pathways, as atmospheric depositions, rivers, industrial discharges, maritime traffic, or submarine groundwaters [3], representing important hazards for the ecosystem's health [1]. These contaminants principally include chemicals (e.g., heavy metals, polycyclic aromatic hydrocarbons (PAHs), pesticides, etc.) and fecal contaminants (e.g., *Escherichia coli*, total coliforms, and enterococci) [3]. Therefore, understanding the fate and distribution of contaminants in the sea and their origins is crucial to assess the ecological and chemical status of water and organisms and to implement management plans [2,5].

Heavy metals (HMs) are a problematic issue for marine ecosystems, ensuring that some of these elements and their derived compounds are listed in the European Water Framework Directive (WFD 2000/60/EC) as priority pollutants. Heavy metals are naturally occurring elements that can be found throughout the earth's crust. Nevertheless, anthropogenic activities (e.g., mining and smelting operations, industrial production and use, metal corrosion, atmospheric deposition, soil erosion of metal ions and leaching of heavy metals, and sediment resuspension) are primarily responsible for the environmental contamination and human exposure to these contaminants [6].

To date, a clear definition of the term "heavy metals" is not yet available. The term has been widely used in chemistry but no authority such as the International Union of Pure and Applied Chemistry (IUPAC) has ever defined it [7]. Even if its use is strongly discouraged, the term is increasingly used in the scientific literature especially in articles pertaining to multidisciplinary environmental issues [8]. On a chemical basis, the group of HMs includes all the transition elements (bloc d of the periodic table), the rare earth elements (the series of lanthanides and the series of actinides, including La and Ac themselves), and all the elements presenting the typical characteristics of metals (bloc p), together with some metalloids (Ge, As, and Te) and some nonmetals (Se) [7,9].

Some HMs (e.g., Fe, Cu, Co, Mn, and Zn) are essential elements for organisms, playing key roles in the functioning of enzyme systems. However, high concentrations induce detrimental effects in organisms and in the environment. Other metals (e.g., Al, As, Ba, Bi, Cd, Pb, Hg, Ni, Pt, and Ag) have no specific biological functions and are considered as nonessential metals [10].

In the marine environment, heavy metals can be found in various chemo-physical forms (metal ions, hydrated ions, charged metal complexes, uncharged inorganic complexes, and organometallic complexes), each of them presenting different bioavailability, toxicity, bioaccumulation, mobility, and biodegradation rates [11]. All aquatic organisms can introduce and/or accumulate HMs in their tissues, taking them up from water or food, and the concentration of these elements can vary between taxonomic levels, from phyla down to species of the same genus [12]. Many studies have been carried out on the bioaccumulation and the effects of the different elements on marine organisms, but mainly on edible species (as bivalves, cephalopods, decapod crustaceans, and fish) e.g., [13,14]. However, other sessile organisms, e.g., polychaetes, tunicates, sponges, and barnacles, can be considered as suitable bioindicators for their physical and physiological characteristics [15,16], and they have been recommended by many authors, e.g., [17–22], by the WFD and by the Marine Strategy Framework Directive (MSFD) [23]. Moreover, these organisms can represent an excellent functional tool since they do not only show detectable concentrations of trace elements [18,24] but the presence of these chemicals in their tissues can also stimulate different physiological responses [25]. Therefore, they can be considered as useful biomonitors of HMs contamination, providing important information on the ecological and chemical status of an area.

The main aim of this systematic review is a spatiotemporal analysis of the literature on the bioaccumulation of HMs at global scale on six sessile marine taxa (Porifera, Cnidaria, Bryozoa, Polychaeta, Cirripedia, and Tunicata), considered as nonconventional biomonitors for chemicals pollution. The current work will provide a schematic summary of the current knowledge on the relationship between the selected taxa and HMs. It will identify possible applications of these organisms as bioremediation tools in contaminated environments, also accordingly to the European MSFD within the achievement of the Good Environmental Status (GES) (MSFD 2008/56/EC).

2. Materials and Methods

The present compilation is the result of an extensive bibliographic research on Elsevier's Scopus database (www.scopus.com), entering different keywords, "heavy metal" AND "taxa" and "heavy metal" AND "common name of the organism" (Figure 1), in the option "Article title, Abstract, Keywords," in all years until the cut-off date of 20 June 2020. The literature considered in this work includes journal articles and grey literature, as congress proceedings. Systematic reviews are not included by choice. All documents found with the aforementioned queries have been screened by reading titles and abstracts, excluding those not matching our criteria (Figure 1), and only the ones eligible for this study have been considered for the analysis. Duplicates, e.g., same documents found using different keywords, were counted as one. Since the Mediterranean province showed the highest number of documents on the topic, we conducted an additional manual research on the references reported in the found articles, in order to include possible documents on the area that could have been missed with the online search (Figure 1). A flow chart of the searching strategy and the eligibility process is given as Figure 1.

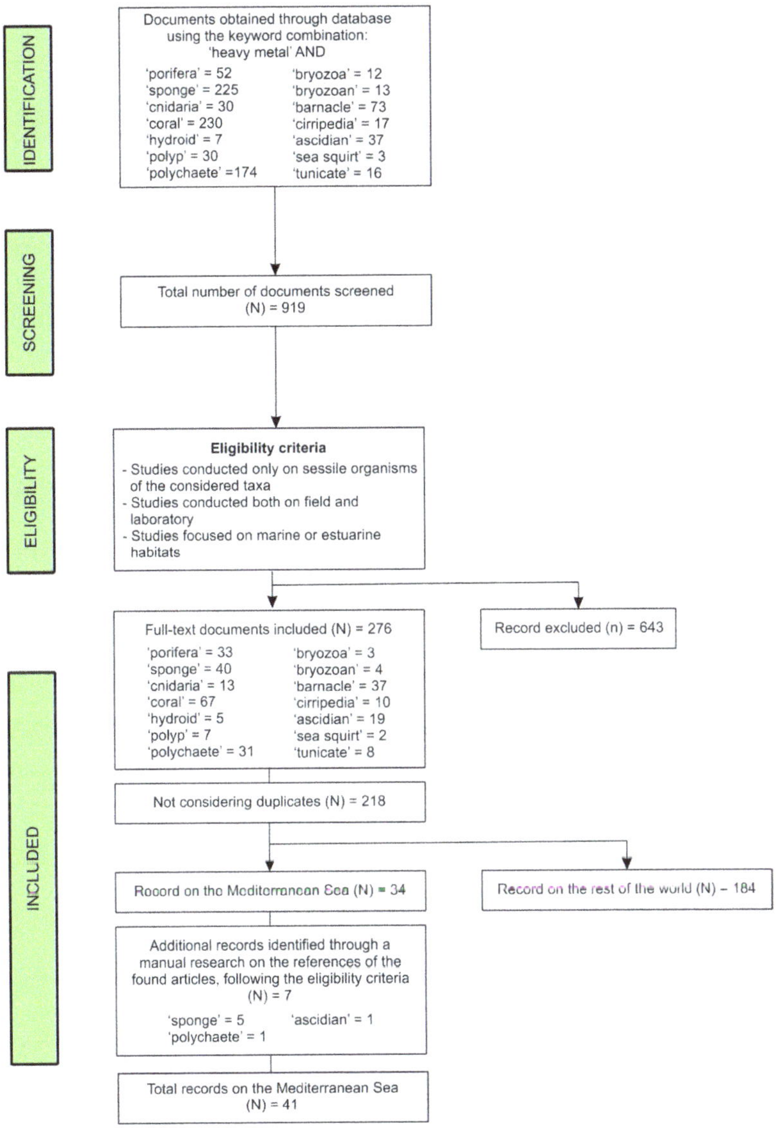

Figure 1. Flow chart illustrating the steps for obtaining documents about the study of heavy metals (HMs) on organisms from Scopus search engine and from the manual research.

Studies from the web-based and manual researches have been analyzed based on the taxa considered (Porifera, Cnidaria, Bryozoa, Polychaeta, Cirripedia, Tunicata), temporal distribution, marine realm of the sampling site(s), habitat (sea, estuarine), analyzed body part (cells and tissues, entire specimen, microbiome, skeleton), science field (biomonitoring, microbiology, physiology), HM(s), work setup (field, laboratory). Moreover, HMs considered in the documents have been divided in two categories, essential (Co, Cr, Cu, Fe, Mn,

Mo, Ni, Se, V, and Zn) and nonessential (Ag, Al, As, Bi, Cd, Ce, Cs, Eu, Ga, Hf, Hg, La, Lu, Nb, Pb, Sb, Sc, Sn, Te, Th, Ti, U, Y, and Zr) elements following [26–28].

Documents related to the Mediterranean Sea province have been analyzed on the basis of the considered taxa (Porifera, Cnidaria, Polychaeta, Cirripedia, and Tunicata), temporal distribution, ecoregion of the sampling site(s), sampling depth(s), HM(s), considered order(s) per taxa. In all the analyses, every document could be included in one or more categories.

The analysis reported in Figure 2B was carried out following the bioregionalization of coastal and shelf areas proposed by Spalding et al. [29]. While, for the creation of the map of documents distribution in the Mediterranean Sea, geographic coordinates were used. In studies showing only a map of the sampling point(s), geographic coordinates were estimated from it. In the absence of a map, coordinates of the sampling point(s) were estimated using the information across the text, if available. When the given coordinates ended up in land, they were adjusted to the nearest coastal points.

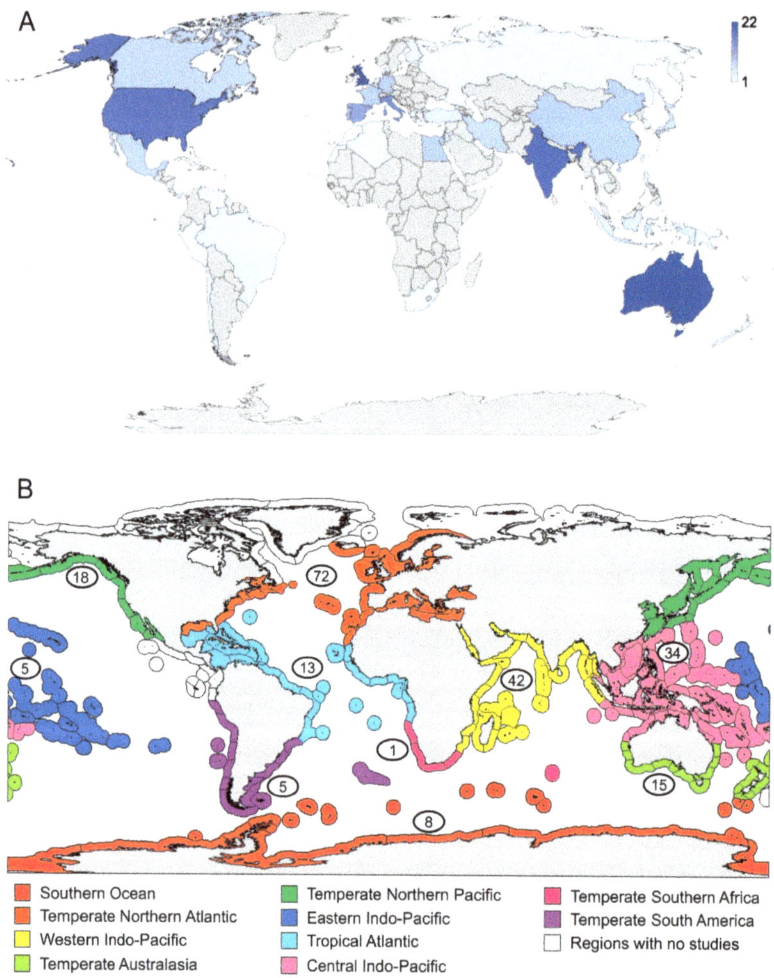

Figure 2. Map showing the number of documents per: (**A**) first authors' affiliation country and (**B**) marine realm.

A list of works divided per taxa with all document's details (publication year, authors, title, DOI/Reference, source, first author affiliation country, realm of the sampling site, habitat, basic physicochemical parameters of water, taxa, application field, analyzed body part, data validation, HM(s), and work setup) is reported in the Supplementary Material as Table S1. The Supplementary Material also included a detailed list of documents (taxa, authors, publication year, marine ecoregion of the sampling site(s), latitude, longitude, depth, application field, order of the analyzed species, analyzed species, HM(s), concentration, and unit) on the Mediterranean Sea province as Table S2. In the case of HMs, concentrations were not reported along the text, they were estimated from graphs. Concentration units, if different, were standardized to make concentrations comparable between documents.

3. Results

3.1. Literature Analysis at Global Scale

A total of 919 documents (Figure 1), including duplicates, have been screened and analyzed by reading the title and abstract, using the aforementioned keywords. Those not following our eligibility criteria were removed (643), leading to 276 documents, spread among the six considered taxa (Figure 1). After excluding duplicates, the final number of analyzed works was 218 (Figure 1; Table S1).

Considering all the literature found, most of the works have been published from researches with an affiliation in the United Kingdom (22), Australia (21), India (19), the United States (18), and Italy (15) (Figure 2A). While in many eastern and northern European countries and various Asian and most of the African countries, only a few studies have been published (Figure 2A). Australia and the United Kingdom were the only two, out of 44 countries, where all six taxa were studied (Table S1). Regarding the sampling area, most of the studies were carried out with organisms collected in the Temperate Northern Atlantic (32.0%) and the Western (18.7%) and Central (15.1%) Indo-Pacific marine realms (Figure 2B). Inside the Temperate Northern Atlantic realm, the Mediterranean Sea province showed the highest number of documents (34). All the other studies are scattered among the other marine realms, with only one work in the Temperate Southern Africa and a total absence of studies in the Tropical Eastern Pacific and the Arctic (Figure 2B). Up to the 12.8% of the total documents have not been taken into account for the creation of Figure 2B in as much as 12 works did not give any information about the sampling site(s), 12 considered marine organisms breed in aquarium, and 5 regarded organisms collected on hydrothermal vents located in areas not included in the division proposed by Spalding et al. [29] (Table S1). Most of the studies included species collected in the marine environment (92.8%), and only an 8.0% focused on estuarine organisms (Table S1).

The heatmap presented as Figure 3A showed that the first document, found using the query "heavy metal" AND "ascidian," was published before 1970 (1956, see Table S1 for citation). After this publication, for all taxa except Bryozoans, for which the first publication is registered in the decade 1991–2000 (1992, see Table S1 for citation), no works could be found until 1971–1980 (Figure 3A). In general, there is an upward trend in the number of documents through time, for almost all the taxa, while for Cirripedia, it is possible to identify a peak in 1991–2000 and a decrease in documents in the following decades (Figure 3A). The taxa of Cnidaria shows the highest number of works in almost all the decades, with the 37.0% of the publications, resulting the most studied group for HMs, followed by Porifera (18.5%), a group presenting a great increase in documents starting from 2001 (Figure 3A).

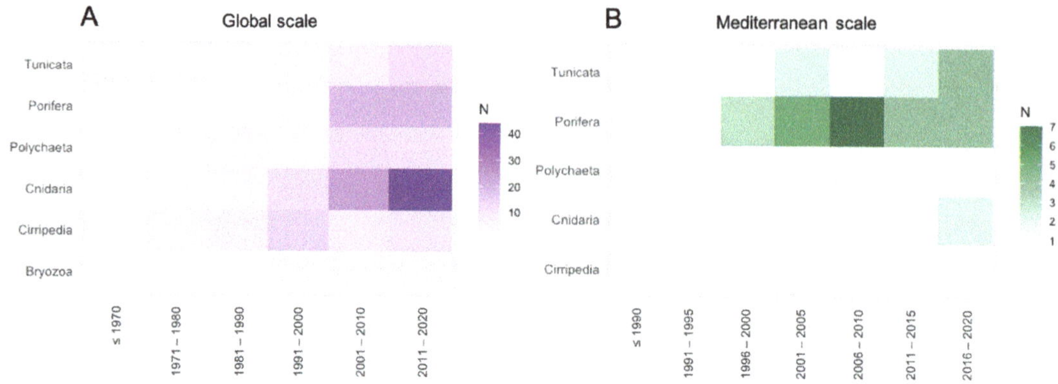

Figure 3. Number of documents per: (**A**) decade, at global-scale and (**B**) 5-year period, at Mediterranean-scale.

From the literature collection, it was evident that most of the documents analyzed both essential and nonessential elements (72.0%), in particular Cd (146), Cu (172), Pb (110), and Zn (142), while the 19.1% analyzed only essential elements and the 8.9% only nonessential ones (Table S1).

Organisms belonging to the six considered taxa have been mainly collected for biomonitoring analysis (Figure 4A), aiming to measure HMs concentrations in organisms and tissues. In particular, a significant proportion of documents in each taxon, for their analysis, processed the entire specimen, followed by the skeleton, especially in the Cnidaria group (Figure 4B). In fact, among the 43 documents on skeletal analysis, 36 examined the one of cnidarians (Table S1). Figure 4A showed also a total absence of physiological studies on Bryozoa, while only 2 have been conducted on Cirripedia (see Table S1 for citations). The remaining documents are spread among the other four taxa (Figure 4A). Works on microbiology represent a small component (15 documents) of the total, most of them (7) on Porifera (Figure 4A). A comparable situation can be observed for studies on cells and tissues and the microbiome, which have not been taken into account very frequently (Figure 4B).

The 59.1% of documents processed samples after their collection on the field (Figure 5A), especially for Cirripedia and Cnidaria (Figure 5B), while the 36.0% conducted laboratory experiments (Figure 5A). Among all taxa, organisms mainly belonging to the Cnidaria group have been used for this aim (Figure 5B). Only a small percentage (4.9%) of documents regarded studies in which collected organisms have been processed immediately to analyze a possible bioaccumulation of HMs and they have been bred to conduct laboratory experiments on the possible physiological effects of elements (Figure 5A). Apart from Cirripedia and Cnidaria, all the other taxa showed studies conducted with this work set up (Figure 5B).

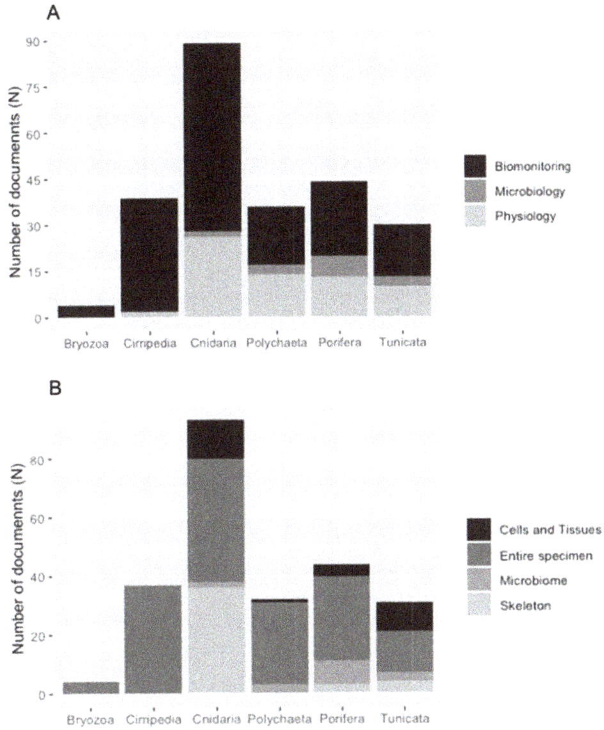

Figure 4. Number of documents per: (**A**) science field and (**B**) analyzed body part.

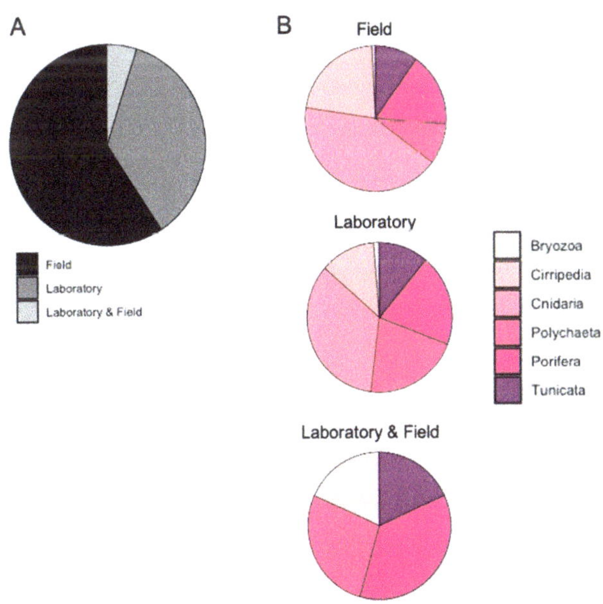

Figure 5. Percentage of documents per: (**A**) work setup and (**B**) taxa in each work setup.

3.2. Literature Analysis Focused on the Mediterranean Sea Province

The literature analysis on documents found within the Elsevier's Scopus database led to the record of 34 works on organisms collected in the Mediterranean Sea province. The manual research carried out on the references reported in these works led to the addition of seven documents. The total 41 documents (Figure 1) regarded 5 of the 6 considered taxa: Porifera (24), Cnidaria (3), Polychaeta (4), Cirripedia (2), and Tunicata (8) (Table S2).

Among all documents, only one work regarding Porifera was published before 1991 (1990, see reference in Table S1) (Figure 3B). The temporal distribution of documents showed how Porifera were considered in every 5-year period, with a peak in 2006–2010 (Figure 3B). Polychaeta were included in a constant number of works from 2001 to 2005, while for Tunicata, there is an increment of documents during time (Figure 3B). On the other hand, the low number of works on Cnidaria and Cirripedia is scattered through the different 5-year periods, from 2001 to 2005 till today (Figure 3B).

Inside the basin, most of the studies collected organisms in the Western Mediterranean ecoregion (22), mainly along the French and Spanish coasts (Figure 6; Table S2), followed by the Adriatic Sea (7), the Ionian Sea (6), the Levantine Sea (4), and the Aegean Sea (2) (Table S2). Figure 6 shows that most of the Porifera have been collected in the Western Mediterranean, while almost all studies on Tunicata and all on Cirripedia are referred to organisms collected in the Levantine and the Aegean Sea. Two-third of documents, moreover, gave information on the sampling depth (Figure 7). Most of the organisms were collected between 0 and 5 m depth, while Porifera is the only taxon collected at all the considered depth ranges (Figure 7).

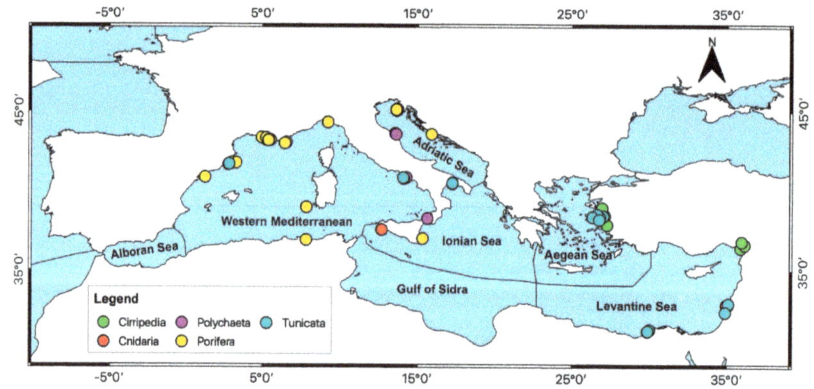

Figure 6. Map of the distribution of documents per taxa in each ecoregion of the Mediterranean Sea.

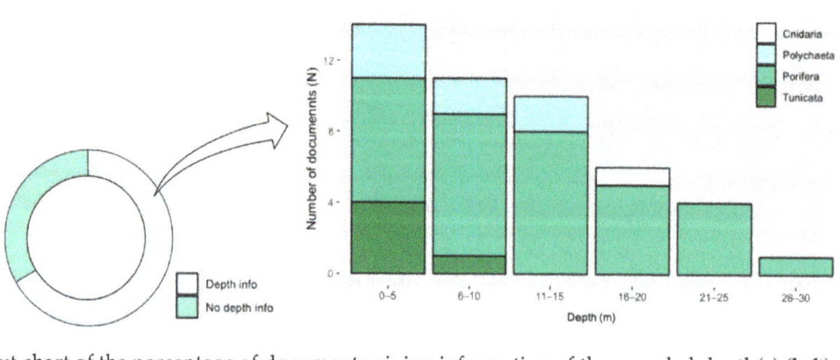

Figure 7. Donut chart of the percentage of documents giving information of the sampled depth(s) (**left**), and number of documents per taxa per depth range (**right**).

Most of the documents studied the role of organisms as bioindicators (28) especially of Cu (28), Pb (23), and Cd (20) contamination. Fourteen works explored the main effects of Cd (10) and Cu (9) on the collected species (Table S2), while one paper highlighted the tolerance of the microbiota extracted from polychaetes to Cd, Cu, and Zn (see Table S1 for citation).

Since Porifera resulted to be the most studied taxa, we considered this group for further considerations. Species studied belong to 11 orders of sponges (Table S2), and Dictyoceratida is undoubtedly the one to which most of works on biomonitoring are related. These organisms showed Cd concentrations ranging between 0.2 and 1.60 mg·kg-1 d.w., Cu between 22 and 300 mg·kg-1 d.w., and Pb between 0.1 and 90 mg·kg-1 d.w. (Table S2). Dictyoceratida has not been investigated for possible effects given by the presence of trace elements, but orders as Poecilosclerida and Scopalinida were considered more suitable for physiological studies and mainly on Cd and Cu (Table S2).

4. Discussion

At present, the sessile taxa considered in this study (Porifera, Cnidaria, Bryozoa, Polychaeta, Cirripedia, and Tunicata) are still little used as biomonitors compared with other groups, despite the fact that their potential application has been highlighted repeatedly during time [12,23], with an increasing annual trend in the last years. For example, searching on Scopus database using the keywords "heavy metal" AND "bivalve" or "heavy metal" AND "fish," without conducting the screening, 1148 and 6800 documents were obtained, respectively, more than all the works screened in our research, conducted using many keywords. In fact, the vast majority of the studies on the topic referred to organisms representing food resources for human consumption, which do not have to exceed specific levels, established by Community legislation or other relevant standards [23,30]. Another factor to take into account is that the considered taxa are not of the same economic importance as bivalves or fish, even though the economic role covered by some of them (e.g., sponges) is more and more often highlighted by the scientific community [31]. Nonetheless, looking at the physical and physiological characteristics a bioindicator should present (see Butler et al. [15] and Haug et al. [16]), these groups fit in many of the listed features. In fact: (i) with the exception of some cnidarians and polychaetes, they are exclusively sessile; (ii) most of them are active filter-feeders; (iii) they present a global distribution, therefore they can be collected and used for this purpose all around the globe, even at the Poles; (iv) these taxa comprehend modular organisms [32], therefore the collection of small portions does not affect their survival and they can rapidly regenerate; (v) some of them present also a gregarious behavior [33–37], allowing to sample just a part of a population and to observe possible changes on long-term monitoring.

From our review, it is clear that, at a global scale, studies are not equally distributed among marine regions [29], with most studies focused the Mediterranean Sea province [29], probably due to its nature of being a semienclosed basin [38], and thus more subjected to anthropogenic stress and pollution [39], as the one caused by HMs (see Table 1 from Danovaro [40], for quantities released annually, and Table 1 from Tovar-Sánchez et al. [41], for the concentration in the Mediterranean water). The basin, moreover, is subjected to the European legislation (e.g., MSFD 2008/56/EC), which suggests biological indicators (e.g., sponges and cnidarians) as potential tools to monitor the environmental status of coastal areas. Therefore, researchers could be more addressed to use unconventional organisms in their biomonitoring studies. However, also in the Mediterranean basin, there is a bias in the studies distribution, since research activities mainly focused on the Western Mediterranean ecoregion (22) [29], while the others are interested only by a low number of studies. This condition should be explained with the affiliation of the researchers studying the topic, which are mostly present in countries facing the Western Mediterranean ecoregion. Moreover, most of studies are limited to the most accessible shallow waters, mainly investigated by scuba diving.

Another strong bias highlighted by the present work is the difference in the number of documents per taxon, most of them concerning Cnidaria (37%) and Porifera (18.5%). Considered cnidarians are especially from tropical areas (Table S1), in which studies principally investigated the concentrations of HMs particularly in the hard skeleton of many scleractinian corals. Their skeleton is, in fact, characterized by the presence of growth bands, used in the determination of coral's age [42,43], and as a proxy of ocean responses to environmental factors, as temperature and salinity [44,45], and also contaminants, such as heavy metals [46,47], hydrocarbons [48], and pesticides [49]. This peculiarity makes the coral skeleton an important biomonitoring tool for marine pollution, allowing to analyze and detect possible variations in the concentrations of HMs over different time periods. However, together with scleractinians, other groups of cnidarians, e.g., octocorals and stylasterids, showed the presence of growth bands in their skeletons [50,51], making these taxa additional organisms suitable for biomonitoring studies.

Regarding the other well-studied group, Porifera have been mainly collected in the Temperate Northern Atlantic realm [29]. Especially interesting are the few studies analyzing the role of their associated bacteria [52–58]. Many sponges, in fact, are characterized by the presence of complex communities of microorganisms composing the associated microbiota, being named "bacteriosponges" or "high microbial abundance sponges" [59]. Since the microbiota can contribute up to 40% of sponge biomass and cover a fundamental role in their secondary metabolism [60,61], it has been suggested that it can actively participate in the bioaccumulation process. Many bacteria isolated from sponges, e.g., *Fasciospongia cavernosa*, have been found to be resistant to different antibiotics and pollutants, including Persistent Organic Pollutants (POPs) and various heavy metals, as Cu, Pb, Co, Cd, Zn, Ni, and Hg together with their organic compounds [55]. Different researches showed how also the bacteria associated with polychaetes and ascidians can be important in the sequestration of HMs from water [62–67]; therefore, increasing study of the effective role of the microbiota is necessary to fully understand the dynamics of the process. Besides, we want to stress the importance of works on the possible negative physiological effects of HMs in organisms (e.g., apoptosis of sponge tissues, bleaching in corals, decreasing of sperm density in polychaetes, etc.) [68–70], but also the "positive" consequences that HMs can trigger, e.g., in the moon jellyfish *Aurelia* spp., which polyps were observed to increase their asexual reproduction if exposed to Ag and Hg [4,71]. However, what we call a "positive" response can have severe environmental consequences. Referring to the example of *Aurelia* spp., an increase in the asexual reproduction can be followed by increase in polyp strobilation and ephyra release and, thus, intense jellyfish blooms [72], with multiple ecological and socio-economic consequences on human health, marine trophic chains, fishery, and aquaculture [73,74]. Moreover, an increase in this kind of studies will help to identify different physiological traits known as "biomarkers." Biomarkers are biological responses of suitable sentinel species to pollutants, which can act as early-warning tools able to detect pollutant-induced stress syndromes in organisms before the occurrence of severe habitat and communities' alterations [75].

From our review emerged clearly that the most studied HMs have been Cd, Cu, Pb, and Zn, both due to their natural toxicity (Cd and Pb) (WFD 2000/60/EC) and both for being used by organisms as micronutrients (Cu and Zn). In fact, essential elements can be considered HMs since they can become toxic, for an organism, above specific thresholds [28], as it was observed, e.g., in *Aurelia aurita* polyps [71] or in the sponge *Crambe crambe* [76]. Many groups of invertebrates are known to synthetize metallothioneins or metallothionein-like proteins, including the taxa considered in this review [77–80], except for Cnidaria and Bryozoa, for which there is still no information available [81–83]. Metallothioneins are cysteine-rich metal-binding proteins, which can sequester metals, being involved in metal resistance and in the detoxification processes [82]. If the detoxification mechanism is not overwhelmed, organisms can increase body metal concentrations without showing toxic effects or signs of suffering. However, if in the environment metal concentrations are elevated, the mechanism overwhelms and metals can bind to sensitive intracellular targets, inducing toxicity [84]. Therefore, to understand the thresholds not to be overcome, for both essential and nonessential elements, results fundamental to set general concentration limits focused on organism wellness, since by now, the actual low limits for HMs concern only edible species (Directive 2002/32/EC) and many are referred to human health [30].

Another important issue emerging from our bibliographic analysis was the lack of a standardization of presented data. First of all, almost half of the studies (generally written in the last century) did not show any information about the quality assurance (Certified Reference Material (CRM), recovery of standard addition, etc.) and, even though the 54% presented quality information, they were provided in an unclear way (no table of data and explanation of the reference material) or inappropriate material has been used (Table S1). In addition, only about the 15% analyzed the physicochemical parameters of water (e.g., pH, conductivity, alkalinity, organic matter, etc.) (Table S1), factors which could influence the speciation of HMs in the marine environment [85] and, thus, modify their bioavailability [11]. Different concentration orders and units are also given, making difficult the data management and the intercomparison among studies. Most of documents expressed data in dry weight (d.w.) while some others expressed results in wet weight (w.w.). If data presented in w.w. do not also provide the water concentration in samples, they are not comparable with those expressed in d.w. These issues arouse the attention of the scientific community since the 1970s [86], and the suggestion for the creation of "quality assurance" protocols for data comparison was given repeatedly in past decades [75,87]. Examples are the MED POL Mediterranean Sea Biomonitoring Programme [88] and the Pollution Effect Network (PEN) [75]. However, these activities did not pursue until today, while the intercomparison problem persists.

5. Conclusions and Future Perspectives

The results of our review showed a growing interest by the scientific community in the identification of sessile taxa as nonconventional bioindicators of HMs. The increasing interest is mainly linked to the higher awareness of researchers to metals contamination in the marine environment and to the international legislations, which are more and more focused on the role of these taxa as fundamental monitors of environmental pollution. Different studies suggested that some sessile organisms can be used also in the bioremediation processes, thanks to their characteristics and the part played by the associated microbiota (bacteria and fungi). This is particularly true for sponges, cnidarians, polychaetes, bivalves, and tunicates [60,64,89–91]. The application of vagile organisms as bioremediation and biomonitoring tools (as fish, cephalopods, or decapods crustaceans) results instead more difficult, since they are not sessile and cannot detect the actual values of a single location. In addition, even if the present work focused only on marine species, some of the considered taxa (Porifera, Cnidaria, and Bryozoa) also includes freshwater species. Therefore, these organisms can be applied as useful tools also near inland farms, industries, and plantations, as it is already done for freshwater bivalves [92].

Nonetheless, researchers should also focus their attention in identifying model organisms, which proved to be more suitable as bioindicators or bioremediation tools compared to other species of the same taxa.

These research activities cannot exclude the consultations of specialized taxonomists, since (1) species belonging to the same genus can bioaccumulate different concentration of HMs [12] or can present different responses if subjected to the same metal (see Figure 1 from Viarengo et al. [75] for the different physiological responses of *Mytilus galloprovincialis*); (2) some species considered by the literature revised for this study (e.g., *Geodia cynodium*, *Spongia (Spongia) lamella*, and *S. (S.) officinalis*) (see Tables S1 and S2 for citations) are included in the Annex II of the SPA/BIO Protocol of Barcelona Convention and, therefore, subjected to protection, due to their status of "threatened" species [93].

Our work highlighted many research gaps, but two aspects need special attention. As future perspectives, we recommend increasing sampling and collection efforts and widening the geographic and bathymetric distributions of the studies not only in the

Mediterranean Sea but also in the other marine realms, provinces and ecoregions, especially in the less investigated ones, integrating the results with the studies currently available. Moreover, data standardization is still a big issue in the field, therefore, we suggest facing the problem again, reopening old programs (e.g., MED POL and PEN) or crating new ones with new possible protocols to standardize methods, validation, analysis, and data presentation.

Supplementary Materials: The following are available online at https://www.mdpi.com/2076-3417/11/2/580/s1, Table S1: List of works divided per taxa with all document's details, Table S2: List of documents on the Mediterranean Sea province.

Author Contributions: Conceptualization, A.A. (Anna Annibaldi) and S.P.; methodology, A.A. (Anna Annibaldi), C.R. and S.P.; software, C.R., C.G., T.P.M., A.A. (Afghan Afghan); formal analysis, C.R.; investigation, C.R.; data curation, C.R. and A.A. (Anna Annibaldi); writing—original draft preparation, C.R., A.A. (Anna Annibaldi) and S.I.; writing—review and editing, all authors; project administration, A.A. (Anna Annibaldi) and S.P.; funding acquisition, S.P. All authors have read and agreed to the published version of the manuscript.

Funding: This study has been conducted with the financial support of PADI FOUNDATION (grant number #32694) and the Università Politecnica delle Marche (RSA—Ricerca Scientifica di Ateneo).

Institutional Review Board Statement: Not applicable.

Informed Consent Statement: Not applicable.

Data Availability Statement: Data is contained within the article and supplementary material.

Acknowledgments: Authors are thankful to Carlo Cerrano for his precious suggestions, which helped to improve our work.

Conflicts of Interest: The authors declare they have no conflict of interest.

References

1. Caeiro, S.; Costa, M.H.; Ramos, T.B.; Fernandes, F.; Silveira, N.; Coimbra, A.; Medeiros, G.; Painho, M. Assessing heavy metal contamination in Sado Estuary sediment: An index analysis approach. *Ecol. Indic.* **2005**, *5*, 151–169. [CrossRef]
2. Orlandi, L.; Bentivoglio, F.; Carlino, P.; Calizza, E.; Rossi, D.; Costantini, M.L.; Rossi, L. δ^{15}N variation in *Ulva lactuca* as a proxy for anthropogenic nitrogen inputs in coastal areas of Gulf of Gaeta (Mediterranean Sea). *Mar. Pollut. Bull.* **2014**, *84*, 76–82. [CrossRef]
3. Tedetti, M.; Guigue, C.; Goutx, M. Utilization of a submersible UV fluorometer for monitoring anthropogenic inputs in the Mediterranean coastal waters. *Mar. Pollut. Bull.* **2010**, *60*, 350–362. [CrossRef] [PubMed]
4. Roveta, C.; Annibaldi, A.; Vagnoni, F.; Mantas, T.P.; Domenichelli, F.; Gridelli, S.; Puce, S. Short-term effects of environmental factors on the asexual reproduction of *Aurelia* sp. polyps. *Chem. Ecol.* **2020**, *36*, 486–492. [CrossRef]
5. Illuminati, S.; Annibaldi, A.; Truzzi, C.; Scarponi, G. Heavy metal distribution in organic and siliceous marine sponge tissues measured by square wave anodic stripping voltammetry. *Mar. Pollut. Bull.* **2016**, *111*, 476–482. [CrossRef] [PubMed]
6. Carnevali, O.; Benedetti, M.; Beolchini, F.; Dell'Anno, A.; Fattorini, D.; Gorbi, S.; Illuminati, S.; Maradonna, F.; Scarponi, G.; Regoli, F. New Insights for Early Warning and Countermeasures to Aquatic Pollution. In *The First Outstanding 50 Years of "Università Politecnica delle Marche"*; Longhi, S., Monteriù, A., Freddi, A., Aquilanti, L., Ceravolo, M.G., Carnelavi., O., Giordano, M., Moroncini, G., Eds.; Springer: Cham, Switzerland, 2020; pp. 431–455. [CrossRef]
7. Duffus, J.H. Heavy metals-a meaningless term? *Pure Appl. Chem.* **2002**, *74*, 793–807. [CrossRef]
8. Pourret, O.; Hursthouse, A. It's Time to Replace the Term "Heavy Metals" with "Potentially Toxic Elements" When Reporting Environmental Research. *Int. J. Environ. Res. Public. Health* **2019**, *16*, 4446. [CrossRef] [PubMed]
9. Appenroth, K.-J. What are "heavy metals" in Plant Sciences? *Acta Physiol. Plant.* **2010**, *32*, 615–619. [CrossRef]
10. Chang, L.W.; Magos, L.; Suzuki, T. *Toxicology of Metals*; CRC Press: Boca Raton, FL, USA, 1996.
11. Simkiss, K.; Taylor, M.G. Transport of metals across membranes. In *Metal Speciation and Bioavailability in Aquatic Systems*; Tessier, A., Turner, D., Eds.; John Wiley & Sons Ltd.: Chichester, UK, 1995; pp. 1–44.
12. Rainbow, P.S. Trace metal concentrations in aquatic invertebrates: Why and so what? *Environ. Pollut.* **2002**, *120*, 497–507. [CrossRef]
13. Annibaldi, A.; Truzzi, C.; Carnevali, O.; Pignalosa, P.; Api, M.; Scarponi, G.; Illuminati, S. Determination of Hg in farmed and wild atlantic bluefin tuna (*Thunnus thynnus* L.) muscle. *Molecules* **2019**, *24*, 1273. [CrossRef]
14. Cammilleri, G.; Galluzzo, P.; Pulvirenti, A.; Giangrosso, I.E.; Lo Dico, G.M.; Montana, G.; Lampisi, N.; Mobilia, M.A.; Lastra, A.; Vazzana, M.; et al. Toxic mineral elements in *Mytilus galloprovincialis* from Sicilian coasts (Southern Italy). *Nat. Prod. Res.* **2020**, *34*, 177–182. [CrossRef] [PubMed]

15. Butler, P.A.; Andren, L.; Bonde, G.J.; Jernelov, A.; Reisch, D.J. Monitoring organisms. In *Food and Agricultural Organization Technical Conference on Marine Pollution and its Effects on Living Resources and Fishing, Rome, 1970. Supplement 1: Methods of Detection, Measurement and Monitoring of Pollutants in the Marine Environment*; Ruivo, M., Ed.; Fishing News Ltd.: London, UK, 1971; pp. 101–112.
16. Haug, A.; Melsom, S.; Omang, S. Estimation of heavy metal pollution in two Norwegian fjord areas by analysis of the brown alga *Ascophyllum nodosum*. *Environ. Pollut.* **1974**, *7*, 179–192. [CrossRef]
17. McKenzie, L.A.; Brooks, R.; Johnston, E.L. Heritable pollution tolerance in a marine invader. *Environ. Res.* **2011**, *111*, 926–932. [CrossRef] [PubMed]
18. Batista, D.; Muricy, G.; Rocha, R.C.; Miekeley, N.F. Marine sponges with contrasting life histories can be complementary biomonitors of heavy metal pollution in coastal ecosystems. *Environ. Sci. Pollut. Res.* **2014**, *21*, 5785–5794. [CrossRef]
19. Colozza, N.; Gravina, M.F.; Amendola, L.; Rosati, M.; Akretche, D.E.; Moscone, D.; Arduini, F. A miniaturized bismuth-based sensor to evaluate the marine organism *Styela plicata* bioremediation capacity toward heavy metal polluted seawater. *Sci. Total Environ.* **2017**, *584*, 692–700. [CrossRef]
20. Giangrande, A.; Licciano, M.; Del Pasqua, M.; Fanizzi, F.P.; Migoni, D.; Stabili, L. Heavy metals in five Sabellidae species (Annelida, Polychaeta): Ecological implications. *Environ. Sci. Pollut. R.* **2017**, *24*, 3759–3768. [CrossRef]
21. Jupp, B.P.; Fowler, S.W.; Dobretsov, S.; van der Wiele, H.; Al-Ghafri, A. Assessment of heavy metal and petroleum hydrocarbon contamination in the Sultanate of Oman with emphasis on harbours, marinas, terminals and ports. *Mar. Pollut. Bull.* **2017**, *121*, 260–273. [CrossRef]
22. Souri, A.; Niyogi, S.; Naji, A. Distribution, source apportionment, bioavailability and ecological risks of metals in reef sediments and corals of the Persian Gulf (Iran): Khark Island, Chirouyeh, and Hendorabi Island. *Mar. Pollut. Bull.* **2019**, *149*, 110654. [CrossRef]
23. Haynes, T.; Bell, J.; Saunders, G.; Irving, R.; Williams, J.; Bell, G. *Marine Strategy Framework Directive Shallow Sublittoral Rock Indicators for Fragile Sponge and Anthozoan Assemblages Part 1: Developing Proposals for Potential Indicators*; JNCC Report No. 524; Nature Bureau and Environment Systems Ltd. for JNCC: Peterborough, UK, 2014; p. 85.
24. Truzzi, C.; Annibaldi, A.; Illuminati, S.; Bassotti, E.; Scarponi, G. Square-wave anodic-stripping voltammetric determination of Cd, Pb, and Cu in a hydrofluoric acid solution of siliceous spicules of marine sponges (from the Ligurian Sea, Italy, and the Ross Sea, Antarctica). *Anal. Bioanal. Chem.* **2008**, *392*, 247–262. [CrossRef]
25. Ledda, F.D.; Ramoino, P.; Ravera, S.; Perino, E.; Bianchini, P.; Diaspro, A.; Gallus, L.; Pronzato, R.; Manconi, R. Tubulin posttranslational modifications induced by cadmium in the sponge *Clathrina clathrus*. *Aquat. Toxicol.* **2013**, *140*, 98–105. [CrossRef]
26. Vincent, J.B. New evidence against chromium as an essential trace element. *J. Nutr.* **2017**, *147*, 2212–2219. [CrossRef] [PubMed]
27. Sobolev, N.; Aksenov, A.; Sorokina, T.; Chashchin, V.; Ellingsen, D.G.; Nieboer, E.; Varakina, Y.; Vaselkina, E.; Kotsur, D.; Thomassen, Y. Essential and non-essential trace elements in fish consumed by indigenous peoples of the European Russian Arctic. *Environ. Pollut.* **2019**, *253*, 966–973. [CrossRef] [PubMed]
28. Zoroddu, M.A.; Aaseth, J.; Crisponi, G.; Medici, S.; Peana, M.; Nurchi, V.M. The essential metals for humans: A brief overview. *J. Inorg. Biochem.* **2019**, *195*, 120–129. [CrossRef] [PubMed]
29. Spalding, M.D.; Fox, H.E.; Allen, G.R.; Davidson, N.; Ferdaña, Z.A.; Finlayson, M.; Halpern, B.S.; Jorge, M.A.; Lombana, A.; Lourie, S.A.; et al. Marine ecoregions of the world: A bioregionalization of coastal and shelf areas. *BioScience* **2007**, *57*, 573–583. [CrossRef]
30. European Parliament; Council of the European Union. *Directive 2002/32/EC of 7 May 2002 on Undesirable Substances in Animal Feed*; Eur-Lex: Brussels, Belgium, 2002; Available online: https://eur-lex.europa.eu/legal-content/EN/TXT/?uri=CELEX%3A32002L0032 (accessed on 4 December 2020).
31. Pham, C.K.; Murillo, F.J.; Lirette, C.; Maldonado, M.; Colaço, A.; Ottaviani, D.; Kenchington, E. Removal of deep sea sponges by bottom trawling in the Flemish Cap area: Conservation, ecology and economic assessment. *Sci. Rep.* **2019**, *9*, 1–13. [CrossRef]
32. Dyrynda, P.E.J. Defensive strategies of modular organisms. *Philos. Trans. R. Soc. Lond. B Biol. Sci.* **1986**, *313*, 227–243. [CrossRef]
33. Rius, M.; Branch, G.M.; Griffiths, C.L.; Turon, X. Larval settlement behaviour in six gregarious ascidians in relation to adult distribution. *Mar. Ecol. Prog. Ser.* **2010**, *418*, 151–163. [CrossRef]
34. Keen, S.L. Recruitment of *Aurelia aurita* (Cnidaria: Scyphozoa) larvae is position-dependent, and independent of conspecific density, within a settling surface. *Mar. Ecol. Prog. Ser.* **1987**, *38*, 151–160. [CrossRef]
35. Miranda, L.S.; Collins, A.G.; Marques, A.C. Molecules Clarify a Cnidarian Life Cycle—The "Hydrozoan" *Microhydrula limopsicola* Is an Early Life Stage of the Staurozoan *Haliclystus antarcticus*. *PLoS ONE* **2010**, *5*, e10182. [CrossRef]
36. Crisp, D.J.; Meadows, P.S. The chemical basis of gregariousness in cirripedes. *Proc. R. Soc. Lond. B Biol. Sci.* **1962**, *156*, 500–520. [CrossRef]
37. Burke, R.D. Pheromones and gregarious settlement of marine invertebrate larvae. *Bull. Mar. Sci.* **1986**, *39*, 323–331.
38. Millot, C.; Taupier-Letage, I. Circulation in the Mediterranean Sea. In *The Mediterranean Sea*; Springer: Berlin/Heidelberg, Germany, 2005; Volume 5K, pp. 29–66. [CrossRef]

39. Bethoux, J.P.; Courau, P.; Nicolas, E.; Ruizpino, D. Trace-metal pollution in the Mediterranean Sea. *Oceanol. Acta* **1990**, *13*, 481–488.
40. Danovaro, R. Pollution threats in the Mediterranean Sea: An overview. *Chem. Ecol.* **2003**, *19*, 15–32. [CrossRef]
41. Tovar-Sánchez, A.; Rodríguez-Romero, A.; Engel, A.; Zäncker, B.; Fu, F.; Marañón, E.; Pérez-Lorenzo, M.; Bressac, M.; Wagener, T.; Triquet, S.; et al. Characterizing the surface microlayer in the Mediterranean Sea: Trace metal concentrations and microbial plankton abundance. *Biogeosciences* **2020**, *17*, 2349–2364. [CrossRef]
42. Adkins, J.F.; Henderson, G.M.; Wang, S.L.; O'Shea, S.; Mokadem, F. Growth rates of the deep-sea scleractinia *Desmophyllum cristagalli* and *Enallopsammia rostrata*. *Earth Planet. Sci. Lett.* **2004**, *227*, 481–490. [CrossRef]
43. Lough, J.M. Coral calcification from skeletal records revisited. *Mar. Ecol. Prog. Ser.* **2008**, *373*, 257–264. [CrossRef]
44. DeLong, K.L.; Quinn, T.M.; Taylor, F.W. Reconstructing twentieth-century sea surface temperature variability in the southwest Pacific: A replication study using multiple coral Sr/Ca records from New Caledonia. *Paleoceanography* **2007**, *22*, PA4212. [CrossRef]
45. Carilli, J.E.; Norris, R.D.; Black, B.; Walsh, S.M.; McField, M. Century-scale records of coral growth rates indicate that local stressors reduce coral thermal tolerance threshold. *Glob. Change Biol.* **2010**, *16*, 1247–1257. [CrossRef]
46. David, C.P. Heavy metal concentrations in growth bands of corals: A record of mine tailings input through time (Marinduque Island, Philippines). *Mar. Pollut. Bull.* **2003**, *46*, 187–196. [CrossRef]
47. Nour, H.E.S.; Nouh, E.S. Using coral skeletons for monitoring of heavy metals pollution in the Red Sea Coast, Egypt. *Arab. J. Geosci.* **2020**, *13*, 341. [CrossRef]
48. Readman, J.W.; Tolosa, I.; Law, A.T.; Bartocci, J.; Azemard, S.; Hamilton, T.; Mee, L.D.; Wagener, A.; Le Tissier, M.; Roberts, C.; et al. Discrete bands of petroleum hydrocarbons and molecular organic markers identified within massive coral skeletons. *Mar. Pollut. Bull.* **1996**, *32*, 437–443. [CrossRef]
49. Yang, T.; Diao, X.; Cheng, H.; Wang, H.; Zhou, H.; Zhao, H.; Chen, C.M. Comparative study of polycyclic aromatic hydrocarbons (PAHs) and heavy metals (HMs) in corals, sediments and seawater from coral reefs of Hainan, China. *Environ. Pollut.* **2020**, *264*, 114719. [CrossRef] [PubMed]
50. Sherwood, O.A.; Edinger, E.N. Ages and growth rates of some deep-sea gorgonian and antipatharian corals of Newfoundland and Labrador. *Can. J. Fish. Aquat. Sci.* **2009**, *66*, 142–152. [CrossRef]
51. Samperiz, A.; Robinson, L.F.; Stewart, J.A.; Strawson, I.; Leng, M.J.; Rosenheim, B.E.; Ciscato, E.R.; Hendry, K.R.; Santodomingo, N. Stylasterid corals: A new paleotemperature archive. *Earth Planet. Sci. Lett.* **2020**, *545*, 116407. [CrossRef]
52. Webster, N.S.; Webb, R.I.; Ridd, M.J.; Hill, R.T.; Negri, A.P. The effects of copper on the microbial community of a coral reef sponge. *Environ. Microbiol.* **2001**, *3*, 19–31. [CrossRef]
53. Mohanty, S.; Bapuji, M.; Mishra, R.K.; Sree, A.; Ray, P.; Mohapatra, S.B.; Rath, C.C. Studies on metal tolerance of bacterial associates of marine sedentary organisms. *Asian J. Microbiol. Biotechnol. Environ. Sci.* **2004**, *6*, 291–296.
54. Selvin, J.; Priya, S.S.; Kiran, G.S.; Bhosle, S. Biomonitoring of heavy metal pollution in the marine environment using indicator organisms. In *Causes and Effects of Heavy Metal Pollution*; Sánchez, M.L., Ed.; Nova Publishers: New York, NY, USA, 2008; Volume 4, pp. 251–264.
55. Selvin, J.; Priya, S.S.; Kiran, G.S.; Thangavelu, T.; Bai, N.S. Sponge-associated marine bacteria as indicators of heavy metal pollution. *Microbiol. Res.* **2009**, *164*, 352–363. [CrossRef]
56. Mangano, S.; Michaud, L.; Caruso, C.; Giudice, A.L. Metal and antibiotic resistance in psychrotrophic bacteria associated with the Antarctic sponge *Hemigellius pilosus* (Kirkpatrick, 1907). *Polar Biol.* **2014**, *37*, 227–235. [CrossRef]
57. Bauvais, C.; Zirah, S.; Piette, L.; Chaspoul, F.; Domart-Coulon, I.; Chapon, V.; Gallice, P.; Rebuffat, S.; Pérez, T.; Bourguet-Kondracki, M.L. Sponging up metals: Bacteria associated with the marine sponge *Spongia officinalis*. *Mar. Environ. Res.* **2015**, *104*, 20–30. [CrossRef]
58. Karimi, E.; Gonçalves, J.M.; Reis, M.; Costa, R. Draft Genome Sequence of *Microbacterium* sp. Strain Alg239_V18, an Actinobacterium Retrieved from the Marine Sponge *Spongia* sp. *Genome Announc.* **2017**, *5*, e01457-16. [CrossRef]
59. Schmitt, S.; Wehrl, M.; Siegl, A.; Hentschel, U. Review article Marine sponges as models for commensal microbe-host interactions. *Symbiosis* **2007**, *44*, 43–50.
60. Santos-Gandelman, J.F.; Giambiagi-deMarval, M.; Oelemann, W.M.R.; Laport, M.S. Biotechnological potential of sponge-associated bacteria. In *Current Pharmaceutical Biotechnology*; Bentham Science Publishers: Valencia, Spain, 2014; Volume 15, pp. 143–155.
61. Thomas, T.; Moitinho-Silva, L.; Lurgi, M.; Björk, J.R.; Easson, C.; Astudillo-García, C.; Olson, J.B.; Erwin, P.M.; Lopez-Legentil, S.; Luter, H.; et al. Diversity, structure and convergent evolution of the global sponge microbiome. *Nat. Commun.* **2016**, *7*, 1–12. [CrossRef] [PubMed]
62. Jeanthon, C.; Prieur, D. Resistance to heavy metals of heterotrophic bacteria isolated from the deep-sea hydrothermal vent polychaete, *Alvinella pompejana*. *Prog. Oceanogr.* **1990**, *24*, 81–88. [CrossRef]
63. Prieur, D.; Chamroux, S.; Durand, P.; Erauso, G.; Fera, P.; Jeanthon, C.; Le Borgne, G.; Mével, G.; Vincent, P. Metabolic diversity in epibiotic microflora associated with the Pompeii worms *Alvinella pompejana* and *A. caudata* (Polychaetae: Annelida) from deep-sea hydrothermal vents. *Mar. Biol.* **1990**, *106*, 361–367. [CrossRef]
64. Rizzo, C.; Michaud, L.; Graziano, M.; De Domenico, E.; Syldatk, C.; Hausmann, R.; Giudice, A.L. Biosurfactant activity, heavy metal tolerance and characterization of *Joostella* strain A8 from the Mediterranean polychaete *Megalomma claparedei* (Gravier, 1906). *Ecotoxicology* **2015**, *24*, 1294–1304. [CrossRef]

65. Ueki, T. Bioaccumulation of vanadium by vanadium-resistant bacteria isolated from the intestine of *Ascidia sydneiensis samea*. *Mar. Biotechnol.* **2016**, *18*, 359–371. [CrossRef]
66. Ueki, T.; Fujie, M.; Satoh, N. Symbiotic bacteria associated with ascidian vanadium accumulation identified by 16S rRNA amplicon sequencing. *Mar. Genom.* **2019**, *43*, 33–42. [CrossRef]
67. Liberti, A.; Bertocci, I.; Pollet, A.; Musco, L.; Locascio, A.; Ristoratore, F.; Spagnuolo, A.; Sordino, P. An indoor study of the combined effect of industrial pollution and turbulence events on the gut environment in a marine invertebrate. *Mar. Environ. Res.* **2020**, *158*, 104950. [CrossRef]
68. Batel, R.; Bihari, N.; Rinkevich, B.; Dapper, J.; Schäcke, H.; Schröder, H.; Müller, W. Modulation of organotin-induced apoptosis by the water pollutant methyl mercury in a human lymphoblastoid tumor cell line and a marine sponge. *Mar. Ecol. Prog. Ser.* **1993**, *93*, 245–251. [CrossRef]
69. Jones, R.J. Testing the 'photoinhibition' model of coral bleaching using chemical inhibitors. *Mar. Ecol. Prog. Ser.* **2004**, *284*, 133–145. [CrossRef]
70. Lockyer, A.; Binet, M.T.; Styan, C.A. Importance of sperm density in assessing the toxicity of metals to the fertilization of broadcast spawners. *Ecotoxicol. Environ. Saf.* **2019**, *172*, 547–555. [CrossRef] [PubMed]
71. Lucas, C.H.; Horton, A.A. Short-term effects of the heavy metals, Silver and copper, on polyps of the common jellyfish, *Aurelia aurita*. *J. Exp. Mar. Biol. Ecol.* **2014**, *461*, 154–161. [CrossRef]
72. Duarte, C.M.; Pitt, K.A.; Lucas, C.H.; Purcell, J.E.; Uye, S.I.; Robinson, K.; Brotz, L.; Decker, M.B.; Sutherland, K.R.; Malej, A.; et al. Is global ocean sprawl a cause of jellyfish blooms? *Front. Ecol. Environ.* **2013**, *11*, 91–97. [CrossRef]
73. Purcell, J.E. Jellyfish and ctenophore blooms coincide with human proliferations and environmental perturbations. *Ann. Rev. Mar. Sci.* **2012**, *4*, 209–235. [CrossRef]
74. Purcell, J.E.; Uye, S.I.; Lo, W.T. Anthropogenic causes of jellyfish blooms and their direct consequences for humans: A review. *Mar. Ecol. Prog. Ser.* **2007**, *350*, 153–174. [CrossRef]
75. Viarengo, A.; Burlando, B.; Giordana, A.; Bolognesi, C.; Gabrielides, G.P. Networking and expert-system analysis: Next frontier in biomonitoring. *Mar. Environ. Res.* **2000**, *49*, 483–486. [CrossRef]
76. Agell, G.; Uriz, M.J.; Cebrian, E.; Martí, R. Does stress protein induction by copper modify natural toxicity in sponges? *Environ. Toxicol. Chem. Int. J.* **2001**, *20*, 2588–2593. [CrossRef]
77. Berthet, B.; Mouneyrac, C.; Pérez, T.; Amiard-Triquet, C. Metallothionein concentration in sponges (*Spongia officinalis*) as a biomarker of metal contamination. *Comp. Biochem. Phys. C Toxicol. Pharmacol.* **2005**, *141*, 306–313. [CrossRef]
78. Amiard, J.C.; Amiard-Triquet, C.; Barka, S.; Pellerin, J.; Rainbow, P.S. Metallothioneins in aquatic invertebrates: Their role in metal detoxification and their use as biomarkers. *Aquat. Toxicol.* **2006**, *76*, 160–202. [CrossRef]
79. Franchi, N.; Boldrin, F.; Ballarin, L.; Piccinni, E. CiMT-1, an unusual chordate metallothionein gene in *Ciona intestinalis* genome: Structure and expression studies. *J. Exp. Zool. A Ecol. Genet. Physiol.* **2011**, *315*, 90–100. [CrossRef]
80. Zanette, J.; Monserrat, J.M.; Bianchini, A. Biochemical biomarkers in barnacles *Balanus improvisus*: Pollution and seasonal effects. *Mar. Environ. Res.* **2015**, *103*, 74–79. [CrossRef] [PubMed]
81. Karntanut, W.; Pascoe, D. The toxicity of copper, cadmium and zinc to four different *Hydra* (Cnidaria: Hydrozoa). *Chemosphere* **2002**, *47*, 1059–1064. [CrossRef]
82. Piola, R.F.; Johnston, E.L. Differential tolerance to metals among populations of the introduced bryozoan *Bugula neritina*. *Mar. Biol.* **2006**, *148*, 997–1010. [CrossRef]
83. Capdevila, M.; Atrian, S. Metallothionein protein evolution: A miniassay. *J. Biol. Inorg. Chem.* **2011**, *16*, 977–989. [CrossRef]
84. Casado-Martinez, M.C.; Smith, B.D.; Luoma, S.N.; Rainbow, P.S. Metal toxicity in a sediment-dwelling polychaete: Threshold body concentrations or overwhelming accumulation rates? *Environ. Pollut.* **2010**, *158*, 3071–3076. [CrossRef]
85. Domingos, R.F.; Gélabert, A.; Carreira, S.; Cordeiro, A.; Sivry, Y.; Benedetti, M.F. Metals in the aquatic environment—Interactions and implications for the speciation and bioavailability: A critical overview. *Aquat. Geochem.* **2015**, *21*, 231–257. [CrossRef]
86. Goldberg, E.D.; Bowen, V.T.; Farrington, J.W.; Harvey, G.; Martin, J.H.; Parker, P.L.; Risebrough, R.W.; Robertson, W.; Schneider, E.; Gamble, E. The Mussel Watch. In *Environmental Conservation*; The Foundation of Environmental Conservation: La Jolla, CA, USA, 1978; Volume 5, pp. 101–105.
87. Roose, P.; Albaigés, J.; Bebianno, M.J.; Camphuysen, C.; Cronin, M.; de Leeuw, J.; Gabrielsen, G.; Hutchinson, T.; Hylland, K.; Jansson, B.; et al. *Chemical Pollution in Europe's Seas: Programmes, Practices and Priorities for Research*; Marine Board Position Paper, 16; Calewaert, J.B., McDonough, N., Eds.; Marine Board-ESF: Ostend, Belgium, 2011.
88. Gabrielides, G. *MED POL Biomonitoring Programme Concerning the Effects of Pollutants on Marine Organisms along the Mediterranean Coasts*; UNEP(OCA)/MED WG 132/3; Athens: Athens, Greece, 1997.
89. Da Silva, M.; Passarini, M.R.Z.; Bonugli, R.C.; Sette, L.D. Cnidarian-derived filamentous fungi from Brazil: Isolation, characterisation and RBBR decolourisation screening. *Environ. Technol.* **2008**, *29*, 1331–1339. [CrossRef]
90. Rosa, I.C.; Costa, R.; Gonçalves, F.; Pereira, J.L. Bioremediation of Metal-Rich Effluents: Could the Invasive Bivalve *Corbicula fluminea* Work as a Biofilter? *J. Environ. Qual.* **2014**, *43*, 1536–1545. [CrossRef]
91. Tamilselvi, M.; Akram, A.S.; Arshan, M.K.; Sivakumar, V. Comparative study on bioremediation of heavy metals by solitary ascidian, *Phallusia nigra*, between Thoothukudi and Vizhinjam ports of India. *Ecotoxicol. Environ. Saf.* **2015**, *121*, 93–99. [CrossRef]

92. Sicuro, B.; Castelar, B.; Mugetti, D.; Pastorino, P.; Chiarandon, A.; Menconi, V.; Galloni, M.; Prearo, M. Bioremediation with freshwater bivalves: A sustainable approach to reducing the environmental impact of inland trout farms. *J. Environ. Manag.* **2020**, *276*, 111327. [CrossRef]
93. Relini, G.; Giaccone, G. Priority habitats according to the SPA/BIO protocol (Barcelona Convention) present in Italy. Identification sheets. *Biol. Mar. Mediterr.* **2009**, *16* (Suppl. 2), 435p.

applied sciences

MDPI

Article

Hg Levels in Marine Porifera of Montecristo and Giglio Islands (Tuscan Archipelago, Italy)

Camilla Roveta *, Daniela Pica, Barbara Calcinai, Federico Girolametti, Cristina Truzzi, Silvia Illuminati, Anna Annibaldi *and Stefania Puce

Department of Life and Environmental Sciences, Polytechnic University of Marche, Via Brecce Bianche, 60131 Ancona, Italy; daniela.pica@gmail.com (D.P.); b.calcinai@staff.univpm.it (B.C.); f.girolametti@pm.univpm.it (F.G.); c.truzzi@staff.univpm.it (C.T.); s.illuminati@staff.univpm.it (S.I.); s.puce@staff.univpm.it (S.P.)

* Correspondence: c.roveta@pm.univpm.it (C.R.); a.annibaldi@staff.univpm.it (A.A.); Tel.: +39-0712204649 (C.R.); +39-0712204981 (A.A.)

Received: 29 May 2020; Accepted: 23 June 2020; Published: 24 June 2020

Abstract: Porifera are filter-feeding organisms known to bioaccumulate different contaminants in their tissues. The presence of mercury (Hg) has been reported in different Mediterranean species, mainly collected in the southern coast of France. In the present study, mercury concentrations in the tissue of the sponges of Montecristo and Giglio, two islands of Tuscany Archipelago National Park (TANP), are presented for the first time. Analyses of total mercury content were performed by Direct Mercury Analyzer. Statistical differences have been reported in the Hg concentrations of species collected in both islands, but they do not appear related to the anthropic impacts of the islands. Among the collected species, a high intra- and inter-variability have been recorded, with *Cliona viridis* showing the lowest concentration (0.0167–0.033 $mg \cdot kg^{-1}$ dry weight), and *Chondrosia reniformis* and *Sarcotragus spinosulus* the highest (0.57 ± 0.15 and 0.64 ± 0.01 $mg \cdot kg^{-1}$ dry weight, respectively). The variability of Hg measured did not allow us to identify sponges as bioindicators of toxic elements. Anyway, these results improve knowledge on the ecosystem of the TANP, underlining the species-specificity of metal concentrations for Porifera, and providing additional data to address the main input of the Marine Strategy guidelines to protect coasts, seas and oceans.

Keywords: toxic element; Mediterranean Sea; sponges; biomonitoring

1. Introduction

Mercury (Hg) is considered to be one of the most toxic heavy metals due to its persistence in the environment, bioaccumulation in organisms and biomagnification in the trophic chain [1,2]. Atmospheric inputs of Hg have tripled during the last 150 years, being two-thirds of its actual concentration from anthropogenic sources [2,3]. Traces of mercury have been found in all the compartments of the ecosphere (atmosphere, hydrosphere, lithosphere and biosphere) [4–6]. In aquatic environments, it is transformed by chemical and biological reactions in organomercury compounds, as methylmercury (MeHg), the most toxic mercury species, which can be bioaccumulated more than other trace elements along the trophic chain [6]. Mercury can have many different effects on a wide range of organisms, both vertebrates, e.g., References [7–9], and invertebrates, e.g., References [10–12]. For all these reasons, Hg is listed in the European Water Framework Directive (WFD 2000/60/EC) as a priority substance representing a risk for the good chemical status for the aquatic environment, with a Maximum Allowable Concentration (MAC) of 0.07 $\mu g \cdot L^{-1}$.

The presence of mercury in the Mediterranean Sea has been documented since the 1970s [13,14]. The principal mercury input (94%) was recognized in rivers' dischargements, while only 5.5% was related to direct industrial wastewater and 0.5% to domestic sewage [15]. The Mediterranean area presents a high number of natural deposits of mercury distributed along the coasts of many countries, containing about 65% of the world's cinnabar (HgS) deposits [15,16]. Particularly, in Tuscany (Italy), the levels of Hg in different environmental matrices are derived from both a natural contribution of the mineralization and the pollution caused by the huge exploitation of the area of Mount Amiata [17]. In fact, Mount Amiata, located in the south of Tuscany, is part of the geologic anomaly of the Mediterranean basin and it is characterized by a large cinnabar deposit [16,18]. Moreover, it is well known from ARPAT (acronym for Regional Agency for Environmental Protection of Tuscany) and ISPRA (acronym for Higher Institute for Environmental Protection and Research) technical reports, and scientific papers [19], the presence of mercury in the waters and organisms of the Tuscany coast and island. From 2012 to 2017, high concentrations of mercury have been reported in water, sediment and biota (as *Posidonia oceanica*, *Mytilus galloprovincialis* and different fish species) in the Tuscan Archipelago islands, e.g., References [20–24].

Many studies have been conducted worldwide on the bioaccumulation and the effects of mercury and its organic compounds in edible species, such as bivalves, cephalopods, decapods crustaceans and fish [6,25–27]. Although, other filter-feeding organisms, such as polychaetes, tunicates, sponges and barnacles, have been proposed as bioindicators in shallow waters [28], but these taxa are still little used for this purpose. Especially, sponges satisfy all the characteristics listed in References [29,30] for a suitable bioindicator and have been recommended by many authors, e.g., References [31,32], and by the WFD as possible monitors for heavy metals. Being filter-feeders, sponges can filter a large amount of water, and they can collect and accumulate many different contaminants (such as hydrocarbons, organochlorinated compounds, heavy metals, etc.) in their tissues, which are present in the water column both in the soluble and particulate phases [10,28]. Moreover, the level of the bioaccumulation in their tissues is a function of the contaminants' concentration in the water [33]. The presence of heavy metals in sponges can affect their physiology and survival [28,34,35]. Only a few authors have investigated the presence of Hg in sponges and most of the studies have been conducted on samples, mainly on the genus *Spongia*, collected in the coast of Marseille [10,36–38].

The current paper presents, for the first time, total mercury content in the tissue of the sponges of Montecristo and Giglio, two islands of Tuscany Archipelago National Park (TANP). The study addresses two hypotheses: (1) sponges collected show detectable total mercury contents, and (2) there are any differences in Hg concentrations between the sponge species collected in both studied islands. Our results point out that sponges have detectable Hg concentrations, showing a high inter- and intra-specific variability. This variability could also be responsible of the lack of trend in the Hg concentrations between the specimens collected from the more anthropic Giglio and the Integral Reserve of Montecristo. Anyway, the results obtained with statistical analyses give new important insights for the area on the differences among Hg concentration in sponges.

2. Materials and Methods

2.1. Study Area

The study was carried out in June 2019 in Montecristo and Giglio islands (Tuscany, Italy) (Figure 1).

The Montecristo Island (42.3317° N, 10.3083° E) (Figure 1B) is an uninhabited and isolated island, sub-circular in shape, fourth in size after Elba, Giglio and Capraia islands, and completely mountainous, reaching a height of 645 m above sea-level with an almost constant slope of 25° [39–42]. Its history and geography distinguish the island from the others: during the Quaternary period, Montecristo remained in contact with the Tuscany's littoral for a shorter period than the other islands, ending up located 63 km from the mainland, on the limit of the continental shelf, closer to Corsica than to the mainland [39].

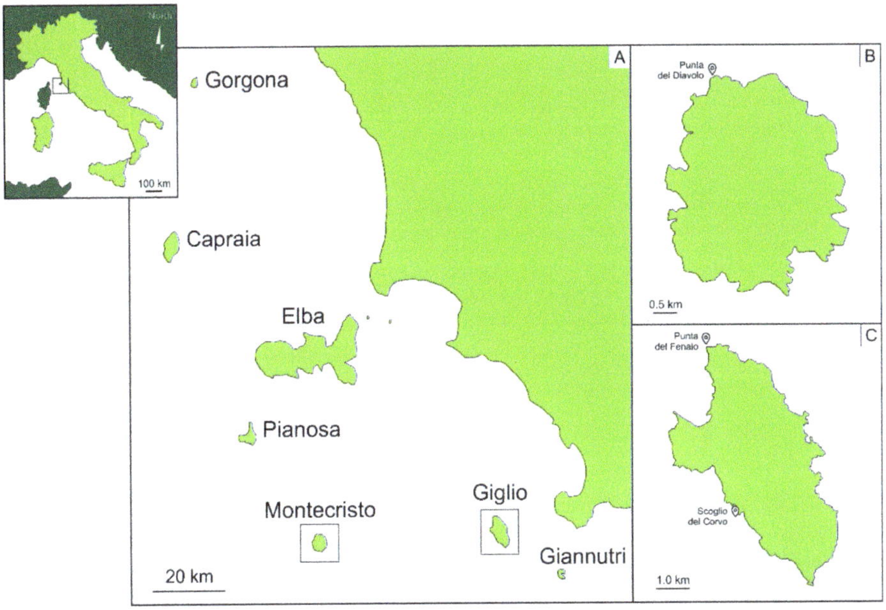

Figure 1. Map of the Tuscan Archipelago National Park (TANP) (**A**), Montecristo Island (**B**) and Giglio Island (**C**). The sampling sites are indicated in the islands' maps.

In 1971, Montecristo was established as an Integral Nature Reserve, and in 1988, was declared a biogenetic nature reserve by the Council of Europe. Moreover, the island has the status of Special Protection Area (SPA, Directive 79/409/EEC). The land and the adjacent waters, up to 1 km offshore, are controlled by the Coast Guard and the Carabinieri Corps [41,43]. Bathing, diving, fishing, mooring and circumnavigating are forbidden, while landing, berthing and scientific activities are allowed, only under specific conditions and with the permission of the Territorial Office for Biodiversity of the Carabinieri Corps of Follonica [41]. Due to the high protection which Montecristo has undergone, the island is considered one of the most pristine and best-preserved sites of the Mediterranean Sea, being previously identified as a reference site for the ecological quality assessment of the western Mediterranean benthic assemblages on rocky bottoms [44,45].

The Giglio Island (42.3603° N, 10.9229° E) (Figure 1C), being the second in size after Elba, is included in the southern group of the Archipelago, only 14 km from the Italian coast [42,46]. It is an oval-shaped island, characterized by a mountain chain along the north–south axis, and it reaches a height of 496 m above the sea-level. Contrary to Montecristo, the island hosts 1500 residents, spread out among the villages of Giglio Castello, Giglio Porto and Campese [42]. Only 40% of the island is included inside the protected area, while adjacent waters and many terrestrial areas do not have any protection (DPR 22 July 1996) [47]. The areas included in the legislation of the TANP are divided into four zones: (1) zone A of strict reserve, including five rocks (Cappa, Corvo, Mezzo Franco, Pietra Bona, and Le Scole), (2) zone B of general reserve, (3) zone C of general protection and (4) zone D of socio-economic promotion [42]. In general, Giglio is considered to have undergone anthropogenic pressures [47], due to the high flow of tourists during Summer [42], and to be more exposed to potential stress deriving from the mainland.

2.2. Samples Collection and Identification

Surveys were conducted by SCUBA diving and eighteen sponge samples were collected between 5 and 40 m, nine on the hard bottom assemblages of Punta del Diavolo (42°21′02.46″ N; 10°17′55.86″ E) in the Montecristo Island (TANP permission #00068010) (Figure 1B), and nine between Punta del Fenaio (42°23′21.54″ N; 10°52′48.18″ E) and Scoglio del Corvo (42°20′17.76″ N; 10°53′21.36″ E) in the Giglio Island (Figure 1C). The low number of samples for this preliminary study is strictly subordinated to protect and respect the island ecosystem, which needs to be protected even in the case of scientific research.

Samples were immediately frozen on dry ice, and then stored at −20 °C until analysis.

Among the samples, fourteen species of Demospongiae have been identified: *Agelas oroides* (Schmidt, 1864), *Axinella damicornis* (Esper, 1794), *Cliona viridis* (Schmidt, 1862), *Haliclona* (*Halichoclona*) *fulva* (Topsent, 1893), *Haliclona* (*Soestella*) *mucosa* (Griessinger, 1971), *Penares euastrum* (Schmidt, 1868) and *Sarcotragus spinosulus* Schmidt, 1862 in Montecristo, and *Chondrosia reniformis* Nardo, 1847, *C. viridis*, *Crambe crambe* (Schmidt, 1862), *H.* (*H.*) *fulva*, *Hemimycale columella* (Bowerbank, 1874), *Hymedesmia* (*Hymedesmia*) *baculifera* (Topsent, 1901) and *Petrosia* (*Petrosia*) *ficiformis* (Poiret, 1789) in Giglio. For the species *P. euastrum* and *C. reniformis*, three samples each were collected; therefore, we denominated them with the name of the species followed by the number 1, 2 or 3 between brackets.

2.3. Samples Treatment and Mercury Analysis

A clean room laboratory ISO 14644-1 Class 6, with areas at ISO Class 5 under laminar flow, was used for all laboratory activities. After the identification, samples were weighted (laboratory analytical balance, AT261 Mettler Toledo Greifensee, Switzerland, readability 0.01 mg, repeatability standard deviation (SD) = 0.015 mg) and cleaned with ultrapure water (A10 Milli-Q system, Merk Millipore, Bedford, MA, USA). The acid-cleaning procedures, used for all the laboratory materials, were performed as described in References [48,49].

After cutting samples into small pieces, sponges were lyophilized (Edwards EF4 modulyo, Crawley, Sussex, England), minced, homogenized and divided in aliquots of about 0.02 g each. Analyses of total mercury content (THg) were performed by Direct Mercury Analyzer (DMA-1 Milestone, Sorisole (BG), Italy), as described in Reference [50]. Briefly, the total mercury content was quantified by thermal decomposition amalgamation atomic absorption spectrometry at 253.7 nm. The calibration curve method was used for the quantification of Hg content. All measurements were replicated at least 4 times.

2.4. Accuracy

Quality assurance and quality control were assessed by processing blank samples and certified reference material (dogfish muscle DORM-2, NRCC; Ottawa, ON, Canada). The experimental values obtained for Hg in blanks are negligible compared with the metal content in sponge tissue (<1%). For DORM-2 analysis (n = 8), Hg content (4.58 ± 0.10 mg·kg^{-1}) is in agreement with the certified value (4.43 ± 0.05 mg·kg^{-1}) and no statistically significant differences were observed (p-value > 0.05, Student's T test, STATGRAPHICS 18 Centurion, 2018).

2.5. Data Analyses

Data are expressed as arithmetic mean ± standard deviation (SD) of the performed replications. Statistical analyses of differences within organisms were performed using the analysis of variance (one-way ANOVA) after testing the homogeneity of the variance with Levene's test [51]. In case of heteroscedasticity, we applied the non-parametric Kruskal–Wallis analysis of variance. Depending on the resulting statistics, post-hoc comparison was eventually performed with the Bonferroni correction, always considering a significant level of 0.05. All graphs and statistical analyses were performed using STATGRAPHICS (STATGRAPHICS Centurion 2018, Statgraphics Technologies Inc., The Plains, VA, USA).

3. Results

THg, expressed on a dry weight (dw) basis, in sponges collected in the area of Montecristo and Giglio islands, are reported in Figure 2A,B, respectively.

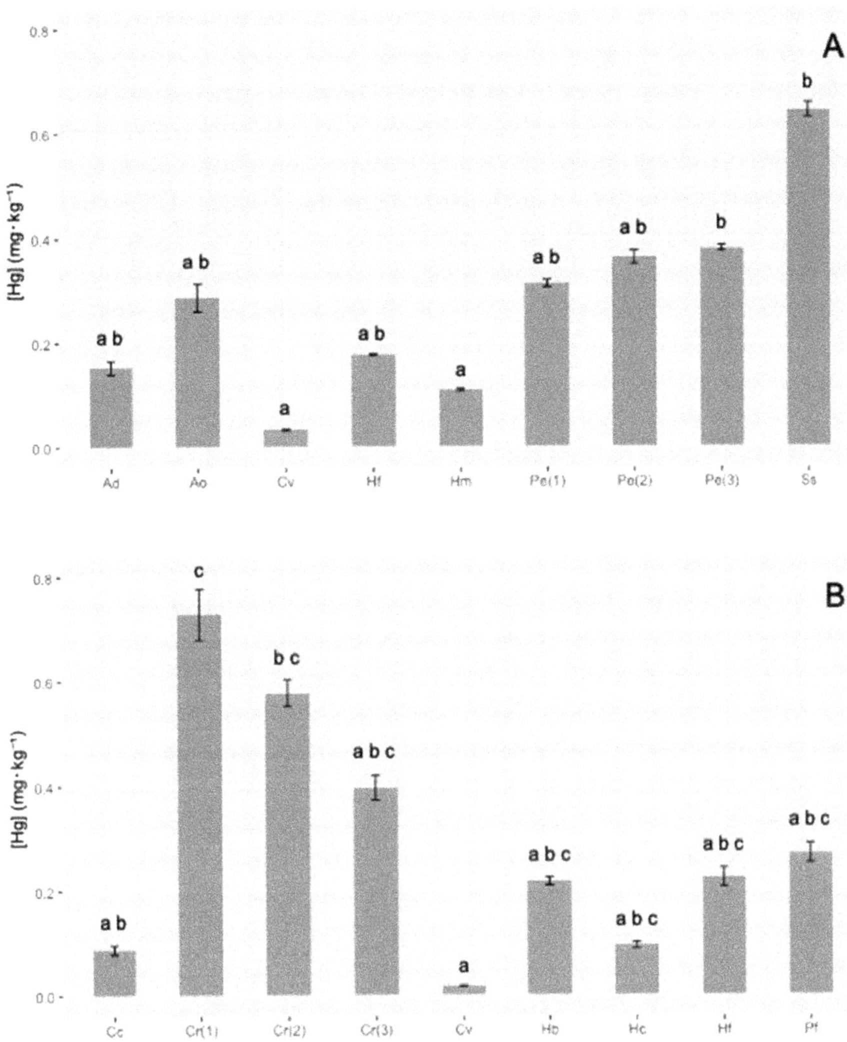

Figure 2. Mercury levels (± standard deviation (SD)) in the sponges collected in Montecristo (**A**) and Giglio (**B**) islands. In each graph, different letters (a, b, c) indicate statistically significant differences among sponges (Kruskal–Wallis test, $p < 0.05$). Ad = *Axinella damicornis;* Ao = *Agelas oroides;* Cv = *Cliona viridis;* Hf = *Haliclona (Halichoclona) fulva;* Hm = *Haliclona (Soestella) mucosa;* Pe = *Penares euastrum;* Ss = *Sarcotragus spinosulus;* Cc = *Crambe crambe;* Cr = *Chondrosia reniformis;* Hb = *Hymedesmia (Hymedesmia) baculifera;* Hc = *Hemimycale columella;* Pf = *Petrosia (Petrosia) ficiformis.*

Total mercury showed high variability in sponges coming from the marine area of Montecristo island, ranging from 0.033 to 0.64 mg·kg^{-1}, with the minimum content in *Cliona viridis* (0.033 ± 0.001 mg·kg^{-1}) and the maximum (20-fold higher) in *Sarcotragus spinosulus* (0.64 ± 0.01 mg·kg^{-1}). Values ranging from about 0.11 to 0.29 mg·kg^{-1} were recorded in *Agelas oroides, Axinella damicornis, Haliclona (Halichoclona) fulva* and *H. (Soestella) mucosa*, whereas slightly higher values were found for the three samples of *Penares euastrum* (1), (2) and (3) (0.35 ± 0.03 mg·kg^{-1}), that showed similar concentration among them. Statistical analysis of the data showed a significant difference (Kruskal–Wallis test, $p < 0.05$) between *C. viridis* vs. *P. euastrum* (3) and *S. spinosulus*, *H. (S.) mucosa* vs. *P. euastrum* (3) and *S. spinosulus* (Figure 2A).

A high variability of THg content was also found in sponges collected in the marine area of Giglio Island: total Hg ranged from 0.017 to 0.73 mg·kg^{-1} dw. Even in this case, the minimum content was measured in *C. viridis* (0.0167 ± 0.0003 mg·kg^{-1}), which showed values from 10 to 70 times lower than other species. On the other hand, the three samples of *Chondrosia reniformis* (1), (2) and (3) showed the highest concentrations (ranging from 0.39 to 0.73 mg·kg^{-1}). In the other sponges, the THg content ranged from ≈0.1 (*Crambe crambe, Hemimycale columella*) to about 0.22 mg·kg^{-1} in *H. (H.) fulva, Hymedesmia baculifera* and *Petrosia (P.) ficiformis*. The Kruskal–Wallis test highlighted significant differences ($p < 0.05$) between *C. reniformis* (1) vs. *C. crambe*, *C. viridis* vs. *C. reniformis* (1) and *C. reniformis* (3) (Figure 2B).

Since samples of *C. viridis* and *H. (H.) fulva* were found in both islands, differences in the THg content between the two islands have been investigated, to compare the possible influence of different sites. *C. viridis* showed a higher THg in Montecristo island (Kruskal–Wallis test, $p < 0.05$), while the concentration of mercury in *H. (H.) fulva* was higher in Giglio (Kruskal–Wallis test, $p < 0.05$).

4. Discussion

Samples show the presence of mercury both in Montecristo and Giglio. Although statistical differences in the Hg concentrations have been detected in the sponge species (*Cliona viridis* and *Haliclona (H.) fulva*) collected in both islands of the Tuscan Archipelago, the lack of a trend excludes a possible influence of the site. In fact, even though since 1971 Montecristo has been an Integral Nature Reserve considered a pristine area not interested by mechanic impacts, such as anchoring and diving disturbances, fishing, etc., its water can still be affected by other different impacts, such as climate change and alien species, e.g., References [52,53], and the presence of different pollutants [20–23] due to water circulation [54].

Currently, no law limit is defined for Hg in sponges and few data are available on Hg concentration in this phylum, most of which regard species and genera different from those collected in this study, e.g., References [28,55,56]. THg values of *Sarcotragus spinosulus*, measured by us, are comparable with the concentrations found in other species belonging to the subclass Keratosa, characterized by a skeleton of spongin fibers [57] such as *Scalarispongia scalaris* (Schmidt, 1862) (cited as *Cacospongia scalaris*) [37], *Spongia (Spongia) lamella* (Schulze, 1879) (cited as *S. agaricina*) [36,37], *S. (S.) nitens* (Schmidt, 1862) (cited as *S. nitens*) [36] and *S. (S.) officinalis* (Linnaeus, 1759) (cited as *S. officinalis*) [10,36–38] (Table 1). THg values of *Chondrosia reniformis* recorded in specimens from Montecristo and Giglio are similar to the ones recorded in Reference [37] for the same species collected in different localities near Marseille (Table 1). On the other hand, results obtained for *Agelas oroides* and *Cliona viridis* in Reference [37] are higher than our THg values for both species (Table 1). Perez et al. [37] collected samples of *Cliona viridis* in polluted and non-polluted areas, and he did not find any differences in the Hg concentrations. Therefore, also the difference reported between our specimen and the one of Reference [37] could be related not only to the species-specificity but also to the individual specificity of sponges, with a high intra- and inter-specific variability [58,59].

Although it is known that some heavy metals (e.g., copper, lead and vanadium) have effects on sponges, increasing their fission frequency, inducing changes in cellular aggregation and reducing growth and filtration rates [34,35], no information is available for Hg. However, some studies showed, from laboratory experiments, that mercury can cause death, inhibition of gemmule formation and malformation in gemmoscleres in the freshwater sponge *Ephydatia fluviatilis* (Linnaeus, 1759) (0.001 to 1.000 mg·kg^{-1}) [60], and it can arrest movement of single sponge cells of *Scopalina lophyropoda*

(Schmidt, 1862) (1 and 5 μg·kg^{-1}), which tended to be rounded without pseudopodia [61], while a high concentration of MeHg (0.6 mg·kg^{-1}) induces apoptosis in tissue of *Geodia cydonium* (Linnaeus, 1767) [62].

Table 1. Selection of literature data for mercury concentrations in sponges.

Species	Location	Methodology	Hg (mg·kg^{-1}) dw	References
Agelas oroides	Montecristo Island, Tuscany (Italy)	Mean ± SD	0.29 ± 0.03	This study
	Coutiou, Marseille (France)	Mean ± SD	1.7 ± 0.2	[37]
Chondrosia reniformis	Giglio Island, Tuscany (Italy)	Mean ± SD	0.73 ± 0.05 0.58 ± 0.02 0.39 ± 0.02	This study
	Coutiou, Marseille (France)	Mean ± SD	0.8 ± 0.1	[37]
Cliona viridis	Montecristo Island, Tuscany (Italy) Giglio Island, Tuscany (Italy)	Mean ± SD	0.033 ± 0.001 0.0167 ± 0.0003	This study
	Coutiou, Marseille (France)	Mean ± SD	0.3 ± 0.4	[37]
Sarcotragus spinosulus	Montecristo Island, Tuscany (Italy)	Mean ± SD	0.64 ± 0.01	This study
Cacospongia scalaris	Coutiou, Marseille (France)	Mean ± SD	0.8 ± 0.0	[37]
Spongia (Spongia) lamella	Marseille and Saint Tropez (France)	Max	0.52	[36]
	Coutiou, Marseille (France)	Mean ± SD	0.3 ± 0.0	[37]
Spongia (Spongia) nitens	Marseille and Saint Tropez (France)	Max	0.52	[36]
Spongia (Spongia) officinalis	Marseille and Saint Tropez (France)	Max	0.52	[36]
	Coutiou, Marseille (France)	Mean ± SD	0.9 ± 0.1	[37]
	Coutiou, Marseille (France) Maire, Marseille (France) Plane, Marseille (France) Jarre, Marseille (France) Riou, Marseille (France) Veyron, Marseille (France) Lavera, Marseille (France) Niolon, Marseille (France) Port-Cros, Saint Tropez (France)	Mean ± SD	0.8 ± 0.1 0.5 ± 0.1 0.9 ± 0.3 0.6 ± 0.1 0.5 ± 0.1 0.6 ± 0.1 0.5 ± 0.1 1.1 ± 0.5 0.6 ± 0.4	[38]
	Coutiou, Marseille (France) Port-Cros, Saint Tropez (France)	Mean ± SD	0.4 ± 0.1 0.3 ± 0.1	[10]

The highest THg values were found in *Chondrosia reniformis* and *Sarcotragus spinosulus*, collected in Giglio and Montecristo islands, respectively. These two species share the absence of a mineral skeleton, showing *C. reniformis*, a dispersal fibrillary collagen, and *S. spinosulus*, an organic skeleton made of spongin [57]. Three papers on Antarctic and Mediterranean species [59,63,64] showed that spicules accumulate only a small part of heavy metals compared to the sponge tissues; while in *Spongia* spp., the skeletal spongin fibers can trap and consequently concentrate metals, such as Fe, Pb, Cr, Zn and V [37,65]. These studies suggest that a collagenous skeleton instead of a mineral one can bioaccumulate a higher concentration of heavy metals. On the other hand, the mercury concentrations recorded for *C. viridis*, which is the only boring species and symbiotic with zooxanthellae [66], were the lowest among all the species collected. It is possible that the association with zooxanthellae influences the accumulation capacity of the sponge, as also pointed out in Reference [37]. It has been demonstrated that *C. viridis* can vary its filtration rates depending on the photosynthetic activities of its zooxanthellae [67], and this could partially explain why our samples present low concentrations of THg in their tissues.

The aforementioned intra- and inter-specific variability suggested by our observations and other authors [37,38] (Table 1) appear to not always be in agreement with the distance from the main source of pollutants. This can be a result of physiological and skeletal differences among sponge species. For example, the diversity of morphology of the aquiferous system could influence the size range of particles that can be filtered as well as the filtration rates [37]. Moreover, sponges are long-living organisms, with variable rate of growth, and it is very difficult to age each individual [36], therefore it is possible that low concentrations of heavy metals found in one organism could be related to its young age. In different sponge species, *Clathria (Clathria) prolifera* (Ellis and Solander, 1786) (cited as *Microciona prolifera*) [68], *Suberites domuncula* (Olivi, 1792) [69] and *Spongia (Spongia) officinalis* [10], authors found the presence of metallothioneins (MT) and metallothionein-like proteins (MTLPs), which are known to be used in the sequestration of metals in some invertebrates [70,71], and a positive correlation between MTLP concentrations and some heavy metals' (Cu, Zn and Hg) concentrations has been established [10].

Moreover, sponges can host complex communities of microorganisms, including bacteria, cyanobacteria and fungi, in their tissues [72]. The relationships between sponges and their microbiota can be defined as a symbiosis, in which microorganism communities can take part in the metabolic cycles and exchange different metabolites with their host [73]. It has been demonstrated that many bacteria, which can contribute up to 40% of the sponge biomass, are resistant to different antibiotics and pollutants, including Persistent Organic Pollutants (POPs) and heavy metals, such as Cu, Pb, Co, Cd, Zn, Ni, Hg and their organic compounds, e.g., References [72,74–76]. Moreover, some studies asserted that sponges and the associated bacteria can potentially be applied in the bioremediation of aquatic environments contaminated by mercury [74,75]. Therefore, the microbiota also seems to play an important role in the bioaccumulation of heavy metals in sponge tissues, suggesting the necessity of deepening the investigation towards this aspect.

5. Conclusions

In conclusion, our results are the first data about total Hg concentration in sponges from the TANP and suggest that these metazoa could accumulate toxic elements in coastal waters. On the other hand, the high variability of concentrations in THg measured in all specimens in both islands did not allow us to identify sponges as bioindicators of toxic elements. However, further studies with a higher number of sponge samples are needed to understand in which compartment (skeleton, tissue or microbiota) they accumulate the heaviest metals, analyzing the main effects on the histology and physiology of this group. Anyway, these results improve the knowledge on the ecosystem of the TANP, pointing out the species/individual-specificity of metal concentrations for Porifera and the key role of the organic skeleton, tissue and microbiota. Moreover, these data provide additional environmental information on the Tuscany Archipelago to address the main input of international guidelines on the Marine Strategy to protect and clean up coasts, seas and oceans.

Author Contributions: Conceptualization, A.A. and S.P.; Formal analysis, A.A.; Funding acquisition, S.P.; Investigation, C.R. and F.G.; Project administration, S.P.; Resources, C.R. and D.P.; Supervision, B.C.; Writing—original draft, C.R. and A.A.; Writing—review and editing, D.P., B.C., C.T., S.I. and S.P. All authors have read and agreed to the published version of the manuscript.

Funding: This study has been conducted with the financial support of PADI FOUNDATION (grant number #32694) and Università Politecnica delle Marche.

Acknowledgments: The authors are grateful to Isla Negra diving, for the logistic assistance during samplings, and to the Tuscan Archipelago National Park, for the authorization of sampling activities.

Conflicts of Interest: The authors declare they have no conflict of interest.

References

1. Ramalhosa, E.C.D.; Pereira, E.; Duarte, A. Mercury behaviour in the water column of an impacted coastal lagoon: Ria de Aveiro (Portugal) as a case study. In *Focus on Water Resource Research*; Heikkinen, E., Ed.; Nova Science Publishers: Hauppauge, NY, USA, 2008; pp. 41–86. ISBN 978-1-60456-093-0.
2. Gomes, R.M. Mercury Effects in Natural Populations of Sea Anemone *Actinia equina*. Ph.D. Thesis, Department of Biology, University of Aveiro, Aveiro, Portugal, 2011.
3. Morel, F.M.; Kraepiel, A.M.; Amyot, M. The chemical cycle and bioaccumulation of mercury. *Annu. Rev. Ecol. Syst.* **1998**, *29*, 543–566. [CrossRef]
4. Ullrich, S.M.; Tanton, T.W.; Abdrashitova, S.A. Mercury in the Aquatic Environment: A Review of Factors Affecting Methylation. *Crit. Rev. Environ. Sci. Technol.* **2001**, *31*, 241–293. [CrossRef]

5. Rajar, R.; ČETINA, M.; Horvat, M.; Žagar, D. Mass balance of mercury in the Mediterranean Sea. *Mar. Chem.* **2007**, *107*, 89–102. [CrossRef]
6. Droghini, E.; Annibaldi, A.; Prezioso, E.; Tramontana, M.; Frapiccini, E.; De Marco, R.; Illuminati, S.; Truzzi, C.; Spagnoli, F. Mercury Content in Central and Southern Adriatic Sea Sediments in Relation to Seafloor Geochemistry and Sedimentology. *Molecules* **2019**, *24*, 4467. [CrossRef] [PubMed]
7. Cardellicchio, N.; DeCataldo, A.; Di Leo, A.; Misino, A. Accumulation and tissue distribution of mercury and selenium in striped dolphins (*Stenella coeruleoalba*) from the Mediterranean Sea (southern Italy). *Environ. Pollut.* **2002**, *116*, 265–271. [CrossRef]
8. Storelli, M.M.; Storelli, A.; D'Addabbo, R.; Marano, C.; Bruno, R.; Marcotrigiano, G.O. Trace elements in loggerhead turtles (*Caretta caretta*) from the eastern Mediterranean Sea: Overview and evaluation. *Environ. Pollut.* **2005**, *135*, 163–170. [CrossRef] [PubMed]
9. DePew, D.; Basu, N.; Burgess, N.; Campbell, L.M.; Devlin, E.W.; Drevnick, P.; Hammerschmidt, C.R.; Murphy, C.A.; Sandheinrich, M.B.; Wiener, J.G. Toxicity of dietary methylmercury to fish: Derivation of ecologically meaningful threshold concentrations. *Environ. Toxicol. Chem.* **2012**, *31*, 1536–1547. [CrossRef]
10. Berthet, B.; Mouneyrac, C.; Perez, T.; Amiard-Triquet, C. Metallothionein concentration in sponges (*Spongia officinalis*) as a biomarker of metal contamination. *Comp. Biochem. Physiol. Part C: Toxicol. Pharmacol.* **2005**, *141*, 306–313. [CrossRef]
11. Carrasco, L.; Díez, S.; Soto, D.X.; Catalan, J.; Bayona, J. Assessment of mercury and methylmercury pollution with zebra mussel (*Dreissena polymorpha*) in the Ebro River (NE Spain) impacted by industrial hazardous dumps. *Sci. Total. Environ.* **2008**, *407*, 178–184. [CrossRef]
12. Roveta, C.; Annibaldi, A.; Vagnoni, F.; Mantas, T.P.; Domenichelli, F.; Gridelli, S.; Puce, S. Short-term effects of environmental factors on the asexual reproduction of *Aurelia* sp. polyps. *Chem. Ecol.* **2020**, *36*, 486–492. [CrossRef]
13. Thibaud, Y. Teneur en mercure dans quelques poissons de consommation courante. *Sci. Peche, Bull. Inst. Peches Marit.* **1971**, *209*, 253–3131.
14. Cumont, G.; Gilles, G.; Bernhard, F.; Briand, M.B.; Stephan, G.; Ramonda, G.; Guillen, G. Bilan de la contamination des poissons de mer par la rnercure al'occasion d'un controle port ant sur 3 annees. *Ann. Hyg. Lang. Fr. Med. Nutr.* **1975**, *11*, 17–25.
15. Cossa, D.; Martin, J.-M. Mercury in the Rhône delta and adjacent marine areas. *Mar. Chem.* **1991**, *36*, 291–302. [CrossRef]
16. Ferrara, R.; Mazzolai, B.; Edner, H.; Svanberg, S.; Wallinder, E. Atmospheric mercury sources in the Mt. Amiata area, Italy. *Sci. Total. Environ.* **1998**, *213*, 13–23. [CrossRef]
17. Ferrara, R.; Maserti, B.; Breder, R. Mercury in abiotic and biotic compartments of an area affected by a geochemical anomaly (Mt. Amiata, Italy). *Water, Air, Soil Pollut.* **1991**, *56*, 219–233. [CrossRef]
18. Cossa, D.; Coquery, M. The Mediterranean Mercury Anomaly, a Geochemical or a Biological Issue. In *The Mediterranean Sea. Handbook of Environmental Chemistry Vol. 5K*; Saliot, A., Ed.; Springer: Berlin/Heidelberg, Germany, 2005; pp. 177–208.
19. Ferrara, R.; Maserti, B.E.; Morelli, M.; Morelli, L.; Nannicini, L.; Scarano, G.; Seritti, A.; Torti, M. Metalli pesanti nelle acque dell'Arcipelago Toscanoe nella Posidonia oceanica dell'Arcipelago Toscano. In *Progetto mare Ricerca sullo Stato Biologico, Chimico e Fisico dell'Alto Tirreno Toscano*; University of Florence: Florence, Italy, 1993; p. 18.
20. Verniani, D. *Monitoraggio Acque Marino Costiere Della Toscana Anno 2012. Proposta Di Classificazione Triennio 2010–2012 (D.Lgs. 152/06)*; Technical report; Agenzia Regionale per la Protezione Ambientale della Toscana (ARPAT): Florence, Italy, 2013; p. 73.
21. Verniani, D. *Monitoraggio Acque Marino Costiere Della Toscana Anno 2012. Proposta Di Classificazione Triennio Anno 2013 (D.Lgs. 152/06)*; Technical report; Agenzia Regionale per la Protezione Ambientale della Toscana (ARPAT): Florence, Italy, 2014; p. 53.
22. Ceccanti, M.; Verniani, D. *Monitoraggio Acque Marino Costiere Della Toscana. Attività Di Monitoraggio 2014. Classificazione Provvisoria II Anno Del Triennio 2013–2015*; Technical report; Agenzia Regionale per la Protezione Ambientale della Toscana (ARPAT): Florence, Italy, 2015; p. 42.
23. Verniani, D.; Mancusi, C. *Monitoraggio Acque Marino Costiere Della Toscana. Attività Di Monitoraggio 2016 e Proposta Di Classificazione*; Technical report; Agenzia Regionale per la Protezione Ambientale della Toscana (ARPAT): Florence, Italy, 2017.

24. Megaletti, E.; Tunesi, L. *Il Report MSFD 2018: Aggiornamento Della Valutazione Ambientale (Art. 8 Del D.Lgs. 190/2010)*; Technical report; Istituto Superiore per laProtezione e la Ricerca Ambientale (ISPRA): Rome, Italy, 2019; p. 44.
25. Denton, G.R.W.; Breck, W.G. Mercury in tropical marine organisms from north Queensland. *Mar. Pollut. Bull.* **1981**, *12*, 116–121. [CrossRef]
26. Schuhmacher, M.; Batiste, J.; Bosque, M.A.; Domingo, J.L.; Corbella, J. Mercury concentrations in marine species from the coastal area of Tarragona Province, Spain. Dietary intake of mercury through fish and seafood consumption. *Sci. Total. Environ.* **1994**, *156*, 269–273. [CrossRef]
27. Perugini, M.; Visciano, P.; Manera, M.; Zaccaroni, A.; Olivieri, V.; Amorena, M. Levels of Total Mercury in Marine Organisms from Adriatic Sea, Italy. *Bull. Environ. Contam. Toxicol.* **2009**, *83*, 244–248. [CrossRef]
28. Batista, D.; Muricy, G.; Rocha, R.C.; Miekeley, N.F. Marine sponges with contrasting life histories can be complementary biomonitors of heavy metal pollution in coastal ecosystems. *Environ. Sci. Pollut. Res.* **2014**, *21*, 5785–5794. [CrossRef]
29. Butler, P.A.; Andren, L.; Bonde, G.J.; Jernelov, A.; Reisch, D.J. Monitoring organisms. In *Food and Agricultural Organization Technical Conference on Marine Pollution and its Effects on Living Resources and Fishing, Rome, 1970. Supplement 1: Methods of Detection, Measurement and Monitoring of Pollutants in the Marine Environment*; Ruivo, M., Ed.; Fishing News Ltd.: London, UK, 1971; pp. 101–112.
30. Haug, A.; Melsom, S.; Omang, S. Estimation of heavy metal pollution in two Norwegian fjord areas by analysis of the brown alga Ascophyllum nodosum. *Environ. Pollut.* **1974**, *7*, 179–192. [CrossRef]
31. Patel, B.; Balani, M.; Patel, S. Sponge 'sentinel' of heavy metals. *Sci. Total. Environ.* **1985**, *41*, 143–152. [CrossRef]
32. Hansen, I.V.; Weeks, J.M.; Depledge, M.H. Accumulation of copper, zinc, cadmium and chromium by the marine sponge *Halichondria panicea* Pallas and the implications for biomonitoring. *Mar. Pollut. Bull.* **1995**, *31*, 133–138. [CrossRef]
33. Capón, R.J.; Elsbury, K.; Butler, M.S.; Lu, C.C.; Hooper, J.N.A.; Rostas, J.A.P.; O'Brien, K.J.; Mudge, L.M.; Sim, A.T.R. Extraordinary levels of cadmium and zinc in a marine sponge, *Tedania charcoti* Topsent: Inorganic chemical defense agents. *Experientia* **1993**, *49*, 263–264. [CrossRef]
34. Cebrian, E.; Martí, R.; Uriz, J.; Turon, X. Sublethal effects of contamination on the Mediterranean sponge *Crambe crambe*: Metal accumulation and biological responses. *Mar. Pollut. Bull.* **2003**, *46*, 1273–1284. [CrossRef]
35. Cebrian, E.; Agell, G.; Marti, R.; Uriz, M.J. Response of the Mediterranean sponge *Chondrosia reniformis* Nardo to copper pollution. *Environ. Pollut.* **2006**, *141*, 452–458. [CrossRef] [PubMed]
36. Verdenal, B.; Diana, C.; Arnoux, A.; Vacelet, J. Pollutant levels in Mediterranean commercial sponges. In *New Perspectives in Sponge Biology, Proceedings of the Third International Conference on Biology of Sponges, Woods Hole, Massachusetts, USA, 17–23 November 1985*; Rützler, K., Ed.; Smithsonian Inst. Press: Washington, DC, USA, 1990; pp. 516–524.
37. Perez, T.; Vacelet, J.; Rebouillon, P. In situ comparative study of several Mediterranean sponges as potential biomonitors of heavy metals. In *Sponge Science in the New Millennium*; Pansini, M., Pronzato, R., Bavestrello, G., Manconi, R., Eds.; Officine Grafiche Canessa Rapallo: Genova, Italy, 2004; pp. 517–525.
38. Perez, T.; Longet, D.; Schembri, T.; Rebouillon, P.; Vacelet, J. Effects of 12 years' operation of a sewage treatment plant on trace metal occurrence within a Mediterranean commercial sponge (*Spongia officinalis*, Demospongiae). *Mar. Pollut. Bull.* **2005**, *50*, 301–309. [CrossRef] [PubMed]
39. Balsamo, M.; Fregni, E.; Tongiorgi, P. Marine and freshwater Gastrotricha from the Island of Montecristo (Tuscan Archipelago, Italy), with the description of new species. *Boll. di Zool.* **1994**, *61*, 217–227. [CrossRef]
40. Innocenti, F.; Westerman, D.S.; Rocchi, S.; Tonarini, S. The Montecristo monzogranite (Northern Tyrrhenian Sea, Italy): A collisional pluton in an extensional setting. *Geol. J.* **1997**, *32*, 131–151. [CrossRef]
41. Angeletti, L.; Ceregato, A.; Ghirelli, M.; Gualandi, B.; Lipparini, E.; Malatesta, D.; Sperotti, A.; Taviani, M. ROV-SCUBA integrated survey of the Montecristo Island Nature Reserve (Tuscan Archipelago National Park, Mediterranean Sea). *Underw. Technol.* **2010**, *29*, 151–154. [CrossRef]
42. IslePark. Available online: http://www.islepark.it (accessed on 14 February 2020).
43. Bo, M.; Canese, S.; Bavestrello, G. Discovering Mediterranean black coral forests: *Parantipathes larix* (Anthozoa: Hexacorallia) in the Tuscan Archipelago, Italy. *Ital. J. Zool.* **2013**, *81*, 112–125. [CrossRef]
44. Cecchi, E.; Gennaro, P.; Piazzi, L.; Ricevuto, E.; Serena, F. Development of a new biotic index for ecological status assessment of Italian coastal waters based on coralligenous macroalgal assemblages. *Eur. J. Phycol.* **2014**, *49*, 298–312. [CrossRef]

45. Turicchia, E.; Abbiati, M.; Sweet, M.; Ponti, M. Mass mortality hits gorgonian forests at Montecristo Island. *Dis. Aquat. Org.* **2018**, *131*, 79–85. [CrossRef]
46. Bavestrello, G.; Bianchi, C.N.; Calcinai, B.; Cattaneo-Vietti, R.; Cerrano, C.; Morri, C.; Puce, S.; Sára, M. Bio-mineralogy as a structuring factor for marine epibenthic communities. *Mar. Ecol. Prog. Ser.* **2000**, *193*, 241–249. [CrossRef]
47. Casoli, E.; Modica, M.V.; Belluscio, A.; Capello, M.; Oliverio, M.; Ardizzone, G.; Ventura, D. A massive ingression of the alien species *Mytilus edulis* L. (Bivalvia: Mollusca) into the Mediterranean Sea following the Costa Concordia cruise-ship disaster. *Mediterr. Mar. Sci.* **2016**, *17*, 404. [CrossRef]
48. Illuminati, S.; Annibaldi, A.; Truzzi, C.; Scarponi, G. Recent temporal variations of trace metal content in an Italian white wine. *Food Chem.* **2014**, *159*, 493–497. [CrossRef]
49. Illuminati, S.; Annibaldi, A.; Truzzi, C.; Libani, G.; Mantini, C.; Scarponi, G. Determination of water-soluble, acid-extractable and inert fractions of Cd, Pb and Cu in Antarctic aerosol by square wave anodic stripping voltammetry after sequential extraction and microwave digestion. *J. Electroanal. Chem.* **2015**, *755*, 182–196. [CrossRef]
50. Truzzi, C.; Illuminati, S.; Girolametti, F.; Antonucci, M.; Scarponi, G.; Ruschioni, S.; Riolo, P.; Annibaldi, A. Influence of Feeding Substrates on the Presence of Toxic Metals (Cd, Pb, Ni, As, Hg) in Larvae of *Tenebrio molitor*: Risk Assessment for Human Consumption. *Int. J. Environ. Res. Public Health* **2019**, *16*, 4815. [CrossRef]
51. Wayne, W.D. Analysis of variance. In *Biostatistics*, 8th ed.; John Wiley & Sons: Hoboken, NJ, USA, 2005; pp. 303–320.
52. Piazzi, L.; Balata, D.; Cecchi, E.; Cinelli, F.; Sartoni, G. Species composition and patterns of diversity of macroalgal coralligenous assemblages in the north-western Mediterranean Sea. *J. Nat. Hist.* **2009**, *44*, 1–22. [CrossRef]
53. Stasolla, G.; Innocenti, G. New records of the invasive crabs *Callinectes sapidus* Rathbun, 1896 and *Percnon gibbesi* (H. Milne Edwards, 1853) along the Italian coasts. *BioInvasions Rec.* **2014**, *3*, 39–43. [CrossRef]
54. Iacono, R.; Napolitano, E.; Marullo, S.; Artale, V.; Vetrano, A. Seasonal Variability of the Tyrrhenian Sea Surface Geostrophic Circulation as Assessed by Altimeter Data. *J. Phys. Oceanogr.* **2013**, *43*, 1710–1732. [CrossRef]
55. Negri, A.P.; Burns, K.; Boyle, S.; Brinkman, D.; Webster, N.S. Contamination in sediments, bivalves and sponges of McMurdo Sound, Antarctica. *Environ. Pollut.* **2006**, *143*, 456–467. [CrossRef] [PubMed]
56. De Mestre, C.; Maher, W.A.; Roberts, D.; Broad, A.; Krikowa, F.; Davis, A.R. Sponges as sentinels: Patterns of spatial and intra-individual variation in trace metal concentration. *Mar. Pollut. Bull.* **2012**, *64*, 80–89. [CrossRef]
57. Pansini, M.; Manconi, R.; Pronzato, R. *Porifera I. Calcarea, Demospongiae (partim), Hexactinellida, Homoscleromorpha. Fauna d'Italia, Vol. 46;* Calderini-Il Sole 24 Ore: Bologna, Italy, 2011; p. 554. ISBN 978-88-506-5395-9.
58. Cebrian, E.; Uriz, M.J.; Turon, X. Sponges as biomonitors of heavy metals in spatial and temporal surveys in northwestern Mediterranean: Multispecies comparison. *Environ. Toxicol. Chem.* **2007**, *26*, 2430–2439. [CrossRef] [PubMed]
59. Illuminati, S.; Annibaldi, A.; Truzzi, C.; Scarponi, G. Heavy metal distribution in organic and siliceous marine sponge tissues measured by square wave anodic stripping voltammetry. *Mar. Pollut. Bull.* **2016**, *111*, 476–482. [CrossRef]
60. Mysing-Gubala, M.; Poirrier, M.A. The effects of cadmium and mercury on gemmule formation and gemmosclere morphology in *Ephydatia fluviatilis* (Porifera: Spongillidae). *Hydrobiologia* **1981**, *76*, 145–148. [CrossRef]
61. Cebrian, E.; Uriz, M.J. Contrasting Effects of Heavy Metals on Sponge Cell Behavior. *Arch. Environ. Contam. Toxicol.* **2007**, *53*, 552–558. [CrossRef] [PubMed]
62. Batel, R.; Bihari, N.; Rinkevich, B.; Dapper, J.; Schäcke, H.; Schröder, H.; Müller, W. Modulation of organotin-induced apoptosis by the water pollutant methyl mercury in a human lymphoblastoid tumor cell line and a marine sponge. *Mar. Ecol. Prog. Ser.* **1993**, *93*, 245–251. [CrossRef]
63. Truzzi, C.; Annibaldi, A.; Illuminati, S.; Bassotti, E.; Scarponi, G. Square-wave anodic-stripping voltammetric determination of Cd, Pb, and Cu in a hydrofluoric acid solution of siliceous spicules of marine sponges (from the Ligurian Sea, Italy, and the Ross Sea, Antarctica). *Anal. Bioanal. Chem.* **2008**, *392*, 247–262. [CrossRef]
64. Annibaldi, A.; Truzzi, C.; Illuminati, S.; Bassotti, E.; Finale, C.; Scarponi, G. First Systematic Voltammetric Measurements of Cd, Pb, and Cu in Hydrofluoric Acid-Dissolved Siliceous Spicules of Marine Sponges: Application to Antarctic Specimens. *Anal. Lett.* **2011**, *44*, 2792–2807. [CrossRef]

65. Vacelet, J.; Verdenal, B.; Perinet, G. The iron mineralization of *Spongia officinalis* L. (Porifera, Dictyoceratida) and its relationships with the collagen skeleton. *Biol. Cell* **1988**, *62*, 189–198. [CrossRef]
66. Rosell, D.; Uriz, M.J. Do associated zooxanthellae and the nature of the substratum affect survival, attachment and growth of *Cliona viridis* (Porifera: Hadromerida)? An experimental approach. *Mar. Biol.* **1992**, *114*, 503–507. [CrossRef]
67. Hanna, C.; Schönberg, L.; De Beer, D.; Lawton, A. oxygen microsensor studies on zooxanthellate clionaid sponges from the Costa Brava, Mediterranean Sea 1. *J. Phycol.* **2005**, *41*, 774–779. [CrossRef]
68. Philp, R.B. Cadmium content of the marine sponge Microciona prolifera, other sponges, water and sediment from the eastern Florida panhandle: Possible effects on Microciona cell aggregation and potential roles of low pH and low salinity. *Comp. Biochem. Physiol. Part C: Pharmacol. Toxicol. Endocrinol.* **1999**, *124*, 41–49. [CrossRef]
69. Schröder, H.; Shostak, K.; Gamulin, V.; Lacorn, M.; Skorokhod, A.; Kavsan, V.; Müller, W. Purification, cDNA cloning and expression of a cadmium-inducible cysteine-rich metallothionein-like protein from the marine sponge *Suberites domuncula*. *Mar. Ecol. Prog. Ser.* **2000**, *200*, 149–157. [CrossRef]
70. Geffard, A.; Amiard-Triquet, C.; Amiard-Triquet, C. Use of metallothionein in gills from oysters (*Crassostrea gigas*) as a biomarker: Seasonal and intersite fluctuations. *Biomarkers* **2002**, *7*, 123–137. [CrossRef] [PubMed]
71. Wallace, W.G.; Lee, B.; Luoma, S. Subcellular compartmentalization of Cd and Zn in two bivalves. I. Significance of metal-sensitive fractions (MSF) and biologically detoxified metal (BDM). *Mar. Ecol. Prog. Ser.* **2003**, *249*, 183–197. [CrossRef]
72. Selvin, J.; Priya, S.S.; Kiran, G.S.; Thangavelu, T.; Bai, N.S. Sponge-associated marine bacteria as indicators of heavy metal pollution. *Microbiol. Res.* **2009**, *164*, 352–363. [CrossRef]
73. Thomas, T.; Moitinho-Silva, L.; Lurgi, M.; Björk, J.R.; Easson, C.; Astudillo-Garcia, C.; Olson, J.B.; Erwin, P.M.; Lopez-Legentil, S.; Luter, H.; et al. Diversity, structure and convergent evolution of the global sponge microbiome. *Nat. Commun.* **2016**, *7*, 11870. [CrossRef]
74. Santos-Gandelman, J.F.; Cruz, K.; Crane, S.; Muricy, G.; Giambiagi-Demarval, M.; Barkay, T.; Laport, M.S. Potential Application in Mercury Bioremediation of a Marine Sponge-Isolated *Bacillus cereus* strain Pj1. *Curr. Microbiol.* **2014**, *69*, 374–380. [CrossRef]
75. Santos-Gandelman, J.F.; Giambiagi-Demarval, M.; Muricy, G.; Barkay, T.; Laport, M.S. Mercury and methylmercury detoxification potential by sponge-associated bacteria. *Antonie van Leeuwenhoek* **2014**, *106*, 585–590. [CrossRef]
76. Stabili, L.; Pizzolante, G.; Morgante, A.; Nonnis Marzano, C.; Longo, C.; Aresta, A.M.; Zambonin, C.; Corriero, G.; Alifano, P. Lindane Bioremediation Capability of Bacteria Associated with the Demosponge *Hymeniacidon perlevis*. *Mar. Drugs* **2017**, *15*, 108. [CrossRef]

applied sciences

MDPI

Review

Methylmercury and Polycyclic Aromatic Hydrocarbons in Mediterranean Seafood: A Molecular Anthropological Perspective

Andrea De Giovanni [1,2,*], Cristina Giuliani [3], Mauro Marini [2,4] and Donata Luiselli [1,2]

1 Department of Cultural Heritage, University of Bologna, Via degli Ariani 1, 48121 Ravenna, Italy; donata.luiselli@unibo.it
2 Fano Marine Center, The Inter-Institute Center for Research on Marine Biodiversity, Resources and Biotechnologies (FMC), Viale Adriatico 1/N, 61032 Fano, Italy; mauro.marini@cnr.it
3 Laboratory of Molecular Anthropology & Centre for Genome Biology, Department of Biological, Geological and Environmental Sciences, University of Bologna, Via Selmi 3, 40126 Bologna, Italy; cristina.giuliani2@unibo.it
4 Institute for Biological Resources and Marine Biotechnologies, National Research Council (IRBIM, CNR), Largo Fiera della Pesca 2, 60125 Ancona, Italy
* Correspondence: andrea.degiovanni5@unibo.it

Abstract: Eating seafood has numerous health benefits; however, it constitutes one of the main sources of exposure to several harmful environmental pollutants, both of anthropogenic and natural origin. Among these, methylmercury and polycyclic aromatic hydrocarbons give rise to concerns related to their possible effects on human biology. In the present review, we summarize the results of epidemiological investigations on the genetic component of individual susceptibility to methylmercury and polycyclic aromatic hydrocarbons exposure in humans, and on the effects that these two pollutants have on human epigenetic profiles (DNA methylation). Then, we provide evidence that Mediterranean coastal communities represent an informative case study to investigate the potential impact of methylmercury and polycyclic aromatic hydrocarbons on the human genome and epigenome, since they are characterized by a traditionally high local seafood consumption, and given the characteristics that render the Mediterranean Sea particularly polluted. Finally, we discuss the challenges of a molecular anthropological approach to this topic.

Keywords: review; DNA methylation; genetic polymorphisms; ecogenetics; anthropology; environmental pollutants; methylmercury; polycyclic aromatic hydrocarbons; seafood

Citation: De Giovanni, A.; Giuliani, C.; Marini, M.; Luiselli, D. Methylmercury and Polycyclic Aromatic Hydrocarbons in Mediterranean Seafood: A Molecular Anthropological Perspective. *Appl. Sci.* **2021**, *11*, 11179. https://doi.org/10.3390/app112311179

Academic Editor: Ibrahim M. Banat

Received: 13 October 2021
Accepted: 23 November 2021
Published: 25 November 2021

1. Introduction

Despite being usually considered a healthy food [1], seafood carries several contaminants that can negatively affect human health [2]. It is recognized that the benefits of fish intake exceed the potential risks, but here we address how contaminants levels in seafood are significantly affected by biological and ecological factors [3–8]. Moreover, seafood habitual intake is a crucial factor in determining contaminants exposure [9,10].

In the present study, we first give a glimpse into the latest findings on the genetic diversity underlying differences in human response to mercury (Hg) and polycyclic aromatic hydrocarbons (PAHs) exposure, and on the impact of these pollutants on human DNA methylation patterns. We decided to include only epidemiological investigations assessing environmental chemical exposures using biomarkers (such as hair and blood mercury, and PAHs urine metabolites). We excluded studies addressing occupational exposure because we were interested in the potential effects on human molecular variability of Hg and PAHs from seafood. As detailed before, Hg in aquatic organisms is mostly found in the form of methylmercury (MeHg), while in occupational exposure, elemental Hg vapor is the major contributor of Hg load in the human body [11,12]. Concerning PAHs, occupational settings are associated with exposure levels much higher than those resulting from diet [13]. Then, we address the ecological evidence that makes Mediterranean coastal communities a potential informative case study to explore this topic.

We decided to focus on MeHg and PAHs because they are two of the most concerning and widespread seafood contaminants, and because of their high levels in Mediterranean seafood.

2. Seafood Contaminants

2.1. MeHg

Hg is a heavy metal found naturally in the Earth's crust. From here, mercury is released into the atmosphere via natural phenomena such as volcanic activity and forest fires, and human activities, such as the burning of coal, oil and wood, and mining. In particular, artisanal and small-scale gold mining in developing countries has recently replaced coal combustion as the largest anthropogenic mercury emission source globally [14]. Once released into the environment, it starts to circulate following what is known as the global mercury cycle, which can last up to 3000 years [15]. When mercury passes into water, it is readily transformed by bacteria in its organic form, methylmercury, which can interact with biological components and eventually biomagnify along aquatic food chains [16]. Many studies have suggested that climate change will increase mercury inputs and methylmercury production and bioaccumulation in aquatic ecosystems [14]. Seafood is recognized as the main source of mercury in the general population, and MeHg accounts for the majority (70–100%) of Hg found in muscle tissue of fishes, molluscs and crustaceans [17].

MeHg is a well-established neurotoxicant, and exposure to MeHg has been associated with nervous system damage in adults and impaired neurological development in infants and children [18]. Decrements in memory, attention, language, and visual–motor skills in childhood have been associated with MeHg biomarkers at birth in populations with moderate MeHg exposure from regular seafood consumption [19]. Even low mercury levels (i.e., levels lower than 4 μg/g in hair; 20 μg/L in cord blood, or approximately 12 μg/L in adult blood) can negatively affect fetal and infant growth and cause neurologic outcomes [20]. Urinary levels of Hg are frequently used to estimate the level of exposure to Hg vapours or inorganic Hg (IHg), whereas blood, hair and toenail [11] Hg predicts MeHg exposure.

Because of the threat that mercury poses to human health, the EU set a maximum level of mercury in seafood of 1 or 0.5 mg/kg, depending on the species, after which seafood shall not be placed on the market [21], while the Joint FAO/WHO Expert Committee on Food Additives (JECFA) established a tolerable intake of 1.6 μg/kg bodyweight per week for methylmercury in order to protect the developing fetus from neurotoxic effects [22]. On the basis of multiple epidemiological studies [23,24] that observed adverse effects in children as consequences of maternal exposures, the European Food Safety Authority (EFSA) eventually decreased this limit to 1.3 μg/kg bodyweight per week [25], corresponding to a Hg level of ~11.5 mg/kg and ~46 μg/L in hair and blood, respectively. This threshold value has been adopted for all classes of consumers, even though adults may be less sensitive to the adverse effects of MeHg [26]. Furthermore, the US-EPA established an oral reference dose (RfD) for MeHg—that is, the maximum acceptable oral dose for this contaminant—of 1×10^{-4} mg/kg day^{-1} (US-EPA, 2010).

2.2. PAHs

PAHs are a class of organic compounds consisting of two or more fused benzene rings, deriving from the incomplete combustion or pyrolysis of organic materials. Natural sources of PAHs include volcanoes, forest fires and petroleum seeps, while the combustion of fossil fuels, oil and wood are among the main anthropogenic sources [27,28]. Due to their physicochemical properties, PAHs are persistent pollutants, in that they can stay in the environment for long periods [29]. They represent the largest share among the main organic contaminants present in the marine environment, due to marine traffic and possible accidents involving oil tankers [30]. Although most PAHs are metabolized a short time after uptake, thanks to their lipophilic nature, a fraction accumulates in lipid-containing tissues such as liver, eggs and muscle [8]. The most important non-occupational source of human exposure to PAHs is the consumption of contaminated food, including seafood [31], especially mollusks and crustaceans [32]. Sixteen PAHs are categorized as priority environmental pollutants, and some of them are deemed to be probable human carcinogens by the US Environmental Protection Agency (US-EPA), with benzo(a)pyrene (B(a)P) arousing more concern because of being the most carcinogenic, teratogenic and toxic compound [33]. The most used biomarkers of PAH exposure are metabolites of PAHs, particularly 1-hydroxypyrene (1-OHP), and PAH–DNA or protein adducts. 1-OHP is the principal product of pyrene metabolism [34], and its urinary excretion has been attributed mainly to the ingestion of PAHs through the diet [35]. Rather, PAH–DNA adducts, which are the products of the Phase I metabolism of PAHs, are deemed a biomarker that integrates multiple BB(a)P exposure routes (including inhalation, dermal absorption, and ingestion) and reflects a biologically effective dose [36]. PAH–DNA adduct formation is significantly influenced by individual susceptibility, which is linked to specific genetic polymorphisms [36,37]. Urine PAHs metabolites and, to a less extent, PAH–DNA adducts are also related to parent air PAH exposures, both at elevated exposures in occupational cohorts, and at low levels of air pollution [34,38].

The EU set a maximum level of B(a)P in seafood to be sold ranging from 2 μg/kg wet weight, for muscle meat of fish (other than smoked fish), to 10 μg/kg for bivalve mollusks [21]. Concerning human exposure, the US-EPA set an RfD for several PAH compounds, including anthracene (0.3 mg/kg day^{-1}), acenaphthene (0.06 mg/kg day^{-1}), fluorene (0.04 mg/kg day^{-1}), fluoranthene (0.04 mg/kg day^{-1}), pyrene (0.03 mg/kg day^{-1}), naphthalene (0.02 mg/kg day^{-1}) and B(a)P (0.0003 mg/kg day^{-1}).

Several findings point to an important role of genetic diversity in shaping individual susceptibility to Hg [39] and PAHs [40] exposure, and consequently some authors claim the urgent need to include this factor in risk assessment and decision making [41]. Moreover, Hg and PAHs, similar to several other environmental toxicants [42], can impact the human epigenome through different mechanisms, and some of the epigenetic alterations driven by these two substances were shown to be associated with adverse health effects [43,44].

Being a semi-enclosed sea, delimited by highly industrialized countries and characterized by large deposits of cinnabar (HgS), the Mediterranean Sea is at a high-risk for contamination by toxic compounds [45], and evidence exists that significant anthropogenic chemical inputs into the Mediterranean began in prehistoric times [46]. In line with this, several studies have shown higher levels of contaminants in marine organisms from the Mediterranean Sea compared to those from other geographic areas [47]. At the same time, the European countries bordering the Mediterranean are among the world's highest seafood consumers, with Spain, Italy and France accounting for more than half of the European expenditure on fish and fishery products, despite having only around a third of the EU's population (EUROSTAT, 2014). Accordingly, high Hg concentrations in the blood and hair of several Mediterranean communities [6] have been found. Given such a traditionally high consumption of contaminated seafood, in our view, it is important to gain insights into the molecular diversity underpinning potential differences in susceptibility to MeHg and PAHs exposure in these communities. Moreover, Mediterranean populations might represent an interesting case study to investigate the potential impact of Hg and PAHs on the human genome and epigenome.

Modern technologies allow us to explore human genomic and epigenomic variability in a cost- and time-efficient way, enabling us, for example, to portray molecular diversity at the populational level [48], and to detect natural selection footprints in genomic regions [49].

3. Human Genetic Diversity

Epidemiological investigations are showing the role of genetics in shaping individual susceptibility to MeHg and PAHs. Through a literature search, we identified 18 (Tables 1 and S1 for further details) and 3 (Tables 2 and S2 for further details) epidemiological studies addressing the role of genetic polymorphisms in MeHg and PAHs toxicokinetics, respectively. Below, we describe some of the main findings of the above studies. Please refer to the tables for the full list of retrieved publications.

Table 1. List of epidemiological studies investigating the influence of genetic polymorphisms on MeHg toxicokinetic. Genes in which the above polymorphisms were identified, biomarkers affected, and samples studied are shown for each study.

Study	Genes	Biomarker	Sample
[50]	GCLC; GSTP1	Erythrocyte Hg	Swedish cases of acute myocardial infarction/stroke and controls
[51]	GSTM1; GSTT1	Hair Hg	Students in Austria
[52]	GCLM; GSTP1	Erythrocyte Hg	Fish-eating Swedish individuals
[53]	GSTP1; MT4; GSTM1; GCLC; GSTT1	Blood and hair Hg	Students in Austria
[54]	GSTM1; GSTT1	Maternal and cord blood Hg	Korean mothers and their infants
[55]	GSTT1; GSS; GSTP1; SEPP1	Hair and urinary Hg	Michigan dental professionals
[56]	MT1M; MT2A; MT1A	Hair and urinary Hg	Michigan dental professionals
[57]	APOE	Cord blood Hg	Children in Taiwan
[58]	GCLM; GSTM1	Blood and hair Hg	Amazonian population in Brazil cronically exposed to MeHg from fish
[59]	TF	Umbilical cord Hg	Children from Bristol
[60]	GCLC; GCLM; GSTM1	Plasmatic Hg and MeHg; whole blood Hg	Fish-eating communities of Brazilian Amazon
[61]	ABCB1; ABCC1; ABCC2	Cord blood Hg	Pregnant women from Greece, Italy and Spain
[62]	APOE	Cord blood Hg	Children in Taiwan
[63]	ABCB1; ABCC1; ABCC2	Hair Hg	Seychellois mother–child pairs with a diet rich in fish of mixed African, European and East Asian origin
[64]	GLRX2; GSTA4; GSTM3; GSTO1; SELS; MT1M; (see Table S1 for the whole list)	Blood and urinary Hg	American dental professionals
[65]	CBS; TXNRD2; SEPHS2; CYP1A2; CBS; MTRR; (see Table S1 for the whole list)	Blood Hg	Inuit from Canada
[66]	GCLC; GCLM; GSTP1	Maternal blood and hair Hg and cord blood Hg	Seychellois mother–child pairs with a diet rich in fish of mixed African, European and East Asian origin
[67]	BDNF; GSTP1	Hair Hg	Children in Valentia

3.1. MeHg Exposure and Human Genetic Diversity

The majority of ingested MeHg passes into the bloodstream, by which route it reaches all tissues. Here, MeHg enters cells thanks to its ability to form water-soluble complexes with the amino acid cysteine. After forming a complex with reduced glutathione (GSH) (Figure 1), MeHg is excreted by the liver cells into the bile. At this point, the glutathione is hydrolyzed, leading to the release of the methylmercury–cysteine complex. The latter is mostly secreted into the intestine tract, where MeHg is demethylated by intestine microflora. The resulting inorganic Hg is then eliminated via the feces [70].

Table 2. List of epidemiological studies investigating the influence of genetic polymorphisms on PAHs toxicokinetics. Genes in which the above polymorphisms were identified, biomarkers affected, and samples studied are shown for each study.

Study	Genes	Biomarker	Sample
[68]	XRCC1	Sperm PAH–DNA adducts	Infertile adult men from Nanjing, China
[69]	MPO; NAT2; ERCC5	Blood PAH–DNA adducts	Non-smoking healthy women from eastern Golestan Province, Iran
[36]	CYP1A1; GSTT2; CYP1B	Cord blood B(a)P–DNA adducts	Mother–infant pairs from Krakow, Poland

B(a)P, Benzo(a)pyrene.

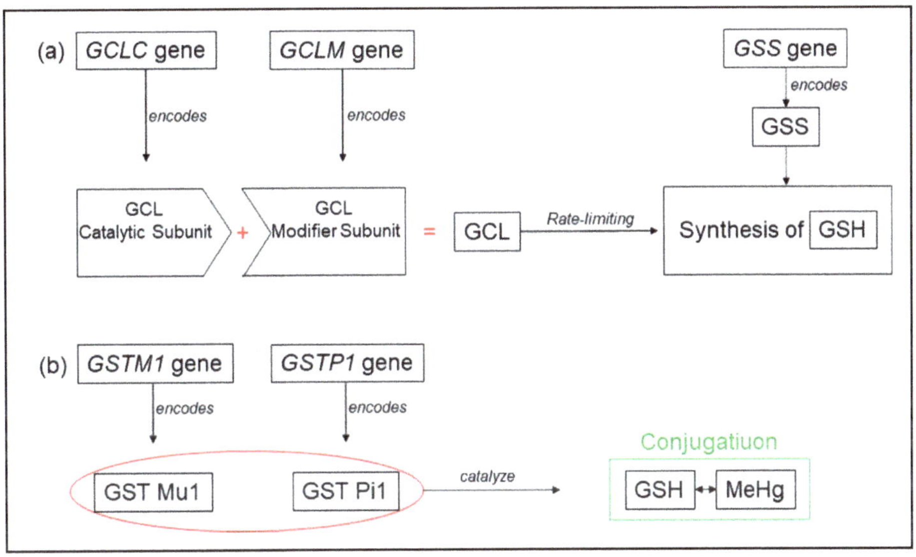

Figure 1. Schematic representation of the role played by several genes in MeHg elimination from the body. (**a**) The GCLC and GCLM genes encode the catalytic and modifier subunits of the glutamate–cysteine ligase (GCL) enzyme, respectively. The GCL is the first rate-limiting enzyme of glutathione (GSH) synthesis. The GSS gene encodes the glutathione synthetase (GSS), another enzyme involved in GSH synthesis. (**b**) The GSTM1 and GSTP1 genes encode the glutathione S transferases (GST) Mu1 and Pi1, respectively, which catalyze the conjugation of GSH to MeHg.

In recent years, epidemiological studies on populations exposed to MeHg have been showing that several genes mediating the toxicokinetics of Hg are polymorphic in humans, and may influence inter-individual variability in Hg exposure biomarker values and health outcomes (Tables 1 and S1 for further details). In line with this, as demonstrated by kinetic studies, MeHg half-life, which is a direct determinant of the Hg body burden [71], can vary widely in humans, which may be due also to a naturally occurring biological basis for the variation in MeHg toxicokinetics.

Goodrich and colleagues [55] analyzed Single-Nucleotide Polymorphisms (SNPs; genetic variants due to a base substitution or the insertion or deletion of a single base) variability in a cohort of dental professionals exposed to inorganic Hg via dental amalgams and to MeHg via seafood consumption, in order to investigate potential associations with Hg levels in hair and urine. In this study, fish consumption as estimated by a self-administered survey was the best predictor of measured hair Hg level, and two SNPs were associated with this biomarker. In particular, SEPP1 3′UTR (rs7579) T allele was associated with lower hair Hg per unit of intake from fish consumption, while the GSS 5′ (rs3761144) minor allele (i.e., the less common allele of a SNP) (G) was associated with increasing hair Hg concentration per unit of fish Hg. SEPP1 encodes a selenoprotein, which combats the oxidative stress created by Hg by binding the toxicant directly via a selenocysteine residue. The latter is an amino acid unique to selenoproteins that can bind Hg–selenium conjugates or MeHg. Interestingly, as demonstrated by previous studies, the 3′UTR T allele is linked to greater SEPP1 expression and Hg-binding capacity. The GSS gene encodes for an enzyme, glutathione synthetase (Figure 1a), that is involved in the synthesis of GSH, to which Hg is conjugated before being eliminated (Figure 1b). The association of the minor allele with increasing hair Hg concentration may be ascribable to a decreased expression of GSS and, thus, to decreased GSH synthesis, which in turn could impact the body's ability to eliminate MeHg as a GSH conjugate, with the higher body burden reflected in hair Hg levels.

The study of de Oliveira and colleagues [60] was the first to investigate the genetic predisposition to mercury accumulation in the plasma, where this pollutant is more bioavailable and therefore potentially harmful to human health. In this study, authors focused on riverside communities of the Brazilian Amazon, for which the only source of Hg exposure was the intake of contaminated fish. They genotyped two glutathione-related genes, GSTM1 and GCLC. The first encodes a glutathione S-transferase, an enzyme that catalyzes the conjugation of GSH to MeHg (Figure 1b), while the second encodes the catalytic subunit of the glutamate-cysteine ligase (GCL), the first rate-limiting enzyme of glutathione synthesis (Figure 1a). What the study found is that null homozygotes for GSTM1, that is, individuals that possess two copies of a non-functional allele for this gene, showed higher plasmatic MeHg levels (MeHgP) compared to subjects with functional GSTM1, which may be related to their lower MeHg-conjugating activity, lower MeHg excretion, and a higher MeHg retention. Moreover, individuals carrying at least one T allele for GCLC (rs17883901) also had significantly higher MeHgP.

As recent findings suggest, apart from being associated with hair mercury level, the SNPs in glutathione-related genes can influence the impact of methylmercury exposure on early child neurodevelopment. Wahlberg and colleagues [66] analyzed GSH-related gene variability in mothers with a diet rich in fish coming from the population of Seychellois. Genotypes of these mothers were analyzed in association with maternal hair and blood Hg, cord blood Hg, and children's mental and motor development, as expressed by the Mental Developmental Index (MDI) and the Psychomotor Developmental Index (PDI), respectively. The authors genotyped SNPs within three genes: GCLC, whose function have been described above; GCLM, encoding the modifier subunit of the GCL (Figure 1a), and GSTP1, which encodes a glutathione S-transferase (Figure 1b). What they found is that individuals with GCLC rs761142 TT genotype showed higher mean maternal hair Hg than AG and GG. Moreover, individuals carrying the combination of GCLC rs761142-TT and GCLM rs41303970-CC genotypes showed higher hair Hg than G plus T carriers. Finally, increasing Hg in maternal and cord blood was associated with lower PDI among GCLC rs761142 TT carriers, while increasing Hg in hair was associated with lower MDI among GSTP1 rs1695 GG carriers.

Another recent study carried out on children from Valentia [67] showed that hair Hg levels were associated with worse neurobehavioral development, and that several SNPs located in the GSTP1 (rs1695) and BDNF (rs1519480, rs7934165, rs7103411) genes modified the association between Hg levels in children's hair samples and two indexes of neurobehavioral function. The brain-derived neurotrophic factor (BDNF), in particular, is a protein that promotes neural survival in adult brains, and is poorly expressed in several diseases, such as Alzheimer's and Parkinson's.

Two recent reviews [39,72] focusing on this topic collectively listed thirty-two genes whose variation is related to Hg body burden and susceptibility to Hg toxicity, and, in particular, twelve of these genes are related to hair Hg level and/or to MeHg exposure outcomes.

3.2. PAHs Exposure and Human Genetic Diversity

PAHs metabolism is a complex process consisting of two major phases (Figure 2): In the first phase, following ingestion or inhalation, the xenobiotic compound is epoxidated by enzymes belonging to the cytochromes P450 family, with the formation of diols and dihydrodiols [31]. Dihydrodiols can thus bind to the DNA, to give rise to DNA adducts, starting the mutagenic processes and eventually leading to cancer [73]. Then, in the second phase, the intermediate diols conjugate with glutathione, thanks to glutathione s-transferase enzymes. This leads to the formation of polar compounds, which can be easily excreted by renal or biliary routes [74]. The liver is the major site of the metabolism of PAHs. However, in the case of ingestion, gut micro flora and intestinal cytochrome P450 enzymes can also contribute to the process [31].

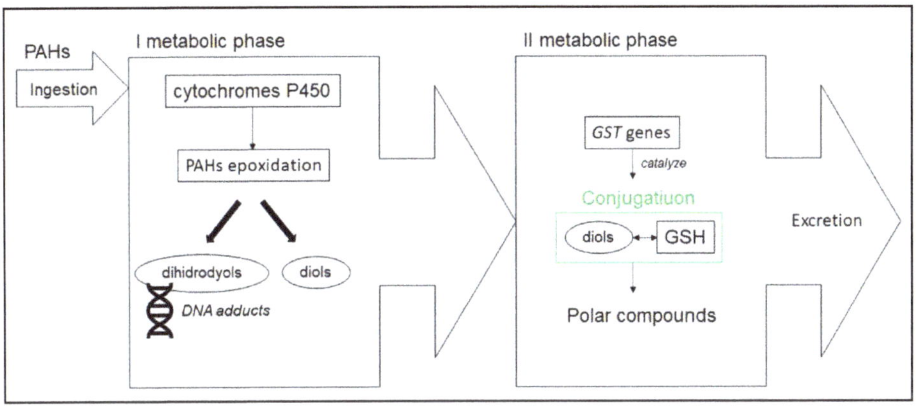

Figure 2. Schematic representation of the metabolic phases following PAHs ingestion. In the first phase, following ingestion or inhalation, PAHs are epoxidated by enzymes belonging to the cytochromes P450 family, with the formation of diols and dihydrodiols. Dihydrodiols can thus bind to the DNA, to give rise to DNA adducts. Then, in the second phase, the intermediate diols conjugate with glutathione, thanks to glutathione s-transferase enzymes. This leads to the formation of polar compounds, which can be easily excreted.

Epidemiological studies showed that polymorphisms at several genes influence levels of biomarkers of exposure to PAHs in several human populations (Tables 2 and S2 for further details).

As suggested by much evidence, PAH–DNA adducts may be a potential source of heritable prezygotic DNA damage in spermatozoa. Ji and colleagues [68] detected PAH–DNA adducts in ejaculated sperm of infertile adults environmentally exposed to low levels of PAHs, showing that the consumption of PAH-rich meals at least three times a week contributed significantly to an increase in DNA adduct formation. Moreover, the authors demonstrated an association between specific XRCC1 polymorphisms and an increase in sperm adduct levels. XRCC1 encodes a protein that is essential to providing an efficient repair of the DNA, and thus polymorphisms at this gene may be useful to identify individuals susceptible to DNA damage resulting from PAHs exposure.

Myeloperoxidase (MPO), an enzyme central to the microbicidal activity of neutrophils, and *N*-acetyltransferase 2 (NAT2), which functions to both activate and deactivate drugs and carcinogens, are involved in Phase I and Phase II of PAH metabolism, respectively, while ERCC5 is a single-strand-specific DNA endonuclease that participates in DNA excision repair. The SNPs in these genes have been shown to affect the PAH-driven formation of the DNA adduct. In a study [69] of more than one hundred healthy female non-smokers from Golestan Province, a region in north-eastern Iran, characterized by very high levels of exposure to PAH probably due to diet and methods of food preparation, the DNA adduct level in blood was significantly lower in homozygotes for NAT2 slow alleles, which are responsible for a less efficient detoxification of carcinogen-reactive metabolites, and the ERCC5 (rs1047768) non-risk-allele genotype. In contrast, DNA adduct level was higher in the MPO (rs2333227) homozygote risk-allele genotype.

In a cohort of non-smoking Polish mothers and newborns, Iyer and colleagues [36] observed a significant interaction between maternal exposure to airborne PAHs, measured by personal air monitoring, and SNPs in selected B(a)P metabolism genes on cord blood B(a)P–DNA adducts. These genes included: maternal CYP1A1 and GSTT2, and newborn CYP1A1 and CYP1B1. CYP1A1 and CYP1B1 are involved in metabolizing the parent B(a)P compound to the reactive B(a)P 7,8-diol-9,10-epoxide (BPDE) metabolite, which is involved in the formation of B(a)P–DNA adducts. In contrast, GSTT2 is involved in shifting B(a)P metabolism so as to prevent the formation of the reactive BPDE. In particular, the authors concluded that the T allele at the GSTT2 SNP position in mothers is protective with regard to cord B(a)P–DNA adduct formation, while the maternal and newborn G allele at the CYP1A1 SNP position, and the newborn G allele at the CYP1B1 SNP position, are not.

4. The Epigenetic Impact of Seafood Contaminants

Several studies have demonstrated the role of the environment in shaping human molecular variability at the epigenetic level in different populations [48,75–77]. Epigenetic changes are defined as any stable changes in the chromatin structure that are heritable from cell to cell, and can result in alteration of gene expression without altering DNA sequences [75]. The epigenome functions as an interface between the inherited genome and the dynamism imposed by the environment [78], and, as such, can be affected by the latter.

DNA methylation is among the most frequently studied epigenetic modifications. It consists in the covalent addition of a methyl group from a methyl group donor, the coenzyme S-adenosylmethionine (SAM), to the fifth carbon atom of a cytosine ring, and it is catalyzed by the DNA methyltransferase (DNMT) enzyme family. In mammals and insects, cytosine methylation is found almost exclusively in the context of CpG dinucleotides [75].

Changes in DNA methylation status influence genes' accessibility, thus altering gene expression, and aberrant DNA methylation has been discovered in a wide range of pathophysiological conditions [79].

Environmental chemicals, such as Hg and PAHs, can interfere with the one-carbon and citric acid metabolism pathways, resulting in anomalous DNA methylation status all over the genome [42]. Hg and PAHs can alter DNA methylation profiles in specific genes [80].

Three reviews address this topic [42,80,81]. Briefly, Ruiz-Hernandez and colleagues retrieve and discuss two and three epidemiologic studies investigating the association between DNA methylation and Hg [82,83] and PAHs [84–86], respectively, in adults. Considering all the strengths and weaknesses of the various studies, e.g., the lack of adjustments for potential confounding, such as sex, age, smoking status and tissue cell heterogeneity, the authors' conclusion is that the evidence they accrued supports the importance of environmental exposures in modulating the epigenome, but is insufficient to support causality because of the heterogeneity among epidemiologic studies in addressing the residual confounding of the associations, differences in DNA methylation assessment methods, and random error. The review of Culbreth and Aschner concludes that, despite some inconsistencies across different studies, dependent on the tissue or species examined, MeHg undoubtedly induces epigenetic modifications, and these modifications can potentially mediate its toxicity. In particular, regarding DNA methylation changes, controlled exposure studies on human and animal in vitro and animal in vivo models reveal that MeHg can lead to hypomethylation of the DNA in brain-derived tissue, but not in the liver, while selected individual genes show exposure-driven DNA hypermethylation.

Through a literature search, we identified seven (Tables 3 and S3 for further details) and eight (Tables 4 and S4 for further details) epidemiological studies assessing the impact of MeHg and PAHs on DNA methylation. Below, we describe some of the main findings of the above studies. Please refer to the tables for the full list of retrieved publications.

Table 3. List of epidemiological studies investigating the impact of MeHg exposure on DNA methylation. Genes in which the differentially methylated CpG dinucleotides were identified, biomarkers measured, tissues from which DNA was extracted, and samples studied are shown for each study.

Study	Genes	Biomarker	Tissue (DNA)	Sample
[82]	GSTM1	Whole blood Hg	Whole blood	Women from San Francisco
[83]	SEPP1	Hair Hg	Buccal mucosa	Michigan dental professionals
[87]	TCEANC2; ANGTP2; PRPF18; FOXD2	Cord whole blood Hg and MeHg	Cord blood	Newborns from Baltimore, USA
[88]	PARM1; PFKFB3; LGMN; CCDC68; LRBA; FBXO31; (see Table S2 for the whole list)	Maternal toenail Hg	Cord blood	Mother–infant pairs from USA
[89]	EMID2	Infant toenail Hg	Placenta	Rhode Island infants
[90]	PON1	Maternal red blood cell Hg	Cord blood and children buffy coat	Mother–children pairs from Massachusetts, USA
[91]	GRIN2B; NR3C1	Maternal hair Hg	Children saliva	Children from Europe and US populations

Table 4. List of epidemiological studies investigating the impact of PAHs exposure on DNA methylation. Genes in which the differentially methylated CpG dinucleotides were identified, biomarkers measured, tissues from which DNA was extracted, and samples studied are shown for each study.

Study	Genes	Biomarker	Tissue (DNA)	Sample
[92]	ACSL3	PAM	Umbilical cord white blood cell	Nonsmoking Dominican and African American mother-infants pairs
[93]	Global DNA methylation	Pyr and B(a)P in PAM	Umbilical cord blood leukocytes	Nonsmoking women from NYC
[94]	IRS2	Nap, Ace, Fl, Phe and Ant in VAT	VAT	Nonsmoking women from Korea with myoma
[95]	LINE1	B(a)P-DNA adducts	Buffy coat	Nonsmoking pregnant women from Tongliang County, China
[96]	239 quality-controlled autosome CpGs	Urinary ΣOH-PAHs, 9-OH-Phe and 1-OH-Pyr	Whole blood	Nonsmoking healthy Chinese individuals
[97]	PAX3	ΣH_PAHs in maternal serum	Fetal neural tissue	Mother–fetus pairs
[44]	ZIC4	ΣH_PAHs in fetal liver tissue	Fetal neural tissue	NTD fetuses
[98]	PLEC1	Urinary ΣOH-PAHs		Nonsmoking healthy Chinese individuals

1-OH-Pyr, 1-hydroxypyrene urinary metabolite; 2-OH-Nap, 9-hydroxynaphthalene urinary metabolite; 9-OH-Phe, 9-hydroxyphenanthrene urinary metabolite; Ace, acenaphthene; Ant, anthracene; B(a)P, benzo(a)pyrene; Fl, fluorine; Nap, naphthalene; NTD, neural tube defects; PAM, personal air monitor; Phe, phenanthrene; PMA, phenylmercuric acetate; Pyr, pyrene; VAT, visceral adipose tissue; ΣH_PAHs, sum of high-molecular weight PAHs including pyrene, benz[a]anthracene, chrysene, benzo[b]fluoranthene; ΣOH-PAHs, total urinary monohydroxy-PAH metabolites.

4.1. The Impact of the Exposure to MeHg on DNA Methylation

Epidemiological investigations on different populations have demonstrated the ability of Hg to impact the DNA methylation pattern in several genes (Tables 3 and S3 for further details).

A study [82] showed increased DNA methylation of the GSTM5 promoter in women with higher Hg levels in whole blood. This gene is a member of th eGSTM gene family, which encodes for enzymes that are involved in the metabolism of several environmental agents.

Hg can also influence the DNA methylation status of genes that are involved in the protection against chemical toxicity. As previously mentioned, the SEPP1 gene encodes a protein known to bind Hg that has antioxidant properties. Its promoter shows a trend of DNA hypomethylation with increasing hair Hg levels, which was predicted by estimated Hg from fish consumption [83].

Some of the neurologic outcomes of the exposure to Hg were associated with DNA methylation changes [43]. The suppressive effect that MeHg exposure has on the expression of the BDNF gene, which is poorly expressed in depressed patients, seem to be mediated also by the hypermethylation of the DNA [99]. As demonstrated by Maccani and colleagues [89], after crossing the placenta, MeHg can disrupt placental DNA methylation patterns, leading to the DNA hypomethylation of the EMID2 gene, and likely to adverse neurobehavioral outcome in infants. Cardenas and collaborators [90] found that, in male children, maternal prenatal blood mercury levels were associated with DNA hypomethylation of the Paraoxonase 1 gene (PON1), a gene involved in drug and fatty acids metabolism, and the DNA methylation pattern of this gene predicted lower cognitive test scores during early childhood. It is important to note that cord blood Hg level has previously been demonstrated to be a more accurate measure of prenatal MeHg exposure than maternal hair Hg level, and so DNA methylation changes associated with this biomarker potentially reflect MeHg effects more accurately [81]. In another very recent study [91], carried out on 406 mother–child pairs from a population who consume large amounts of fish and who are characterized by hair Hg levels that are higher than those in European and US populations, the authors found a positive association between prenatal MeHg exposure and DNA methylation in two nervous system-related genes, GRIN2B and NR3C1, measured in children's saliva. GRIN2B encodes a subunit of receptors that are important for the regulation of neural morphology, learning and memory, while NR3C1 is a receptor that is crucial to the stress responses in the brain. As stated by the authors, the observed DNA methylation changes associated with MeHg prenatal exposure at these two genes are predicted to lead to lower gene expression, and are likely to influence neurodevelopment and mental health.

4.2. The Impact of the Exposure to PAHs on DNA Methylation

Both in vitro and in vivo analyses have revealed the ability of these substances to disrupt human DNA methylation patterns [42] (Tables 4 and S4 for further details).

The developing fetus is particularly susceptible to PAH-induced DNA damage, and studies support the hypothesis that this may be due also to epigenetic dysregulations caused by these chemicals. In a study of non-smoking African-American and Dominican women from New York City [93], the authors found that prenatal exposure to PAHs measured using a personal air monitor was associated with lower global DNA methylation levels measured in umbilical cord blood DNA.

Kim and colleagues [94] analyzed samples of visceral adipose tissue of non-smoking female patients with myoma. They showed that the DNA methylation level of IRS2 gene increased as the concentrations of PAHs in adipose tissue increased. Interestingly, the IRS2 gene mediates the effects of insulin on various cellular processes, and it has been associated with several diseases, such as type 2 diabetes. Furthermore, promoter methylation of the IRS2 gene turned out to mediate the transcriptional silencing of this gene in the same study, and this led the authors to suggest that exposure to PAHs might contribute to the pathogenesis of insulin resistance through the methylation-mediated suppression of IRS2.

PAHs can accelerate human aging through epigenetic modifications. In their study of Chinese and Caucasian populations, Li and collaborators [96] first developed a DNA methylation age predictor based on the methylation status of many CpG sites across the genome, and then defined two aging indicators: Δage, defined as methylation age minus chronological age; and aging rate, defined as the ratio between methylation age and chronological age. Evaluating the association of PAHs exposure biomarkers with the above-defined aging indicators, the authors found that the increase in several urine PAHs metabolites was associated with an increase in both Δage and aging rate.

Possible hints of PAHs-mediated DNA methylation changes that may affect neurodevelopment emerged also from a study of pregnant women living close to a coal-fired power plant in China [95]. In that study, the authors analyzed cord blood samples for PAH–DNA adducts and assessed global DNA methylation by measuring genomic long interspersed nuclear elements (LINE1) methylation. LINE1 is one of the transposable repetitive elements, repetitive DNA sequences scattered across the genome and found in most eukaryotic organisms, which can change their position. Changes in LINE1 methylation can disrupt gene expression, and have been associated with birth defects, such as NTDs. In Lee and collaborators' study, a significant inverse relationship was observed between PAH–DNA adducts and LINE1 DNA methylation. Interestingly, the latter was a positive predictor of IQ (Intelligence Quotient) scores at 5 years of age in women enrolled before the closure of the power plant.

Neural tube defects (NTDs) are common and severe congenital malformations that arise from a failed or disordered closure of the neural tube during embryogenesis. Studies have linked NTDs to abnormal genome-wide DNA methylation. Authors found that PAX3, a gene encoding a transcription factor involved in development, is hypermethylated in NTD cases, and that the mean DNA methylation level of this gene in fetal neural tissue is positively correlated with median concentrations of PAHs in maternal serum [97]. Moreover, mean DNA methylation levels in the promoter region and 5′ UTR of ZIC4 gene tended to be inversely associated with levels of HMW-PAHs in the livers of NTD fetuses in a recent survey [44]. ZIC4 encodes a zinc finger protein whose absence can hamper cerebellum development in both humans and mice.

Concerning potential mechanisms underlying DNA methylation changes driven by Hg and PAHs, several findings support different hypotheses. Evidence exists that MeHg exposure is associated with the reduced expression or biochemical activity of DNMT, but Hg may also affect the methionine cycle, thus influencing the availability of SAM for DNA methylation [41]. Moreover, various studies support the hypothesis that oxidative stress mediates the effects of PAHs exposure on DNA methylation, via both the suppression of DNMT and excessive SAM consumption [44].

Even if not exhaustive, given that they report all the evidence on the subject beyond the scope of this paper, the above-described results demonstrate the ability of these important seafood pollutants to impact the human epigenome and, in particular, the DNA methylation profiles.

5. An Anthropological Perspective

Human populations that traditionally consume seafood are at an increased risk of MeHg exposure and bioaccumulation. This is supported by recent data that demonstrate that populations consuming more fish or marine mammals have greater blood MeHg values than those consuming marine foods less than once a week [17]. Moreover, evidence exists that fish-eating populations tend to show the typical symptoms associated with Hg exposure at a high rate [19]. One of the first studies addressing this topic was carried out on a cohort of 1022 consecutive singleton births from the Faroe Islands [100], where maternal exposure to MeHg is derived from the consumption of pilot whale meat. This study found a statistically significant relationship between higher prenatal Hg exposure and poorer scores on tests of neurologic function [101]. In a cross-sectional study conducted on the adults of six fishing villages of the Pantanal region of Brazil, Hg exposures associated with fish consumption, as measured by hair mercury levels, were associated with detectable alterations in performance in tests of fine motor speed, dexterity, and concentration, and the magnitude of the effects increased with hair mercury concentration, consistent with a dose-dependent effect [102].

At present, unlike the case of MeHg, there is no direct evidence that populations that consume high amounts of seafood are more exposed to PAHs. Nonetheless, evidence exists that traditional fish smoking methods can introduce potentially harmful combustion by-products into the smoked fillets, leading to concentrations of PAHs that pose a threat to human health [103,104]. Moreover, human exposure to PAHs in seafood may date back to ancient times: with fish, shellfish and sea mammals being rich in fats, and considering the high lipophilicity of PAHs, these foods may have had absorbed substantial amount of PAHs from the bitumen used for prehistoric container production [105]. Finally, as detailed below, several investigations revealed high levels of PAHs in several commercially relevant marine species, with concentrations sometimes exceeding legal limits.

Modern technologies allow us to explore human molecular variation, both at a locus-specific and at a genome-wide level, enabling us to answer several questions about population evolutionary history and the relationship between environment and human biodiversity.

The first evidence of human adaptation to a toxic chemical was reported in arsenic-exposed women from the northern Argentinean Andes [106]. The inhabitants of this region, which is characterized by elevated arsenic concentrations in available drinking water, show a uniquely efficient arsenic metabolism. Accordingly, the authors found that the AS3MT gene, which encodes the arsenite methyltransferase and functions as the major gene for arsenic metabolism in humans, strongly differentiates the Argentinean Andes population from a highly related Peruvian population much less exposed to this environmental toxicant. Then, they confirmed that SNPs mapping in that gene was positively selected.

Similar results were obtained from investigations on another Andean community. Through analyses of ancient human remains from the Camarones Valley, it has been shown that the inhabitants of the area have been exposed to arsenic-contaminated drinking water for the last 7000 years [107]. Interestingly, a decreasing trend has been detected in the average hair and bone arsenic levels, starting from Archaic hunter–gatherers and leading to the current populations, and this evidence has been interpreted as the potential result of an adaptive increasingly efficient metabolic detoxification [108]. In support of the above scenario, analyses carried out through polymerase chain reaction-restriction fragment length polymorphism (PCR-RFLP), targeting SNPs strongly associated with arsenic metabolization, showed that, contrary to alleles associated with increased toxicity risk, protective variants are much more frequent in exposed populations compared to a southern Chilean community [109].

Potential hints of human adaptation to PAHs exposure come from the comparison between the exome sequence (i.e., the sequence of nucleotides that make up the protein-coding portion of the genome) of the aryl hydrocarbon receptor (AHR) gene in Neanderthal, Denisovan and modern human individuals [110]. Once activated by endogenous or exogenous ligands, such as diet-derived metabolites and PAHs, respectively, the complex made up of AHR and other proteins passes from the cytoplasm to the nucleus. Here, following further biochemical mechanisms, the AHR regulates the CYP1A1/1A2/1B1 genes expression, thus initiating PAHs metabolism, with the resultant production of PAHs' reactive metabolites and DNA adducts. Hubbard and colleagues found that modern humans carry the same allele at a codon of the AHR gene that is unique to our species, and which is associated with a reduced AHR activation by PAHs, specifically 2,3,7,8-tetrachlorodibenzofuran (TCDF), B(a)P and benz(a)anthracene, compared to Neanderthal and other primates receptors. On the other hand, modern humans and Neanderthal AHR showed similar levels of activation by endogenous ligands. Based on the above results, the authors postulate that exposure to potentially toxic environmental AHR ligands, such as PAHs derived from controlled fires in caves, may have driven the selection of genetic variants conferring a reduced sensitivity to AHR exogenous ligands, and thus a lower DNA adduct synthesis [110].

As demonstrated also by the above-mentioned studies, molecular anthropologists are only starting to depict the role that environmental toxicants could have had in human evolution, and this is thanks to modern genomic technologies. Moreover, environmental toxicants can impact human biological variability at multiple levels, as shown by above-mentioned studies on the DNA methylation.

Environmental pressures can shape DNA methylation variability across human groups, and methods have been developed to explore the epigenetic side of human diversity at different levels. In a study analyzing the DNA methylation profiles of three human populations at 450,000 CpG sites, the authors found that DNA methylation differences contribute to the phenotypic variability of these populations, and that 68% of differentially methylated CpG sites were significantly related to underlying genetic variation, while the remaining 32% was probably related to external stimuli able to induce epigenetic changes with an impact on subsequent generations (e.g., toxic xenobiotics, differences in dietary or hormone exposure, or stress response) [111].

Many toxicants are sources of differences within human populations [80]. In Giuliani and colleagues' [112] investigation, for example, the authors compared the DNA methylation of individuals living in areas that were heavily sprayed with Agent Orange during the Vietnam War with that of individuals from non-contaminated areas. What they found is that past exposure to dioxin, the main ingredient of Agent Orange, led to DNA methylation changes among Vietnamese individuals from areas heavily sprayed, and in those whose parents participated in the war in sprayed zones.

6. Mediterranean Coastal Communities as an Informative Case Study

Certain Mediterranean communities may be more exposed to the negative effects of seafood contaminants, for both environmental and cultural reasons.

The Mediterranean Sea is a semi-enclosed basin, surrounded by countries highly industrialized and with high agricultural development, and, as such, is of particular concern with respect to contamination by toxic compounds. The Mediterranean Sea is generally considered a geological hot spot for mercury [26], as it is characterized by large deposits of HgS that account for about 65% of the global mercury reserves [113]. Moreover, it should not be surprising that several studies have shown higher levels of contaminants in marine organisms from the Mediterranean Sea compared to those from other geographic areas [6,47,114–116], with levels of PAHs and especially Hg often exceeding recommended limits for human consumption [6,30,117,118]

Concerning Hg, in particular, the highest concentrations in Europe tend to be found in fish caught in the Mediterranean Sea [16]. Data showed a more marked Hg bioavailability in the Tyrrhenian and the Adriatic coastal waters compared to the rest of the Mediterranean [119], and MeHg levels higher than the legal limit have been discovered in seafood caught in both areas [117,118,120–124], as well as in the Ionian Sea (Sidimar 2018), and in different classes of marine organisms, including fishes, crustaceans, and mollusks. MeHg contamination hotspots are represented by the Trieste gulf [125], the coastal waters between Cattolica and Rimini, in the Central Adriatic Sea [124], and those between Anzio and Civitavecchia, in the Central Tyrrhenian Sea [126]. Turning to PAHs, potential contamination hotspots are generally deemed to be located near Taranto, Trieste [125], Naples [127,128], Genoa and Palermo [127]. It is also worth reporting that some studies revealed possible risks to human health arising from the consumption of PAHs-contaminated seafood from the Mediterranean. The above conclusions were drawn from the calculation of risk indexes, such as the Excess Lifetime Cancer Risk (ELCR) and the Target Hazard Quotient (THQ) [30,73,129] The authors calculated the ELCR according to the following equation:

$$\text{ELCR} = \text{EF (day/y)} \times \text{ED (y)} \times \text{IR (Kg/day)} \times \text{CSF (mg Kg}^{-1}\text{ day)} \div \text{BW(Kg)} \times \text{AT}$$

where EF is the exposure frequency (365 days/year), ED is the exposure duration (y), IR is the ingestion rate, which is equal to PAH concentration times the mean ingestion rate of the species, CSF is the Cancer Slope Factor for each of the analyzed PAHs (OEHHA, 2009), BW is the body weight, and AT is the averaging time, which is equal to EF × ED.

A CR above the acceptable lifetime risk (ALR) of 10^{-5} [130] indicates a probability greater than 1 in 100,000 of developing cancer [30].

THQ indicates the ratio between exposure and the reference dose, and is calculated through the formula:

$$THQ = EF\ (day/y) \times ED\ (y) \times IR\ (Kg/day) \times C\ (mg\ Kg^{-1}) \div RfD\ (mg\ kg^{-1}\ day) \times BW(Kg) \times AT$$

In Europe, coastal populations consume greater amounts of seafood compared to inland populations [11]. Moreover, Mediterranean coastal populations are characterized by food habits based on local seafood consumption [26], and European countries bordering the Mediterranean are among the world's highest seafood consumers, with Spain, Italy and France accounting for more than half of the European expenditure on fish and fishery products, despite having only around a third of the EU's population (EUROSTAT, 2014). In Italy, apparent consumption of fish and seafood products amounted to 28.4 kg per capita, a share significantly higher than the EU average (EUROFISH, 2015). The traditionally high consumption of local seafood can lead to high MeHg and PAHs exposure levels among Mediterranean communities.

The above scenario is supported by several lines of evidence. Assuming an exposure frequency of 365 days a year, an exposure duration of 80 years (equivalent to the average lifetime in Italy in 2011), an ingestion rate of 18 g per day, and a body weight of 70 kg, after measuring Hg levels in various species of demersal fish commonly consumed in Italy, Storelli and Barone calculated a high target hazard quotient (THQ) and estimated weekly intake (EWI) for larger fish specimens caught in the Adriatic Sea [123]. In a study taking into account various commercially relevant marine species from the Ionian Sea, assuming a consumption rate greater than once per week, the authors found a possible risk for chronic systemic effects derived from Hg content [131]. Similar findings were derived from several other studies on seafood from various parts of the Mediterranean Sea, focusing on levels of Hg [124,132] and PAHs [129].

Given all the above, and considering the high average seafood consumption in the Mediterranean regions, we highlight the need for policymakers to take into account evidence of potentially high exposure to seafood contaminants among Mediterranean communities, especially as regards MeHg, and to consider if it is reasonable to revise law limits and/or recommendations.

In line with the above-mentioned evidence, studies on newborns and preschool children from Mediterranean populations have shown high Hg concentrations in blood and hair [6]. Analyzing data collected from over 200 cross-sectional studies measuring Hg biomarkers in human populations, Basu and collaborators [12] found geographic differences in Hg exposure, with pooled central median blood mercury concentrations being higher in general background populations (i.e., those with no particular or significant exposure to mercury) living in certain geographic areas, including the Eastern Mediterranean. They also found that subpopulations that consume high amounts of seafood are approximately four times more exposed than the general background population. In particular, exposures were higher in Indigenous people in many regions of the world, in populations living in proximity to water bodies or associated with marine ecosystems, among which were populations living along the Mediterranean Sea.

Višnjevec and colleagues [11] compared the results of several investigations on Hg exposure in Europe countries, and found that the highest hair Hg levels were found in Madeira fishermen, habitual tuna consumers in Sardinia, and Greenlandic children and mothers.

A study on the adult population of Naples, Italy, found a strong correlation between total mercury concentrations (THg) in hair and fish consumption, while almost no association was found between THg and number, surface and area of dental amalgam fillings, another possible source of MeHg in the general population, thus confirming previous findings of the major role of seafood consumption in human exposure to this pollutant [133]. Moreover, the same study found THg levels higher than the reference dose adopted by the U.S. Environmental Protection Agency (0.1 μg per kg body weight) in 5.9% of the samples.

Finally, it is interesting to note that two of the three Mediterranean countries taken into consideration in the EU-funded human biomonitoring study named DEMOCOPHES [134], i.e., Spain and Cyprus, with the fourth being Slovenia, are respectively the first and the third countries as regards mercury levels in the hair of mothers compared to all the others, with Spain being also the second European country for per capita seafood consumption in 2016 [135], as well as the Mediterranean country with the second highest share of domestic wild capture consumption compared to imported seafood (FAOSTAT, 2018), after Croatia, which was not included in the DEMOCOPHES study.

Some subgroups of the Mediterranean coastal population may be more exposed than others to the harmful action of seafood pollutants. As shown by the above-described review from Basu and colleagues [12], coastal communities are more exposed to MeHg. A study [136] of mother–infant pairs from Croatia found higher levels of Hg and selenium (Se) in hair, blood, placenta and cord blood of mothers from the coast compared to those living in continental areas, due to higher fish consumption. Interestingly, in this study, the authors also evaluated the relationship between Hg and Se levels and a polymorphism (rs28366003) in the MT2A-5 gene, which encodes a protein that plays a role in the detoxification of heavy metals, but they did not find any association.

Fishermen, in particular, have shown a tendency for a greater accumulation of Hg derived from fish, and this is related to their higher mean consumption of this food compared to the general population. Evidence in this sense comes from a fishing community on the Mediterranean coast of Morocco. In this community, researchers measured hair Hg levels, and found that these were closely related to fish intake, and that fishermen and their families were the most exposed population subgroup [137]. Additionally, in Sicily, greater Hg accumulation has been proven in fishermen, as they showed significantly higher mean hair Hg levels (6.45 ± 7.03 μg g^{-1} vs. 0.23 ± 0.4 μg g^{-1} in the control group) [138]. Similar results have been derived from investigations carried out on Italian coastal communities from the northern Adriatic Sea [139].

What may emerge from all the above is that Mediterranean fishing communities could represent an informative case study to gain insight into the potential impact of Hg and PAHs on the human genome and epigenome.

Additionally, to fulfil the need for an "ecogenetic approach" to the study of the health effects of environmental chemicals stressed by Basu and colleagues [41], what we suggest is to extend the research on Mediterranean seafood contamination by Hg and PAHs by including information about the genomic and epigenomic backgrounds of the exposed communities. Such an approach would involve the following main steps: identification of communities that are particularly exposed to seafood contaminants, in terms of both cultural (i.e., traditional high consumption of local seafood) and environmental factors (i.e., subsisting on resources caught from pollution hotspots), and the selection of communities that would represent the control group, for example inland communities, characterized by a very low fish intake; simultaneous collection of biological samples (e.g., buccal mucosa cells) and information on the family history, diet and lifestyle of participants (e.g., through a Food Frequency Questionnaire); analysis of the genetic variability and DNA methylation profiles of those genes implicated, for example, in fatty acids metabolism and in susceptibility to environmental chemicals. Such an approach would enable us to answer different questions, such as, are there any biological differences between fishing and non-fishing communities that could have been caused by different seafood intakes? Are there any differences in the biological predisposition of Mediterranean communities to the health effects of seafood intake?

7. Challenges

The main challenges in such a study refer to the epigenetic investigation, and this is for several reasons. First of all, DNA methylation may be influenced by several factors [80], including many dietary components (e.g., folate, vitamin B6, vitamin B12, betaine, methionine and choline) [75], other environmental chemicals [42], pathogens load, various environmental and climatic conditions [140], sex and age [42], socioeconomic status [141], and genetic background [48,78]. Consequently, it is difficult to tell which is the actual correlative factor underlying the observed patterns, even when an association between a given factor or biomarker has been detected.

The simultaneous analysis of genetic and epigenetic data, coupled with information on eating habits and lifestyle and personal details of participants, would allow us to account for several potential confounders, possibly distorting the association between estimated exposure and epigenetic changes. To this end, it is important to gather information that is crucial to depict the whole set of environmental stimuli affecting the individual's methylome (e.g., smoking status, diet, occupation), as well as to identify, sample and to compare populations that differ markedly when it comes to seafood consumption rate and/or levels of Hg and PAHs in fish consumed.

However, it is worth noting that the potential simultaneous exposure of individuals to a plethora of chemicals constitutes one of the main challenges in the field.

Seafood, along with other food items, contains several nutrients and contaminants, which can impact human biology at a molecular level, and this may confound the association between MeHg and/or PAHs exposure and DNA methylation.

Among seafood contaminants, the heavy metals arsenic (As), cadmium (Cd) and lead (Pb) also constitute an emerging issue due to their concentrations often exceeding regulatory limits [123,142,143] and studies have demonstrated their ability to elicit DNA methylation changes [80]. This is quite expected, as comparisons of the mechanisms of action reported similar biological pathways of these metals inducing toxicity, such as ROS generation, weakening of the antioxidant defense, enzyme inactivation, and oxidative stress (for a detailed review see [144]).

Moreover, epidemiological studies showed a general hypomethylation of LINE-1 elements after PAHs, As, Cd and Pb exposure [145–149].

In vitro and animal studies performed under rigorous experimental conditions constitute a powerful method to identify the impact of single chemicals on DNA methylation. In this respect, molecular anthropological investigations could help to make a list of candidate genes to be tested through functional studies, or vice versa, could constitute a method to evaluate the real effect.

Another crucial aspect to consider is the tissue-specificity of DNA methylation [42]. Most of the retrieved epidemiological studies on MeHg and PAHs epigenetic effects measured DNA methylation in blood, with only one study using buccal mucosa [83], another study sampling saliva [91], one study measuring adipose tissue DNA methylation [94], and two studies using neural tissue [44,97]. Molecular anthropologists, on the other hand, often collect saliva or buccal mucosa cells as an alternative DNA source, because whole blood is difficult to collect during fieldwork [150,151]. It is also important to note that, despite the risk of discordant results due to potential tissue-specific DNA methylation changes (and especially to heterogeneity in cell composition), several studies demonstrated high correlations between the DNA methylation profiles of blood and saliva [152–154], pointing to the suitability of saliva as a source for genomic DNA in cohort studies. In the same way, recent investigations have shown that DNA methylation also correlates well between saliva and the brain [155]. Additionally, buccal cells also offer potential advantages to human epigenetic studies, as they represent a better surrogate tissue for brain tissues, with both being ectodermal tissues, and because they can be collected via a non-invasive method (buccal swabs) [156]. Finally, it should be noted that statistical methods to account for cell composition in DNA methylation assays are implemented and available [157].

Seafood is not the only source of exposure to Hg and PAHs. Working with dental amalgam fillings and working or residing among artisanal and small-scale gold mining sites result in elevated exposures to elemental and inorganic Hg [12], which may lead to DNA methylation changes [83,158]. In the same way, several occupations, such as coke oven manufacturing, chimney sweeping, paving and roofing, entail relevant exposure to PAH mixtures [159,160], with consequent impacts on the DNA methylation status [86,161].

Asking participants about their occupation is therefore of fundamental importance in order to address potential confounders of the association between exposure level to MeHg or PAHs via seafood consumption and DNA methylation changes.

As regards the influence of genetics on the individual's biological response to MeHg and PAHs exposure, it should be considered that, sometimes, the same genes are involved in the toxicokinetic of and/or susceptibility to different substances. This, obviously, complicates the detection of genes that may be subject to natural selection driven by a specific chemical. This is the case of GSTP1, MT4 and ALAD genes, whose variation can influence the toxicokinetic of Hg (Tables 1 and S1), but also of Cd [162] and Pb [163–165]. The same is true for NAT2 gene polymorphisms, which influence the toxicokinetics of both Hg [65] and PAHs [69] (Tables 3 and 4).

A further level of complexity is related to the fact that concentrations of these chemicals in edible tissues of aquatic organisms are influenced by several factors.

Several studies have highlighted the role of trophic level, habitat and size of the organism in determining the level of MeHg in sea animals. In particular, despite discordant evidence [121,166], MeHg concentration in fishes often increases with trophic level [126,167], size and age [47], with MeHg uptake being a process of bioaccumulation during the whole life [118], and other studies show that MeHg concentrations are affected also by changes in feeding habits during fish lifespan [132]. The relationship between size and MeHg concentration in marine invertebrates is rather less clear: while some results point to a positive correlation between these two variables in bivalves and crustaceans [124,168], others show a negative [120,169] or no significant correlation [124] in the same taxonomic classes, or even in the same species. As regards PAHs, only few studies have tried to assess the influence of biological factors on their accumulation in aquatic organisms [8], and these led to discordant results. Several studies found no significant correlation between PAHs concentrations and fish size or age [170–173], while Frapiccini and colleagues [28] found a negative correlation between body size and PAH concentrations in the liver and gills of common soles caught in the Po Delta and off Chioggia. Additionally, the sex [170,173,174] and the reproductive stage [8] of the organism seem to affect PAHs accumulation and metabolism in fish.

MeHg and PAHs concentration in fish also vary with seasons [8,171,175–178] and with the geographic origin of the fished specimen, with some areas of the Mediterranean being more polluted than others.

Addressing the above factors through a questionnaire is not easy, if not impossible, which means that a proper estimate of the habitual seafood intake or, more generally, of the eating habits of the sampled individuals, does not always correspond to a precise estimate of their habitual MeHg and/or PAHs intake [179].

To overcome these limitations, the most effective solution is the use of biomarkers of exposition. In particular, as already mentioned, hair mercury level has proven to predict exposition to MeHg [136,180], whose primary route in the general population is seafood consumption [181], while urinary excretion of the metabolite 1-OHP has been attributed mainly to the ingestion of PAHs through the diet [35,182].

However, it is important to note that, unlike MeHg, seafood is not always the main dietary contributor to PAHs intake, and that the concentrations of PAHs in food are also influenced by cooking procedures [183]; as a consequence, it would be difficult to tell whether eventual epigenetic modifications correlated with 1-OHP urinary level are actually driven by PAHs in seafood, unless a very detailed questionnaire on eating habits is collected.

As regards the uncertainties on the geographic origin of seafood consumed, sampling fishermen could help trace the origin of the fish they eat, as fishermen tend to consume their own catch (unpublished data). Moreover, as already mentioned, fishermen represent an interesting case study, given their exposure to high levels of MeHg and, potentially, PAHs, due to their traditionally high seafood consumption. Studying fishermen, however, implies the inclusion of a potential further modifier of the effect of MeHg and PAHs on

DNA methylation, that is, the typical lifestyle of fishermen. Fishing is strongly demanding, both physically and psychologically, implying, among the several challenges, working long hours, frequent night shifts, the unpredictability of the sea, and prolonged separation from the family [184]. Moreover, several investigations have linked the above factors to health conditions and harmful habits that are common among fishermen from different parts of the world, including the Mediterranean Sea [185], which comprise tobacco smoking, alcohol abuse, sleep deprivation, chronic stress, and so on [186,187]. Such habits are known to impact human DNA methylation [188–191], and hence must be taken into consideration when asking sampled individuals about their daily life.

8. Conclusions

The Mediterranean Sea is considered a pollution hotspot for both natural and anthropogenic factors. As a consequence, Mediterranean communities may be particularly exposed to MeHg and PAHs through ingestion, due to their traditional high consumption rate of local seafood, and much evidence supports the above scenario. MeHg and PAHs can impact DNA methylation patterns in humans, even at low doses. Moreover, some of the epigenetic changes associated with MeHg and PAHs exposure are in turn associated with their known health outcomes. Finally, increasing evidence points to a significant contribution of human genetic variability in determining individual susceptibility to the chronic exposure to these chemicals, which, in certain cases, may be the results of population adaptation to certain ecological settings. In this framework, and also considering the growing concern about MeHg pollution due to climate change, we highlighted the benefit of an integrated approach, including molecular anthropologists and environmental and marine chemists, to the investigation of the relationship between the molecular diversity of Mediterranean communities and the exposure to MeHg and PAHs through seafood intake. Such an approach will help us to cope with uncertainties when it comes to risk assessment and decision-making about contaminant limits in seafood.

Supplementary Materials: The following are available online at https://www.mdpi.com/article/10.3390/app112311179/s1, Table S1: List of epidemiological studies investigating the influence of genetic polymorphisms on MeHg toxicokinetic. Genes in which above polymorphisms were identified, polymorphisms, alleles/genotypes effect on biomarkers, and samples studied are shown for each study. Table S2: List of epidemiological studies investigating the impact of MeHg exposure on DNA methylation. Genes in which differentially methylated CpG dinucleotides were identified, technology used for methylation assay, biomarkers measured, tissues in which biomarkers were measured, tissues from which DNA was extracted, and samples (with sample sizes) studied are shown for each study. Table S3: List of epidemiological studies investigating the influence of genetic polymorphisms on PAHs toxicokinetic. Genes in which above polymorphisms were identified, polymorphisms, alleles/genotypes effect on biomarkers, and samples studied are shown for each study. Table S4: List of epidemiological studies investigating the impact of PAHs exposure on DNA methylation. Genes in which differentially methylated CpG dinucleotides were identified, technology used for methylation assay, biomarkers measured, tissues in which biomarkers were measured, tissues from which DNA was extracted, and samples (with sample sizes) studied are shown for each study.

Author Contributions: Conceptualization, A.D.G., C.G., M.M. and D.L.; methodology, A.D.G., C.G. and D.L.; formal analysis, A.D.G. and C.G.; investigation, A.D.G.; resources, D.L.; data curation, A.D.G. and C.G.; writing—original draft preparation, A.D.G., C.G., M.M. and D.L.; writing—review and editing, A.D.G. and C.G.; visualization, A.D.G. and C.G.; supervision, A.D.G., C.G., M.M. and D.L.; project administration, D.L. and M.M.; funding acquisition, D.L. All authors have read and agreed to the published version of the manuscript.

Funding: This research received no external funding.

Institutional Review Board Statement: Not applicable.

Informed Consent Statement: Not applicable.

Data Availability Statement: Not applicable.

Acknowledgments: The research leading to these results was conceived under the collaboration between the University of Bologna and the National Research Council for the implementation of the International PhD Program "Innovative Technologies and Sustainable Use of Mediterranean Sea Fishery and Biological Resources" (www.FishMed-PhD.org).

Conflicts of Interest: The authors declare no conflict of interest.

References

1. Weichselbaum, E.; Coe, S.; Buttriss, J.; Stanner, S. Fish in the Diet: A Review: Fish in the Diet. *Nutr. Bull.* **2013**, *38*, 128–177. [CrossRef]
2. Thompson, L.A.; Darwish, W.S. Environmental Chemical Contaminants in Food: Review of a Global Problem. *J. Toxicol.* **2019**, *2019*, 1–14. [CrossRef]
3. Marini, M.; Frapiccini, E. Persistence of Polycyclic Aromatic Hydrocarbons in Sediments in the Deeper Area of the Northern Adriatic Sea (Mediterranean Sea). *Chemosphere* **2013**, *90*, 1839–1846. [CrossRef]
4. Marini, M.; Frapiccini, E. Do Lagoon Area Sediments Act as Traps for Polycyclic Aromatic Hydrocarbons? *Chemosphere* **2014**, *111*, 80–88. [CrossRef]
5. Ferrante, M.; Copat, C.; Mauceri, C.; Grasso, A.; Schilirò, T.; Gilli, G. The Importance of Indicators in Monitoring Water Quality According to European Directives. *Epidemiol. Prev.* **2015**, *5*, 71–75.
6. Junqué, E.; Garí, M.; Llull, R.M.; Grimalt, J.O. Drivers of the Accumulation of Mercury and Organochlorine Pollutants in Mediterranean Lean Fish and Dietary Significance. *Sci. Total Environ.* **2018**, *634*, 170–180. [CrossRef]
7. Schartup, A.T.; Thackray, C.P.; Qureshi, A.; Dassuncao, C.; Gillespie, K.; Hanke, A.; Sunderland, E.M. Climate Change and Overfishing Increase Neurotoxicant in Marine Predators. *Nature* **2019**, *572*, 648–650. [CrossRef]
8. Frapiccini, E.; Panfili, M.; Guicciardi, S.; Santojanni, A.; Marini, M.; Truzzi, C.; Annibaldi, A. Effects of Biological Factors and Seasonality on the Level of Polycyclic Aromatic Hydrocarbons in Red Mullet (Mullus Barbatus). *Environ. Pollut.* **2020**, *258*, 113742. [CrossRef] [PubMed]
9. Castaño, A.; Cutanda, F.; Esteban, M.; Pärt, P.; Navarro, C.; Gómez, S.; Rosado, M.; López, A.; López, E.; Exley, K.; et al. Fish Consumption Patterns and Hair Mercury Levels in Children and Their Mothers in 17 EU Countries. *Environ. Res.* **2015**, *141*, 58–68. [CrossRef] [PubMed]
10. Domingo, J.L. Nutrients and Chemical Pollutants in Fish and Shellfish. Balancing Health Benefits and Risks of Regular Fish Consumption. *Crit. Rev. Food Sci. Nutr.* **2016**, *56*, 979–988. [CrossRef]
11. Višnjevec, A.M.; Kocman, D.; Horvat, M. Human Mercury Exposure and Effects in Europe: Human Mercury Exposure and Effects in Europe. *Environ. Toxicol. Chem.* **2014**, *33*, 1259–1270. [CrossRef] [PubMed]
12. Basu, N.; Horvat, M.; Evers, D.C.; Zastenskaya, I.; Weihe, P.; Tempowski, J. A State-of-the-Science Review of Mercury Biomarkers in Human Populations Worldwide between 2000 and 2018. *Environ. Health Perspect.* **2018**, *126*, 106001. [CrossRef] [PubMed]
13. Jacob, J.; Seidel, A. Biomonitoring of Polycyclic Aromatic Hydrocarbons in Human Urine. *J. Chromatogr. B* **2002**, *778*, 31–47. [CrossRef]
14. Krabbenhoft, D.P.; Sunderland, E.M. Global Change and Mercury. *Science* **2013**, *341*, 1457–1458. [CrossRef]
15. Selin, N.E. Global Biogeochemical Cycling of Mercury: A Review. *Annu. Rev. Environ. Resour.* **2009**, *28*, 43–63. [CrossRef]
16. EEA EEA Report No 11/2018. *Mercury in Europe's Environment. A Priority for European and Global Action*; Publications Office of the European Union: Luxembourg, 2018; Available online: https://www.eea.europa.eu/publications/mercury-in-europe-s-environment (accessed on 13 October 2021). [CrossRef]
17. Szefer, P. Safety Assessment of Seafood with Respect to Chemical Pollutants in European Seas. *Oceanol. Hydrobiol. Stud.* **2013**, *42*, 110–118. [CrossRef]
18. Rice, K.M.; Walker, E.M.; Wu, M.; Gillette, C.; Blough, E.R. Environmental Mercury and Its Toxic Effects. *J. Prev. Med. Pub. Health* **2014**, *47*, 74–83. [CrossRef]
19. Zahir, F.; Rizwi, S.J.; Haq, S.K.; Khan, R.H. Low Dose Mercury Toxicity and Human Health. *Environ. Toxicol. Pharmacol.* **2005**, *20*, 351–360. [CrossRef]
20. Karagas, M.R.; Choi, A.L.; Oken, E.; Horvat, M.; Schoeny, R.; Kamai, E.; Cowell, W.; Grandjean, P.; Korrick, S. Evidence on the Human Health Effects of Low-Level Methylmercury Exposure. *Environ. Health Perspect.* **2012**, *120*, 799–806. [CrossRef]
21. EEA Commission Regulation (EC) No 1881/2006 of 19 December 2006 Setting Maximum Levels for Certain Contaminants in Foodstuffs (1). *Off. J. Eur. Union* **2006**, *49*, 5–24.
22. WHO Evaluation of Certain Food Additives and Contaminants. *Sixty-First Report of the Joint FAO/WHO Expert Committee on Food Additives*; WHO: Geneva, Switzerland, 2004.
23. Davidson, P.W.; van Wijngaarden, E.; Shamlaye, C.; Strain, J.J.; Myers, G.J. Putting Findings from the Seychelles Child Development Study into Perspective: The Importance of a Historical Special Issue of the Seychelles Medical and Dental Journal. *NeuroToxicology* **2020**, *76*, 111–113. [CrossRef] [PubMed]
24. Weihe, P.; Grandjean, P. Cohort Studies of Faroese Children Concerning Potential Adverse Health Effects after the Mothers' Exposure to Marine Contaminants during Pregnancy. *Acta Vet. Scand.* **2012**, *54*, S7, 1751-0147-54-S1–S7. [CrossRef]

25. EFSA Panel on Contaminants in the Food Chain (CONTAM). Scientific Opinion on the Risk for Public Health Related to the Presence of Mercury and Methylmercury in Food. *EFSA J.* **2012**, *10*, 2985. [CrossRef]
26. Brambilla, G. Mercury Occurrence in Italian Seafood from the Mediterranean Sea and Possible Intake Scenarios of the Italian Coastal Population. *Regul. Toxicol. Pharmacol.* **2013**, *9*, 269–277. [CrossRef]
27. Cocci, P.; Mosconi, G.; Bracchetti, L.; Nalocca, J.M.; Frapiccini, E.; Marini, M.; Caprioli, G.; Sagratini, G.; Palermo, F.A. Investigating the Potential Impact of Polycyclic Aromatic Hydrocarbons (PAHs) and Polychlorinated Biphenyls (PCBs) on Gene Biomarker Expression and Global DNA Methylation in Loggerhead Sea Turtles (Caretta Caretta) from the Adriatic Sea. *Sci. Total Environ.* **2018**, *619–620*, 49–57. [CrossRef]
28. Frapiccini, E.; Annibaldi, A.; Betti, M.; Polidori, P.; Truzzi, C.; Marini, M. Polycyclic Aromatic Hydrocarbon (PAH) Accumulation in Different Common Sole (Solea Solea) Tissues from the North Adriatic Sea Peculiar Impacted Area. *Mar. Pollut. Bull.* **2018**, *137*, 61–68. [CrossRef]
29. US-EPA Polycyclic Aromatic Hydrocarbons (PAHs) Fact Sheet. US EPA ARCHIVE DOCUMENT 2008. Available online: https://archive.epa.gov/epawaste/hazard/wastemin/web/pdf/pahs.pdf (accessed on 12 October 2021).
30. Ferrante, M.; Zanghì, G.; Cristaldi, A.; Copat, C.; Grasso, A.; Fiore, M.; Signorelli, S.S.; Zuccarello, P.; Oliveri Conti, G. PAHs in Seafood from the Mediterranean Sea: An Exposure Risk Assessment. *Food Chem. Toxicol.* **2018**, *115*, 385–390. [CrossRef]
31. Ramesh, A.; Walker, S.A.; Hood, D.B.; Guillén, M.D.; Schneider, K.; Weyand, E.H. Bioavailability and Risk Assessment of Orally Ingested Polycyclic Aromatic Hydrocarbons. *Int. J. Toxicol.* **2004**, *23*, 301–333. [CrossRef]
32. Balcıoğlu, E.B. Potential Effects of Polycyclic Aromatic Hydrocarbons (PAHs) in Marine Foods on Human Health: A Critical Review. *Toxin Rev.* **2016**, *35*, 98–105. [CrossRef]
33. Haritash, A.K. A Comprehensive Review of Metabolic and Genomic Aspects of PAH-Degradation. *Arch. Microbiol.* **2020**, *26*, 2033–2058.
34. Castano-Vinyals, G. Biomarkers of Exposure to Polycyclic Aromatic Hydrocarbons from Environmental Air Pollution. *Occup. Environ. Med.* **2004**, *61*, e12. [CrossRef] [PubMed]
35. Pruneda-Álvarez, L.G.; Pérez-Vázquez, F.J.; Ruíz-Vera, T.; Ochoa-Martínez, Á.C.; Orta-García, S.T.; Jiménez-Avalos, J.A.; Pérez-Maldonado, I.N. Urinary 1-Hydroxypyrene Concentration as an Exposure Biomarker to Polycyclic Aromatic Hydrocarbons (PAHs) in Mexican Women from Different Hot Spot Scenarios and Health Risk Assessment. *Environ. Sci. Pollut. Res.* **2016**, *23*, 6816–6825. [CrossRef]
36. Iyer, S.; Wang, Y.; Xiong, W.; Tang, D.; Jedrychowski, W.; Chanock, S.; Wang, S.; Stigter, L.; Mróz, E.; Perera, F. Significant Interactions between Maternal PAH Exposure and Single Nucleotide Polymorphisms in Candidate Genes on B[*a*]P–DNA Adducts in a Cohort of Non-Smoking Polish Mothers and Newborns. *Carcinogenesis* **2016**, *37*, 1110–1115. [CrossRef]
37. Gu, A.; Ji, G.; Jiang, T.; Lu, A.; You, Y.; Liu, N.; Luo, C.; Yan, W.; Zhao, P. Contributions of Aryl Hydrocarbon Receptor Genetic Variants to the Risk of Glioma and PAH-DNA Adducts. *Toxicol. Sci.* **2012**, *128*, 357–364. [CrossRef]
38. Nethery, E.; Wheeler, A.J.; Fisher, M.; Sjödin, A.; Li, Z.; Romanoff, L.C.; Foster, W.; Arbuckle, T.E. Urinary Polycyclic Aromatic Hydrocarbons as a Biomarker of Exposure to PAHs in Air: A Pilot Study among Pregnant Women. *J. Expo. Sci. Environ. Epidemiol.* **2012**, *22*, 70–81. [CrossRef]
39. Joneidi, Z.; Mortazavi, Y.; Memari, F.; Roointan, A.; Chahardouli, B.; Rostami, S. The Impact of Genetic Variation on Metabolism of Heavy Metals: Genetic Predisposition? *Biomed. Pharmacother.* **2019**, *113*, 108642. [CrossRef]
40. Shen, H.; Tao, S.; Liu, J.; Huang, Y.; Chen, H.; Li, W.; Zhang, Y.; Chen, Y.; Su, S.; Lin, N.; et al. Global Lung Cancer Risk from PAH Exposure Highly Depends on Emission Sources and Individual Susceptibility. *Sci. Rep.* **2015**, *4*, 6561. [CrossRef] [PubMed]
41. Basu, N.; Goodrich, J.M.; Head, J. Ecogenetics of Mercury: From Genetic Polymorphisms and Epigenetics to Risk Assessment and Decision-Making: Ecogenetics of Mercury. *Environ. Toxicol. Chem.* **2014**, *33*, 1248–1258. [CrossRef]
42. Ruiz-Hernandez, A.; Kuo, C.-C.; Rentero-Garrido, P.; Tang, W.-Y.; Redon, J.; Ordovas, J.M.; Navas-Acien, A.; Tellez-Plaza, M. Environmental Chemicals and DNA Methylation in Adults: A Systematic Review of the Epidemiologic Evidence. *Clin. Epigenetics* **2015**, *7*, 55. [CrossRef]
43. Ijomone, O.M.; Ijomone, O.K.; Iroegbu, J.D.; Ifenatuoha, C.W.; Olung, N.F.; Aschner, M. Epigenetic Influence of Environmentally Neurotoxic Metals. *NeuroToxicology* **2020**, *81*, 51–65. [CrossRef]
44. Huang, Y.; Lin, S.; Wang, C.; Pi, X.; Jin, L.; Li, Z.; Wang, L.; Ren, A. Neural Tube Defects and ZIC4 Hypomethylation in Relation to Polycyclic Aromatic Hydrocarbon Exposure. *Front. Cell Dev. Biol.* **2020**, *8*, 582661. [CrossRef]
45. Esposito, M.; De Roma, A.; Sansone, D.; Capozzo, D.; Iaccarino, D.; di Nocera, F.; Gallo, P. Non-Essential Toxic Element (Cd, As, Hg and Pb) Levels in Muscle, Liver and Kidney of Loggerhead Sea Turtles (Caretta Caretta) Stranded along the Southwestern Coasts of Tyrrhenian Sea. *Comp. Biochem. Physiol. Part C Toxicol. Pharmacol.* **2020**, *231*, 108725. [CrossRef]
46. Serrano, O.; Martínez-Cortizas, A.; Mateo, M.A.; Biester, H.; Bindler, R. Millennial Scale Impact on the Marine Biogeochemical Cycle of Mercury from Early Mining on the Iberian Peninsula: Early Hg Pollution in the Iberian Region. *Glob. Biogeochem. Cycles* **2013**, *27*, 21–30. [CrossRef]
47. Ancora, S.; Mariotti, G.; Ponchia, R.; Fossi, M.C.; Leonzio, C.; Bianchi, N. Trace Elements Levels in Muscle and Liver of a Rarely Investigated Large Pelagic Fish: The Mediterranean Spearfish Tetrapturus Belone (Rafinesque, 1810). *Mar. Pollut. Bull.* **2020**, *151*, 110878. [CrossRef]

48. Giuliani, C.; Sazzini, M.; Bacalini, M.G.; Pirazzini, C.; Marasco, E.; Fontanesi, E.; Franceschi, C.; Luiselli, D.; Garagnani, P. Epigenetic Variability across Human Populations: A Focus on DNA Methylation Profiles of the *KRTCAP3, MAD1L1* and *BRSK2* Genes. *Genome Biol. Evol.* **2016**, *8*, 2760–2773. [CrossRef] [PubMed]
49. Sazzini, M.; Abondio, P.; Sarno, S.; Gnecchi-Ruscone, G.A.; Ragno, M.; Giuliani, C.; De Fanti, S.; Ojeda-Granados, C.; Boattini, A.; Marquis, J.; et al. Genomic History of the Italian Population Recapitulates Key Evolutionary Dynamics of Both Continental and Southern Europeans. *BMC Biol.* **2020**, *18*, 51. [CrossRef]
50. Custodio, H.M.; Broberg, K.; Wennberg, M.; Jansson, J.-H.; Vessby, B.; Hallmans, G.; Stegmayr, B.; Skerfving, S. Polymorphisms in Glutathione-Related Genes Affect Methylmercury Retention. *Arch. Environ. Health Int. J.* **2004**, *59*, 588–595. [CrossRef]
51. Gundacker, C.; Komarnicki, G.; Jagiello, P.; Gencikova, A.; Dahmen, N.; Wittmann, K.; Gencik, M. Glutathione-S-Transferase Polymorphism, Metallothionein Expression, and Mercury Levels among Students in Austria. *Sci. Total Environ.* **2007**, *385*, 37–47. [CrossRef]
52. Engström, K.S.; Strömberg, U.; Lundh, T.; Johansson, I.; Vessby, B.; Hallmans, G.; Skerfving, S.; Broberg, K. Genetic Variation in Glutathione-Related Genes and Body Burden of Methylmercury. *Environ. Health Perspect.* **2008**, *116*, 734–739. [CrossRef]
53. Gundacker, C.; Wittmann, K.J.; Kukuckova, M.; Komarnicki, G.; Hikkel, I.; Gencik, M. Genetic Background of Lead and Mercury Metabolism in a Group of Medical Students in Austria. *Environ. Res.* **2009**, *109*, 786–796. [CrossRef] [PubMed]
54. Lee, B.-E.; Hong, Y.-C.; Park, H.; Ha, M.; Koo, B.S.; Chang, N.; Roh, Y.-M.; Kim, B.-N.; Kim, Y.-J.; Kim, B.-M.; et al. Interaction between *GSTM1 / GSTT1* Polymorphism and Blood Mercury on Birth Weight. *Environ. Health Perspect.* **2010**, *118*, 437–443. [CrossRef]
55. Goodrich, J.M.; Wang, Y.; Gillespie, B.; Werner, R.; Franzblau, A.; Basu, N. Glutathione Enzyme and Selenoprotein Polymorphisms Associate with Mercury Biomarker Levels in Michigan Dental Professionals. *Toxicol. Appl. Pharmacol.* **2011**, *257*, 301–308. [CrossRef]
56. Wang, Y.; Goodrich, J.M.; Gillespie, B.; Werner, R.; Basu, N.; Franzblau, A. An Investigation of Modifying Effects of Metallothionein Single-Nucleotide Polymorphisms on the Association between Mercury Exposure and Biomarker Levels. *Environ. Health Perspect.* **2012**, *120*, 530–534. [CrossRef]
57. Ng, S.; Lin, C.-C.; Hwang, Y.-H.; Hsieh, W.-S.; Liao, H.-F.; Chen, P.-C. Mercury, APOE, and Children's Neurodevelopment. *NeuroToxicology* **2013**, *37*, 85–92. [CrossRef]
58. Barcelos, G.R.M.; Grotto, D.; de Marco, K.C.; Valentini, J.; van Helvoort Lengert, A.; de Oliveira, A.Á.S.; Garcia, S.C.; Braga, G.Ú.L.; Schläwicke Engström, K.; de Syllos Cólus, I.M.; et al. Polymorphisms in Glutathione-Related Genes Modify Mercury Concentrations and Antioxidant Status in Subjects Environmentally Exposed to Methylmercury. *Sci. Total Environ.* **2013**, *463–464*, 319–325. [CrossRef]
59. Julvez, J.; Smith, G.D.; Golding, J.; Ring, S.; Pourcain, B.S.; Gonzalez, J.R.; Grandjean, P. Prenatal Methylmercury Exposure and Genetic Predisposition to Cognitive Deficit at Age 8 Years. *Epidemiology* **2013**, *24*, 643–650. [CrossRef]
60. De Oliveira, A.Á.S.; de Souza, M.F.; van Lengert, A.H.; de Oliveira, M.T.; de Camargo, R.B.O.G.; Braga, G.Ú.L.; de Cólus, I.M.S.; Barbosa, F.; Barcelos, G.R.M. Genetic Polymorphisms in Glutathione (GSH-) Related Genes Affect the Plasmatic Hg/Whole Blood Hg Partitioning and the Distribution between Inorganic and Methylmercury Levels in Plasma Collected from a Fish-Eating Population. *BioMed Res. Int.* **2014**, *2014*, 1–8. [CrossRef]
61. Llop, S.; Pisa, F.; Tratnik, J.S.; Mazej, D.; Murcia, M.; Rebagliato, M.; Bustamante, M.; Sunyer, J.; Antonopoulou, E.; Antoniadou, I.; et al. Polymorphisms in ABC Transporter Genes and Concentrations of Mercury in Newborns – Evidence from Two Mediterranean Birth Cohorts. *PLoS ONE* **2014**, *9*, 9. [CrossRef]
62. Ng, S.; Lin, C. C.; Jeng, S.-F.; Hwang, Y.-H.; Hsieh, W.-S.; Chen, P.-C. Mercury, APOE, and Child Behavior. *Chemosphere* **2015**, *120*, 123–130. [CrossRef]
63. Engström, K.; Love, T.M.; Watson, G.E.; Zareba, G.; Yeates, A.; Wahlberg, K.; Alhamdow, A.; Thurston, S.W.; Mulhern, M.; McSorley, E.M.; et al. Polymorphisms in ATP-Binding Cassette Transporters Associated with Maternal Methylmercury Disposition and Infant Neurodevelopment in Mother-Infant Pairs in the Seychelles Child Development Study. *Environ. Int.* **2016**, *94*, 224–229. [CrossRef]
64. Parajuli, R.P.; Goodrich, J.M.; Chou, H.-N.; Gruninger, S.E.; Dolinoy, D.C.; Franzblau, A.; Basu, N. Genetic Polymorphisms Are Associated with Hair, Blood, and Urine Mercury Levels in the American Dental Association (ADA) Study Participants. *Environ. Res.* **2016**, *149*, 247–258. [CrossRef]
65. Parajuli, R.P. Genetic Polymorphisms Are Associated with Exposure Biomarkers for Metals and Persistent Organic Pollutants among Inuit from the Inuvialuit Settlement Region, Canada. *Sci. Total Environ.* **2018**, *10*, 569–578. [CrossRef]
66. Wahlberg, K.; Love, T.M.; Pineda, D.; Engström, K.; Watson, G.E.; Thurston, S.W.; Yeates, A.J.; Mulhern, M.S.; McSorley, E.M.; Strain, J.J.; et al. Maternal Polymorphisms in Glutathione-Related Genes Are Associated with Maternal Mercury Concentrations and Early Child Neurodevelopment in a Population with a Fish-Rich Diet. *Environ. Int.* **2018**, *115*, 142–149. [CrossRef]
67. Lozano, M.; Murcia, M.; Soler-Blasco, R.; González, L.; Iriarte, G.; Rebagliato, M.; Lopez-Espinosa, M.-J.; Esplugues, A.; Ballester, F.; Llop, S. Exposure to Mercury among 9-Year-Old Children and Neurobehavioural Function. *Environ. Int.* **2021**, *146*, 106173. [CrossRef]
68. Ji, G.; Gu, A.; Zhou, Y.; Shi, X.; Xia, Y.; Long, Y.; Song, L.; Wang, S.; Wang, X. Interactions between Exposure to Environmental Polycyclic Aromatic Hydrocarbons and DNA Repair Gene Polymorphisms on Bulky DNA Adducts in Human Sperm. *PLoS ONE* **2010**, *5*, e13145. [CrossRef]

69. Etemadi, A.; Islami, F.; Phillips, D.H.; Godschalk, R.; Golozar, A.; Kamangar, F.; Malekshah, A.F.-T.; Pourshams, A.; Elahi, S.; Ghojaghi, F.; et al. Variation in PAH-Related DNA Adduct Levels among Non-Smokers: The Role of Multiple Genetic Polymorphisms and Nucleotide Excision Repair Phenotype. *Int. J. Cancer* **2013**, *132*, 2738–2747. [CrossRef]
70. Clarkson, T.W.; Magos, L. The Toxicology of Mercury and Its Chemical Compounds. *Crit. Rev. Toxicol.* **2006**, *36*, 609–662. [CrossRef]
71. Caito, S.W.; Jackson, B.P.; Punshon, T.; Scrimale, T.; Grier, A.; Gill, S.R.; Love, T.M.; Watson, G.E.; van Wijngaarden, E.; Rand, M.D. Editor's Highlight: Variation in Methylmercury Metabolism and Elimination Status in Humans Following Fish Consumption. *Toxicol. Sci.* **2018**, *161*, 443–453. [CrossRef]
72. Andreoli, V.; Sprovieri, F. Genetic Aspects of Susceptibility to Mercury Toxicity: An Overview. *Int. J. Environ. Res. Public. Health* **2017**, *14*, 93. [CrossRef]
73. Conte, F.; Copat, C.; Longo, S.; Conti, G.O.; Grasso, A.; Arena, G.; Dimartino, A.; Brundo, M.V.; Ferrante, M. Polycyclic Aromatic Hydrocarbons in Haliotis Tuberculata (Linnaeus, 1758) (Mollusca, Gastropoda): Considerations on Food Safety and Source Investigation. *Food Chem. Toxicol.* **2016**, *94*, 57–63. [CrossRef]
74. Rehman, M.Y.A.; Taqi, M.M.; Hussain, I.; Nasir, J.; Rizvi, S.H.H.; Syed, J.H. Elevated Exposure to Polycyclic Aromatic Hydrocarbons (PAHs) May Trigger Cancers in Pakistan: An Environmental, Occupational, and Genetic Perspective. *Environ. Sci. Pollut. Res.* **2020**, *27*, 42405–42423. [CrossRef]
75. Feil, R.; Fraga, M.F. Epigenetics and the Environment: Emerging Patterns and Implications. *Nat. Rev. Genet.* **2012**, *13*, 97–109. [CrossRef]
76. Fagny, M.; Patin, E.; MacIsaac, J.L.; Rotival, M.; Flutre, T.; Jones, M.J.; Siddle, K.J.; Quach, H.; Harmant, C.; McEwen, L.M.; et al. The Epigenomic Landscape of African Rainforest Hunter-Gatherers and Farmers. *Nat. Commun.* **2015**, *6*, 10047. [CrossRef]
77. Natri, H.M.; Bobowik, K.S.; Kusuma, P.; Crenna Darusallam, C.; Jacobs, G.S.; Hudjashov, G.; Lansing, J.S.; Sudoyo, H.; Banovich, N.E.; Cox, M.P.; et al. Genome-Wide DNA Methylation and Gene Expression Patterns Reflect Genetic Ancestry and Environmental Differences across the Indonesian Archipelago. *PLoS Genet.* **2020**, *16*, e1008749. [CrossRef]
78. Carja, O.; MacIsaac, J.L.; Mah, S.M.; Henn, B.M.; Kobor, M.S.; Feldman, M.W.; Fraser, H.B. Worldwide Patterns of Human Epigenetic Variation. *Nat. Ecol. Evol.* **2017**, *1*, 1577–1583. [CrossRef]
79. Hodjat, M.; Rahmani, S.; Khan, F.; Niaz, K.; Navaei–Nigjeh, M.; Mohammadi Nejad, S.; Abdollahi, M. Environmental Toxicants, Incidence of Degenerative Diseases, and Therapies from the Epigenetic Point of View. *Arch. Toxicol.* **2017**, *91*, 2577–2597. [CrossRef]
80. Martin, E.M.; Fry, R.C. Environmental Influences on the Epigenome: Exposure- Associated DNA Methylation in Human Populations. *Annu. Rev. Public Health* **2018**, *39*, 309–333. [CrossRef]
81. Culbreth, M.; Aschner, M. Methylmercury Epigenetics. *Toxics* **2019**, *7*, 56. [CrossRef]
82. Hanna, C.W.; Bloom, M.S.; Robinson, W.P.; Kim, D.; Parsons, P.J.; vom Saal, F.S.; Taylor, J.A.; Steuerwald, A.J.; Fujimoto, V.Y. DNA Methylation Changes in Whole Blood Is Associated with Exposure to the Environmental Contaminants, Mercury, Lead, Cadmium and Bisphenol A, in Women Undergoing Ovarian Stimulation for IVF. *Hum. Reprod.* **2012**, *27*, 1401–1410. [CrossRef]
83. Goodrich, J.M.; Basu, N.; Franzblau, A.; Dolinoy, D.C. Mercury Biomarkers and DNA Methylation among Michigan Dental Professionals: Mercury and DNA Methylation. *Environ. Mol. Mutagen.* **2013**, *54*, 195–203. [CrossRef]
84. Pavanello, S.; Bollati, V.; Pesatori, A.C.; Kapka, L.; Bolognesi, C.; Bertazzi, P.A.; Baccarelli, A. Global and Gene-Specific Promoter Methylation Changes Are Related to *Anti* -B[*a*]PDE-DNA Adduct Levels and Influence Micronuclei Levels in Polycyclic Aromatic Hydrocarbon-Exposed Individuals. *Int. J. Cancer* **2009**, *125*, 1692–1697. [CrossRef]
85. Yang, P.; Ma, J.; Zhang, B.; Duan, H.; He, Z.; Zeng, J.; Zeng, X.; Li, D.; Wang, Q.; Xiao, Y.; et al. CpG Site–Specific Hypermethylation of $P16^{INK4\alpha}$ in Peripheral Blood Lymphocytes of PAH-Exposed Workers. *Cancer Epidemiol. Biomark. Prev.* **2012**, *21*, 182–190. [CrossRef]
86. Alegría-Torres, J.A.; Barretta, F.; Batres-Esquivel, L.E.; Carrizales-Yáñez, L.; Pérez-Maldonado, I.N.; Baccarelli, A.; Bertazzi, P.A. Epigenetic Markers of Exposure to Polycyclic Aromatic Hydrocarbons in Mexican Brickmakers: A Pilot Study. *Chemosphere* **2013**, *91*, 475–480. [CrossRef]
87. Bakulski, K.M.; Lee, H.; Feinberg, J.I.; Wells, E.M.; Brown, S.; Herbstman, J.B.; Witter, F.R.; Halden, R.U.; Caldwell, K.; Mortensen, M.E.; et al. Prenatal Mercury Concentration Is Associated with Changes in DNA Methylation at *TCEANC2* in Newborns. *Int. J. Epidemiol.* **2015**, *44*, 1249–1262. [CrossRef]
88. Cardenas, A.; Koestler, D.C.; Houseman, E.A.; Jackson, B.P.; Kile, M.L.; Karagas, M.R.; Marsit, C.J. Differential DNA Methylation in Umbilical Cord Blood of Infants Exposed to Mercury and Arsenic *in Utero*. *Epigenetics* **2015**, *10*, 508–515. [CrossRef]
89. Maccani, J.Z.J.; Koestler, D.C.; Lester, B.; Houseman, E.A.; Armstrong, D.A.; Kelsey, K.T.; Marsit, C.J. Placental DNA Methylation Related to Both Infant Toenail Mercury and Adverse Neurobehavioral Outcomes. *Environ. Health Perspect.* **2015**, *123*, 723–729. [CrossRef]
90. Cardenas, A.; Rifas-Shiman, S.L.; Agha, G.; Hivert, M.-F.; Litonjua, A.A.; DeMeo, D.L.; Lin, X.; Amarasiriwardena, C.J.; Oken, E.; Gillman, M.W.; et al. Persistent DNA Methylation Changes Associated with Prenatal Mercury Exposure and Cognitive Performance during Childhood. *Sci. Rep.* **2017**, *7*, 288. [CrossRef]
91. Cediel Ulloa, A.; Gliga, A.; Love, T.M.; Pineda, D.; Mruzek, D.W.; Watson, G.E.; Davidson, P.W.; Shamlaye, C.F.; Strain, J.J.; Myers, G.J.; et al. Prenatal Methylmercury Exposure and DNA Methylation in Seven-Year-Old Children in the Seychelles Child Development Study. *Environ. Int.* **2021**, *147*, 106321. [CrossRef]

92. Perera, F.; Tang, W.; Herbstman, J.; Tang, D.; Levin, L.; Miller, R.; Ho, S. Relation of DNA Methylation of 5′-CpG Island of ACSL3 to Transplacental Exposure to Airborne Polycyclic Aromatic Hydrocarbons and Childhood Asthma. *PLoS ONE* **2009**, *4*, e4488. [CrossRef]
93. Herbstman, J.B.; Tang, D.; Zhu, D.; Qu, L.; Sjödin, A.; Li, Z.; Camann, D.; Perera, F.P. Prenatal Exposure to Polycyclic Aromatic Hydrocarbons, Benzo[*a*]Pyrene–DNA Adducts, and Genomic DNA Methylation in Cord Blood. *Environ. Health Perspect.* **2012**, *120*, 733–738. [CrossRef]
94. Kim, Y.H.; Lee, Y.S.; Lee, D.H.; Kim, D.S. Polycyclic Aromatic Hydrocarbons Are Associated with Insulin Receptor Substrate 2 Methylation in Adipose Tissues of Korean Women. *Environ. Res.* **2016**, *150*, 47–51. [CrossRef]
95. Lee, J.; Kalia, V.; Perera, F.; Herbstman, J.; Li, T.; Nie, J.; Qu, L.R.; Yu, J.; Tang, D. Prenatal Airborne Polycyclic Aromatic Hydrocarbon Exposure, LINE1 Methylation and Child Development in a Chinese Cohort. *Environ. Int.* **2017**, *99*, 315–320. [CrossRef]
96. Li, J.; Zhu, X.; Yu, K.; Jiang, H.; Zhang, Y.; Wang, B.; Liu, X.; Deng, S.; Hu, J.; Deng, Q.; et al. Exposure to Polycyclic Aromatic Hydrocarbons and Accelerated DNA Methylation Aging. *Environ. Health Perspect.* **2018**, *126*, 067005. [CrossRef]
97. Lin, S.; Ren, A.; Wang, L.; Santos, C.; Huang, Y.; Jin, L.; Li, Z.; Greene, N.D.E. Aberrant Methylation of Pax3 Gene and Neural Tube Defects in Association with Exposure to Polycyclic Aromatic Hydrocarbons. *Clin. Epigenetics* **2019**, *11*, 13. [CrossRef]
98. Liu, K.; Jiang, J.; Lin, Y.; Liu, W.; Zhu, X.; Zhang, Y.; Jiang, H.; Yu, K.; Liu, X.; Zhou, M.; et al. Exposure to Polycyclic Aromatic Hydrocarbons, DNA Methylation and Heart Rate Variability among Non-Current Smokers. *Environ. Pollut.* **2021**, *288*, 117777. [CrossRef]
99. Onishchenko, N.; Karpova, N.; Sabri, F.; Castrn, E.; Ceccatelli, S. Long-Lasting Depression-like Behavior and Epigenetic Changes of BDNF Gene Expression Induced by Perinatal Exposure to Methylmercury. *J. Neurochem.* **2008**, *106*, 1378–1387. [CrossRef]
100. Grandjean, P.; Weihe, P.; White, R.F.; Debes, F.; Araki, S.; Yokoyama, K.; Murata, K.; Sørensen, N.; Dahl, R.; Jørgensen, P.J. Cognitive Deficit in 7-Year-Old Children with Prenatal Exposure to Methylmercury. *Neurotoxicol. Teratol.* **1997**, *19*, 417–428. [CrossRef]
101. Axelrad, D.A.; Bellinger, D.C.; Ryan, L.M.; Woodruff, T.J. Dose–Response Relationship of Prenatal Mercury Exposure and IQ: An Integrative Analysis of Epidemiologic Data. *Environ. Health Perspect.* **2007**, *115*, 609–615. [CrossRef]
102. Yokoo, E.M.; Valente, J.G.; Grattan, L.; Schmidt, S.L.; Platt, I.; Silbergeld, E.K. Low Level Methylmercury Exposure Affects Neuropsychological Function in Adults. *Environ. Health* **2003**, *2*, 8. [CrossRef]
103. Akpambang, V.; Purcaro, G.; Lajide, L.; Amoo, I.; Conte, L.; Moret, S. Determination of Polycyclic Aromatic Hydrocarbons (PAHs) in Commonly Consumed Nigerian Smoked/Grilled Fish and Meat. *Food Addit. Contam.* **2009**, *22*, 1096–1103. [CrossRef]
104. Forsberg, N.D.; Stone, D.; Harding, A.; Harper, B.; Harris, S.; Matzke, M.M.; Cardenas, A.; Waters, K.M.; Anderson, K.A. Effect of Native American Fish Smoking Methods on Dietary Exposure to Polycyclic Aromatic Hydrocarbons and Possible Risks to Human Health. *J. Agric. Food Chem.* **2012**, *60*, 6899–6906. [CrossRef]
105. Sholts, S.B.; Smith, K.; Wallin, C.; Ahmed, T.M.; Wärmländer, S.K.T.S. Ancient Water Bottle Use and Polycyclic Aromatic Hydrocarbon (PAH) Exposure among California Indians: A Prehistoric Health Risk Assessment. *Environ. Health* **2017**, *16*, 61. [CrossRef]
106. Schlebusch, C.M.; Gattepaille, L.M.; Engström, K.; Vahter, M.; Jakobsson, M.; Broberg, K. Human Adaptation to Arsenic-Rich Environments. *Mol. Biol. Evol.* **2015**, *32*, 1544–1555. [CrossRef]
107. Arriaza, B.; Amarasiriwardena, D.; Cornejo, L.; Standen, V.; Byrne, S.; Bartkus, L.; Bandak, B. Exploring Chronic Arsenic Poisoning in Pre-Columbian Chilean Mummies. *J. Archaeol. Sci.* **2010**, *37*, 1274–1278. [CrossRef]
108. Apata, M.; Pfeifer, S.P. Recent Population Genomic Insights into the Genetic Basis of Arsenic Tolerance in Humans: The Difficulties of Identifying Positively Selected Loci in Strongly Bottlenecked Populations. *Heredity* **2020**, *124*, 253–262. [CrossRef]
109. Apata, M.; Arriaza, B.; Llop, E.; Moraga, M. Human Adaptation to Arsenic in Andean Populations of the Atacama Desert. *Am. J. Phys. Anthropol.* **2017**, *163*, 192–199. [CrossRef]
110. Hubbard, T.D.; Murray, I.A.; Bisson, W.H.; Sullivan, A.P.; Sebastian, A.; Perry, G.H.; Jablonski, N.G.; Perdew, G.H. Divergent Ah Receptor Ligand Selectivity during Hominin Evolution. *Mol. Biol. Evol.* **2016**, *33*, 2648–2658. [CrossRef]
111. Heyn, H.; Moran, S.; Hernando-Herraez, I.; Sayols, S.; Gomez, A.; Sandoval, J.; Monk, D.; Hata, K.; Marques-Bonet, T.; Wang, L.; et al. DNA Methylation Contributes to Natural Human Variation. *Genome Res.* **2013**, *23*, 1363–1372. [CrossRef]
112. Giuliani, C.; Biggs, D.; Nguyen, T.T.; Marasco, E.; De Fanti, S.; Garagnani, P.; Le Phan, M.T.; Nguyen, V.N.; Luiselli, D.; Romeo, G. First Evidence of Association between Past Environmental Exposure to Dioxin and DNA Methylation of CYP1A1 and IGF2 Genes in Present Day Vietnamese Population. *Environ. Pollut.* **2018**, *242*, 976–985. [CrossRef]
113. Droghini, E.; Annibaldi, A.; Prezioso, E.; Tramontana, M.; Frapiccini, E.; De Marco, R.; Illuminati, S.; Truzzi, C.; Spagnoli, F. Mercury Content in Central and Southern Adriatic Sea Sediments in Relation to Seafloor Geochemistry and Sedimentology. *Molecules* **2019**, *24*, 4467. [CrossRef]
114. Koenig, S.; Solé, M.; Fernández-Gómez, C.; Díez, S. New Insights into Mercury Bioaccumulation in Deep-Sea Organisms from the NW Mediterranean and Their Human Health Implications. *Sci. Total Environ.* **2013**, *442*, 329–335. [CrossRef]
115. Cammilleri, G.; Vazzana, M.; Arizza, V.; Giunta, F.; Vella, A.; Lo Dico, G.; Giaccone, V.; Giofrè, S.V.; Giangrosso, G.; Cicero, N.; et al. Mercury in Fish Products: What's the Best for Consumers between Bluefin Tuna and Yellowfin Tuna? *Nat. Prod. Res.* **2018**, *32*, 457–462. [CrossRef]

116. Cinnirella, S.; Bruno, D.E.; Pirrone, N.; Horvat, M.; Živković, I.; Evers, D.C.; Johnson, S.; Sunderland, E.M. Mercury Concentrations in Biota in the Mediterranean Sea, a Compilation of 40 Years of Surveys. *Sci. Data* **2019**, *6*, 205. [CrossRef]
117. Gallo, P.; De Carlo, E.; Marigliano, L.; Maglio, P.; Amato, A.; Improta, A.; Caruso, C.; De Roma, A. Food Safety Assessment of Heavy Metals in Uncommon and Abyssal Fish and Cephalopod from the Tyrrhenian Sea. *J. Consum. Prot. Food Saf.* **2018**, *13*, 399–402. [CrossRef]
118. Annibaldi, A.; Truzzi, C.; Carnevali, O.; Pignalosa, P.; Api, M.; Scarponi, G.; Illuminati, S. Determination of Hg in Farmed and Wild Atlantic Bluefin Tuna (*Thunnus Thynnus* L.) Muscle. *Molecules* **2019**, *24*, 1273. [CrossRef]
119. Copat, C.; Conti, G.O.; Fallico, R.; Sciacca, S.; Ferrante, M. Heavy Metals in Fish from the Mediterranean Sea. In *The Mediterranean Diet*; Elsevier: Amsterdam, The Netherlands, 2015; pp. 547–562. ISBN 978-0-12-407849-9.
120. Perugini, M.; Visciano, P.; Manera, M.; Abete, M.C.; Gavinelli, S.; Amorena, M. Contamination of Different Portions of Raw and Boiled Specimens of Norway Lobster by Mercury and Selenium. *Environ. Sci. Pollut. Res.* **2013**, *20*, 8255–8262. [CrossRef]
121. Perugini, M.; Visciano, P.; Manera, M.; Zaccaroni, A.; Olivieri, V.; Amorena, M. Heavy Metal (As, Cd, Hg, Pb, Cu, Zn, Se) Concentrations in Muscle and Bone of Four Commercial Fish Caught in the Central Adriatic Sea, Italy. *Environ. Monit. Assess.* **2014**, *186*, 2205–2213. [CrossRef]
122. Perugini, M.; Zezza, D.; Tulini, S.M.R.; Abete, M.C.; Monaco, G.; Conte, A.; Olivieri, V.; Amorena, M. Effect of Cooking on Total Mercury Content in Norway Lobster and European Hake and Public Health Impact. *Mar. Pollut. Bull.* **2016**, *109*, 521–525. [CrossRef]
123. Storelli, M.M.; Barone, G. Toxic Metals (Hg, Pb, and Cd) in Commercially Important Demersal Fish from Mediterranean Sea: Contamination Levels and Dietary Exposure Assessment: Toxic Metals in Fish. *J. Food Sci.* **2013**, *78*, T362–T366. [CrossRef]
124. Di Lena, G.; Casini, I.; Caproni, R.; Orban, E. Total Mercury Levels in Crustacean Species from Italian Fishery. *Food Addit. Contam. Part B* **2018**, *11*, 175–182. [CrossRef]
125. Bajt, O.; Ramšak, A.; Milun, V.; Andral, B.; Romanelli, G.; Scarpato, A.; Mitrić, M.; Kupusović, T.; Kljajić, Z.; Angelidis, M.; et al. Assessing Chemical Contamination in the Coastal Waters of the Adriatic Sea Using Active Mussel Biomonitoring with Mytilus Galloprovincialis. *Mar. Pollut. Bull.* **2019**, *141*, 283–298. [CrossRef]
126. Di Lena, G.; Casini, I.; Caproni, R.; Fusari, A.; Orban, E. Total Mercury Levels in Commercial Fish Species from Italian Fishery and Aquaculture. *Food Addit. Contam. Part B* **2017**, *10*, 118–127. [CrossRef]
127. Galgani, F.; Martínez-Gómez, C.; Giovanardi, F.; Romanelli, G.; Caixach, J.; Cento, A.; Scarpato, A.; BenBrahim, S.; Messaoudi, S.; Deudero, S.; et al. Assessment of Polycyclic Aromatic Hydrocarbon Concentrations in Mussels (Mytilus Galloprovincialis) from the Western Basin of the Mediterranean Sea. *Environ. Monit. Assess.* **2011**, *172*, 301–317. [CrossRef]
128. Arienzo, M.; Toscanesi, M.; Trifuoggi, M.; Ferrara, L.; Stanislao, C.; Donadio, C.; Grazia, V.; Gionata, D.V.; Carella, F. Contaminants Bioaccumulation and Pathological Assessment in Mytilus Galloprovincialis in Coastal Waters Facing the Brownfield Site of Bagnoli, Italy. *Mar. Pollut. Bull.* **2019**, *140*, 341–352. [CrossRef]
129. Conti, G.O.; Copat, C.; Ledda, C.; Fiore, M.; Fallico, R.; Sciacca, S.; Ferrante, M. Evaluation of Heavy Metals and Polycyclic Aromatic Hydrocarbons (PAHs) in Mullus Barbatus from Sicily Channel and Risk-Based Consumption Limits. *Bull. Environ. Contam. Toxicol.* **2012**, *88*, 946–950. [CrossRef]
130. US-EPA Guidance for Assessing Chemical Contamination Data for Use in Fish Advisories, Vol. II. Risk Assessment and Fish Consumption Limits EPA/823-B94-004 2000. Available online: https://www.epa.gov/sites/default/files/2015-06/documents/volume2.pdf (accessed on 12 October 2021).
131. Copat, C.; Vinceti, M.; D'Agati, M.G.; Arena, G.; Mauceri, V.; Grasso, A.; Fallico, R.; Sciacca, S.; Ferrante, M. Mercury and Selenium Intake by Seafood from the Ionian Sea: A Risk Evaluation. *Ecotoxicol. Environ. Saf.* **2014**, *100*, 87–92. [CrossRef]
132. Barone, G.; Storelli, A.; Garofalo, R.; Busco, V.P.; Quaglia, N.C.; Centrone, G.; Storelli, M.M. Assessment of Mercury and Cadmium via Seafood Consumption in Italy: Estimated Dietary Intake (EWI) and Target Hazard Quotient (THQ). *Food Addit. Contam. Part A* **2015**, *32*, 1277–1286. [CrossRef]
133. Díez, S.; Montuori, P.; Pagano, A.; Sarnacchiaro, P.; Bayona, J.M.; Triassi, M. Hair Mercury Levels in an Urban Population from Southern Italy: Fish Consumption as a Determinant of Exposure. *Environ. Int.* **2008**, *6*, 162–167. [CrossRef]
134. Den Hond, E.; Govarts, E.; Willems, H.; Smolders, R.; Casteleyn, L.; Kolossa-Gehring, M.; Schwedler, G.; Seiwert, M.; Fiddicke, U.; Castaño, A.; et al. First Steps toward Harmonized Human Biomonitoring in Europe: Demonstration Project to Perform Human Biomonitoring on a European Scale. *Environ. Health Perspect.* **2015**, *123*, 255–263. [CrossRef]
135. EUMOFA The EU Fish Market. European Market Observatory for Fisheries and Aquaculture Products. European Commission, Directorate-General for Maritime Affairs and Fisheries, Director-General, 2018. Available online: http://agricultura.gencat.cat/web/.content/de_departament/de02_estadistiques_observatoris/27_butlletins/02_butlletins_nd/documents_nd/fitxers_estatics_nd/2018/0218_2018_Pesca_Productes-pesquers-mercats-UE-2018-exportacions.pdf (accessed on 13 October 2021). [CrossRef]
136. Sekovanic, A.; Piasek, M.; Orct, T.; Grgec, A.S.; Saric, M.M.; Stasenko, S.; Jurasovic, J. Mercury Exposure Assessment in Mother–Infant Pairs from Continental and Coastal Croatia. *Biomolecules* **2020**, *27*, 821. [CrossRef]
137. Elhamri, H.; Idrissi, L.; Coquery, M.; Azemard, S.; Abidi, A.E.; Benlemlih, M.; Saghi, M.; Cubadda, F. Hair Mercury Levels in Relation to Fish Consumption in a Community of the Moroccan Mediterranean Coast. *Food Addit. Contam.* **2007**, *24*, 1236–1246. [CrossRef]

138. Giangrosso, G.; Cammilleri, G.; Macaluso, A.; Vella, A.; D'Orazio, N.; Graci, S.; Lo Dico, G.M.; Galvano, F.; Giangrosso, M.; Ferrantelli, V. Hair Mercury Levels Detection in Fishermen from Sicily (Italy) by ICP-MS Method after Microwave-Assisted Digestion. *Bioinorg. Chem. Appl.* **2016**, *2016*, 1–5. [CrossRef] [PubMed]
139. Barbone, F.; Valent, F.; Pisa, F.; Daris, F.; Fajon, V.; Logar, M.; Horvat, M. Prenatal low-level methyl mercury exposure and child development in an Italian coastal area. *Seychelles Med Dent. J.* **2004**, *7*, 149–154. [CrossRef]
140. Giuliani, C.; Bacalini, M.G.; Sazzini, M.; Pirazzini, C.; Franceschi, C.; Garagnani, P.; Luiselli, D. The Epigenetic Side of Human Adaptation: Hypotheses, Evidences and Theories. *Ann. Hum. Biol.* **2015**, *42*, 1–9. [CrossRef]
141. Lam, L.L.; Emberly, E.; Fraser, H.B.; Neumann, S.M.; Chen, E.; Miller, G.E.; Kobor, M.S. Factors Underlying Variable DNA Methylation in a Human Community Cohort. *Proc. Natl. Acad. Sci. USA* **2012**, *109*, 17253–17260. [CrossRef]
142. Bosch, A.C.; O'Neill, B.; Sigge, G.O.; Kerwath, S.E.; Hoffman, L.C. Heavy Metals in Marine Fish Meat and Consumer Health: A Review: Heavy Metals in Marine Fish Meat. *J. Sci. Food Agric.* **2016**, *96*, 32–48. [CrossRef]
143. Maggi, C.; Berducci, M.T.; Di Lorenzo, B.; Dattolo, M.; Cozzolino, A.; Mariotti, S.; Fabrizi, V.; Spaziani, R.; Virno Lamberti, C. Temporal Evolution of the Environmental Quality of the Vallona Lagoon (Northern Mediterranean, Adriatic Sea). *Mar. Pollut. Bull.* **2017**, *125*, 45–55. [CrossRef] [PubMed]
144. Balali-Mood, M.; Naseri, K.; Tahergorabi, Z.; Khazdair, M.R.; Sadeghi, M. Toxic Mechanisms of Five Heavy Metals: Mercury, Lead, Chromium, Cadmium, and Arsenic. *Front. Pharmacol.* **2021**, *12*, 643972. [CrossRef]
145. Alegría-Torres, J.A.; Carrizales-Yánez, L.; Díaz-Barriga, F.; Rosso-Camacho, F.; Motta, V.; Tarantini, L.; Bollati, V. DNA Methylation Changes in Mexican Children Exposed to Arsenic from Two Historic Mining Areas in San Luis Potosí: DNA Methylation Changes. *Environ. Mol. Mutagen.* **2016**, *57*, 717–723. [CrossRef]
146. Bandyopadhyay, A.K.; Paul, S.; Adak, S.; Giri, A.K. Reduced LINE-1 Methylation Is Associated with Arsenic-Induced Genotoxic Stress in Children. *BioMetals* **2016**, *29*, 731–741. [CrossRef]
147. Lambrou, A.; Baccarelli, A.; Wright, R.O.; Weisskopf, M.; Bollati, V.; Amarasiriwardena, C.; Vokonas, P.; Schwartz, J. Arsenic Exposure and DNA Methylation among Elderly Men. *Epidemiology* **2012**, *23*, 668–676. [CrossRef]
148. Hossain, M.B.; Vahter, M.; Concha, G.; Broberg, K. Low-Level Environmental Cadmium Exposure Is Associated with DNA Hypomethylation in Argentinean Women. *Environ. Health Perspect.* **2012**, *120*, 879–884. [CrossRef]
149. Li, C.; Yang, X.; Xu, M.; Zhang, J.; Sun, N. Epigenetic Marker (LINE-1 Promoter) Methylation Level Was Associated with Occupational Lead Exposure. *Clin. Toxicol.* **2013**, *51*, 225–229. [CrossRef]
150. Non, A.L.; Thayer, Z.M. Epigenetics for Anthropologists: An Introduction to Methods: Epigenetic Methods for Anthropologists. *Am. J. Hum. Biol.* **2015**, *27*, 295–303. [CrossRef]
151. Samson, C.A.; Whitford, W.; Snell, R.G.; Jacobsen, J.C.; Lehnert, K. Contaminating DNA in Human Saliva Alters the Detection of Variants from Whole Genome Sequencing. *Sci. Rep.* **2020**, *10*, 19255. [CrossRef] [PubMed]
152. Wu, H.-C.; Wang, Q.; Chung, W.K.; Andrulis, I.L.; Daly, M.B.; John, E.M.; Keegan, T.H.; Knight, J.; Bradbury, A.R.; Kappil, M.A.; et al. Correlation of DNA Methylation Levels in Blood and Saliva DNA in Young Girls of the LEGACY Girls Study. *Epigenetics* **2014**, *9*, 929–933. [CrossRef]
153. Langie, S.A.S.; Szarc vel Szic, K.; Declerck, K.; Traen, S.; Koppen, G.; Van Camp, G.; Schoeters, G.; Vanden Berghe, W.; De Boever, P. Whole-Genome Saliva and Blood DNA Methylation Profiling in Individuals with a Respiratory Allergy. *PLoS ONE* **2016**, *11*, e0151109. [CrossRef]
154. Murata, Y.; Fujii, A.; Kanata, S.; Fujikawa, S.; Ikegame, T.; Nakachi, Y.; Zhao, Z.; Jinde, S.; Kasai, K.; Bundo, M.; et al. Evaluation of the Usefulness of Saliva for DNA Methylation Analysis in Cohort Studies. *Neuropsychopharmacol. Rep.* **2019**, *39*, 301–305. [CrossRef] [PubMed]
155. Braun, P.R.; Han, S.; Hing, B.; Nagahama, Y.; Gaul, L.N.; Heinzman, J.T.; Grossbach, A.J.; Close, L.; Dlouhy, B.J.; Howard, M.A.; et al. Genome-Wide DNA Methylation Comparison between Live Human Brain and Peripheral Tissues within Individuals. *Transl. Psychiatry* **2019**, *9*, 47. [CrossRef]
156. Van Dongen, J.; Ehli, E.A.; Jansen, R.; van Beijsterveldt, C.E.M.; Willemsen, G.; Hottenga, J.J.; Kallsen, N.A.; Peyton, S.A.; Breeze, C.E.; Kluft, C.; et al. Genome-Wide Analysis of DNA Methylation in Buccal Cells: A Study of Monozygotic Twins and MQTLs. *Epigenetics Chromatin* **2018**, *11*, 54. [CrossRef] [PubMed]
157. McGregor, K.; Bernatsky, S.; Colmegna, I.; Hudson, M.; Pastinen, T.; Labbe, A.; Greenwood, C.M.T. An Evaluation of Methods Correcting for Cell-Type Heterogeneity in DNA Methylation Studies. *Genome Biol.* **2016**, *17*, 84. [CrossRef]
158. Narváez, D.M.; Groot, H.; Diaz, S.M.; Palma, R.M.; Muñoz, N.; Cros, M.-P.; Hernández-Vargas, H. Oxidative Stress and Repetitive Element Methylation Changes in Artisanal Gold Miners Occupationally Exposed to Mercury. *Heliyon* **2017**, *3*, e00400. [CrossRef] [PubMed]
159. Olsson, A.C.; Fevotte, J.; Fletcher, T.; Cassidy, A.; 't Mannetje, A.; Zaridze, D.; Szeszenia-Dabrowska, N.; Rudnai, P.; Lissowska, J.; Fabianova, E.; et al. Occupational Exposure to Polycyclic Aromatic Hydrocarbons and Lung Cancer Risk: A Multicenter Study in Europe. *Occup. Environ. Med.* **2010**, *67*, 98–103. [CrossRef]
160. Patel, A.B.; Shaikh, S.; Jain, K.R.; Desai, C.; Madamwar, D. Polycyclic Aromatic Hydrocarbons: Sources, Toxicity, and Remediation Approaches. *Front. Microbiol.* **2020**, *11*, 562813. [CrossRef]
161. Liu, Y.; Li, X.; Zhang, B.; Fu, Y.; Yang, A.; Zhang, H.; Zhang, H.; Niu, Y.; Nie, J.; Yang, J. CYP1A1 Methylation Mediates the Effect of Smoking and Occupational Polycyclic Aromatic Hydrocarbons Co-Exposure on Oxidative DNA Damage among Chinese Coke-Oven Workers. *Environ. Health* **2019**, *18*, 69. [CrossRef]

162. Khansakorn, N.; Wongwit, W.; Tharnpoophasiam, P.; Hengprasith, B.; Suwannathon, L.; Chanprasertyothin, S.; Sura, T.; Kaojarern, S.; Sritara, P.; Sirivarasai, J. Genetic Variations of Glutathione S-Transferase Influence on Blood Cadmium Concentration. *J. Toxicol.* **2012**, *2012*, 1–6. [CrossRef]
163. Chen, H.-I.; Chiu, Y.-W.; Hsu, Y.K.; Li, W.-F.; Chen, Y.-C.; Chuang, H.-Y. The Association of Metallothionein-4 Gene Polymorphism and Renal Function in Long-Term Lead-Exposed Workers. *Biol. Trace Elem. Res.* **2010**, *137*, 55–62. [CrossRef]
164. Nariya, A.; Pathan, A.; Shah, N.; Patel, A.; Chettiar, S.; Vyas, J.; Shaikh, I.; Jhala, D. Association of Delta-Aminolevulinic Acid Dehydratase Polymorphism with Blood Lead and Hemoglobin Level in Lead Exposed Workers. *Annu. Res. Rev. Biol.* **2017**, *17*, 1–7. [CrossRef]
165. Yang, Y.; Wu, J.; Sun, P. Effects of Delta-Aminolevulinic Acid Dehydratase Polymorphisms on Susceptibility to Lead in Han Subjects from Southwestern China. *Int. J. Environ. Res. Public. Health* **2012**, *9*, 2326–2338. [CrossRef] [PubMed]
166. Miniero, R.; Beccaloni, E.; Carere, M.; Ubaldi, A.; Mancini, L.; Marchegiani, S.; Cicero, M.R.; Scenati, R.; Lucchetti, D.; Ziemacki, G.; et al. Mercury (Hg) and Methyl Mercury (MeHg) Concentrations in Fish from the Coastal Lagoon of Orbetello, Central Italy. *Mar. Pollut. Bull.* **2013**, *76*, 365–369. [CrossRef]
167. Storelli, M.M.; Giacominelli-Stuffler, R.; Storelli, A.; D'Addabbo, R.; Palermo, C.; Marcotrigiano, G.O. Survey of Total Mercury and Methylmercury Levels in Edible Fish from the Adriatic Sea. *Food Addit. Contam.* **2003**, *20*, 1114–1119. [CrossRef]
168. Saavedra, Y.; González, A.; Fernández, P.; Blanco, J. The Effect of Size on Trace Metal Levels in Raft Cultivated Mussels (Mytilus Galloprovincialis). *Sci. Total Environ.* **2004**, *318*, 115–124. [CrossRef]
169. Puel, D.; Zsürger, N.; Breittmayer, J.P.H. Statistical Assessment of a Sampling Pattern for Evaluation of Changes in Mercury and Zinc Concentrations InPatella Coerulea. *Bull. Environ. Contam. Toxicol.* **1987**, *38*, 700–706. [CrossRef] [PubMed]
170. Vives, I.; Grimalt, J.; Fernandez, P.; Rosseland, B. Polycyclic Aromatic Hydrocarbons in Fish from Remote and High Mountain Lakes in Europe and Greenland. *Sci. Total Environ.* **2004**, *324*, 67–77. [CrossRef] [PubMed]
171. Perugini, M.; Visciano, P.; Giammarino, A.; Manera, M.; Di Nardo, W.; Amorena, M. Polycyclic Aromatic Hydrocarbons in Marine Organisms from the Adriatic Sea, Italy. *Chemosphere* **2007**, *66*, 1904–1910. [CrossRef] [PubMed]
172. Storelli, M.M.; Barone, G.; Perrone, V.G.; Storelli, A. Risk Characterization for Polycyclic Aromatic Hydrocarbons and Toxic Metals Associated with Fish Consumption. *J. Food Compos. Anal.* **2013**, *31*, 115–119. [CrossRef]
173. Rahmanpour, S.; Farzaneh Ghorghani, N.; Lotfi Ashtiyani, S.M. Polycyclic Aromatic Hydrocarbon (PAH) in Four Fish Species from Different Trophic Levels in the Persian Gulf. *Environ. Monit. Assess.* **2014**, *186*, 7047–7053. [CrossRef] [PubMed]
174. Vuorinen, P.J.; Keinänen, M.; Vuontisjärvi, H.; Baršienė, J.; Broeg, K.; Förlin, L.; Gercken, J.; Kopecka, J.; Köhler, A.; Parkkonen, J.; et al. Use of Biliary PAH Metabolites as a Biomarker of Pollution in Fish from the Baltic Sea. *Mar. Pollut. Bull.* **2006**, *53*, 479–487. [CrossRef] [PubMed]
175. Piccardo, M.T.; Coradeghini, R.; Valerio, F. Polycyclic Aromatic Hydrocarbon Pollution in Native and Caged Mussels. *Mar. Pollut. Bull.* **2001**, *42*, 951–956. [CrossRef]
176. Perugini, M.; Visciano, P.; Manera, M.; Turno, G.; Lucisano, A.; Amorena, M. Polycyclic Aromatic Hydrocarbons in Marine Organisms from the Gulf of Naples, Tyrrhenian Sea. *J. Agric. Food Chem.* **2007**, *55*, 2049–2054. [CrossRef]
177. Guerranti, C.; Grazioli, E.; Focardi, S.; Renzi, M.; Perra, G. Levels of Chemicals in Two Fish Species from Four Italian Fishing Areas. *Mar. Pollut. Bull.* **2016**, *111*, 449–452. [CrossRef] [PubMed]
178. Bajc, Z.; Kirbiš, A. Trace Element Concentrations in Mussels (Mytilus Galloprovincialis) from the Gulf of Trieste, Slovenia. *J. Food Prot.* **2019**, *82*, 429–434. [CrossRef] [PubMed]
179. Noisel, N.; Bouchard, M.; Carrier, G.; Plante, M. Comparison of a Toxicokinetic and a Questionnaire-Based Approach to Assess Methylmercury Intake in Exposed Individuals. *J. Expo. Sci. Environ. Epidemiol.* **2011**, *21*, 328–335. [CrossRef]
180. Branco, V.; Caito, S.; Farina, M.; Teixeira da Rocha, J.; Aschner, M.; Carvalho, C. Biomarkers of Mercury Toxicity: Past, Present, and Future Trends. *J. Toxicol. Environ. Health Part B* **2017**, *20*, 119–154. [CrossRef] [PubMed]
181. Kusanagi, E.; Takamura, H.; Chen, S.-J.; Adachi, M.; Hoshi, N. Children's Hair Mercury Concentrations and Seafood Consumption in Five Regions of Japan. *Arch. Environ. Contam. Toxicol.* **2018**, *74*, 259–272. [CrossRef]
182. Suzuki, K.; Yoshinaga, J. Inhalation and Dietary Exposure to Polycyclic Aromatic Hydrocarbons and Urinary 1-Hydroxypyrene in Non-Smoking University Students. *Int. Arch. Occup. Environ. Health* **2007**, *81*, 115–121. [CrossRef]
183. Domingo, J.L.; Nadal, M. Human Dietary Exposure to Polycyclic Aromatic Hydrocarbons: A Review of the Scientific Literature. *Food Chem. Toxicol.* **2015**, *86*, 144–153. [CrossRef]
184. Doddamani, A.; Ballala, A.B.K.; Madhyastha, S.P.; Kamath, A.; Kulkarni, M.M. A Cross-Sectional Study to Identify the Determinants of Non-Communicable Diseases among Fishermen in Southern India. *BMC Public Health* **2021**, *21*, 414. [CrossRef]
185. Zytoon, M.A. Occupational Injuries and Health Problems in the Egyptian Mediterranean Fisheries. *Saf. Sci.* **2012**, *50*, 113–122. [CrossRef]
186. Doza, S.; Bovbjerg, V.E.; Vaughan, A.; Nahorniak, J.S.; Case, S.; Kincl, L.D. Health-Related Exposures and Conditions among US Fishermen. *J. Agromedicine* **2021**, 1–8. [CrossRef]
187. Wu, Y.; Li, S.; Hu, K.; Yang, J. Evidence of the Moderating Role of Hair Cortisol and Hair Cortisone in the Relationship between Work Stress and Depression Symptoms among Chinese Fishermen. *J. Affect. Disord.* **2021**, *294*, 868–875. [CrossRef] [PubMed]
188. Skuladottir, G.V.; Nilsson, E.K.; Mwinyi, J.; Schiöth, H.B. One-Night Sleep Deprivation Induces Changes in the DNA Methylation and Serum Activity Indices of Stearoyl-CoA Desaturase in Young Healthy Men. *Lipids Health Dis.* **2016**, *15*, 137. [CrossRef]

189. Matosin, N.; Cruceanu, C.; Binder, E.B. Preclinical and Clinical Evidence of DNA Methylation Changes in Response to Trauma and Chronic Stress. *Chronic Stress* **2017**, *1*, 247054701771076. [CrossRef] [PubMed]
190. Mahna, D.; Puri, S.; Sharma, S. DNA Methylation Signatures: Biomarkers of Drug and Alcohol Abuse. *Mutat. Res. Mutat. Res.* **2018**, *777*, 19–28. [CrossRef] [PubMed]
191. Christiansen, C.; Castillo-Fernandez, J.E.; Domingo-Relloso, A.; Zhao, W.; El-Sayed Moustafa, J.S.; Tsai, P.-C.; Maddock, J.; Haack, K.; Cole, S.A.; Kardia, S.L.R.; et al. Novel DNA Methylation Signatures of Tobacco Smoking with Trans-Ethnic Effects. *Clin. Epigenetics* **2021**, *13*, 36. [CrossRef]

applied sciences

MDPI

Article

Meta-Analysis of a New Georeferenced Database on Polycyclic Aromatic Hydrocarbons in Western and Central Mediterranean Seafood

Andrea De Giovanni [1,2,*], Paolo Abondio [1,3], Emanuela Frapiccini [4], Donata Luiselli [1,2] and Mauro Marini [2,4]

[1] Department of Cultural Heritage, University of Bologna, Via degli Ariani 1, 48121 Ravenna, Italy; paolo.abondio2@unibo.it (P.A.); donata.luiselli@unibo.it (D.L.)
[2] Fano Marine Center, The Inter-Institute Center for Research on Marine Biodiversity, Resources and Biotechnologies (FMC), Viale Adriatico 1/N, 61032 Fano, Italy; mauro.marini@cnr.it
[3] Laboratory of Molecular Anthropology, Department of Biological, Geological and Environmental Sciences, Centre for Genome Biology, University of Bologna, Via Selmi 3, 40126 Bologna, Italy
[4] Institute for Biological Resources and Marine Biotechnologies, National Research Council (IRBIM, CNR), Largo Fiera della Pesca 2, 60125 Ancona, Italy; emanuela.frapiccini@cnr.it
* Correspondence: andrea.degiovanni5@unibo.it

Featured Application: The database featured in the present work aims to support researchers and decisionmakers in planning future investigations and in evaluating current knowledge on polycyclic aromatic hydrocarbons when it comes to fixing the limits to PAH levels in fishery products.

Abstract: The aim of this work was to collect and harmonize the results of several studies achieved over the years, in order to obtain a database of georeferenced observations on polycyclic aromatic hydrocarbons (PAHs) in Western and Central Mediterranean seafood. For each observation, some information on the taxonomy and the ecology of the sampled species are reported, as well as details on the investigated hydrocarbon, and spatial and temporal information on sampling. Moreover, two health risk indexes were calculated for each record and included in the database. Through several statistical methods, we conducted a meta-analysis of the data on some of the species in this database, identifying trends that could be related to the biology of the investigated organisms, as well as to the physico-chemical properties of each hydrocarbon and to the oceanographic characteristic of this part of the Mediterranean. The analysis of the data showed that, at a consumption rate like the one typical of the Italian population, seafood caught from the area considered in the present work seems to pose a minimal risk to health. However, we also found evidence of an increasing trend of PAH concentrations in Mediterranean mussels, pointing to the need for constant monitoring.

Keywords: environmental pollutants; polycyclic aromatic hydrocarbons; seafood; Mediterranean Sea; Adriatic Sea; Ionian Sea; Tyrrhenian Sea; database; meta-analysis; contaminants

Citation: De Giovanni, A.; Abondio, P.; Frapiccini, E.; Luiselli, D.; Marini, M. Meta-Analysis of a New Georeferenced Database on Polycyclic Aromatic Hydrocarbons in Western and Central Mediterranean Seafood. *Appl. Sci.* **2022**, *12*, 2776. https://doi.org/10.3390/app12062776

Academic Editors: Raffaele Marotta and Anna Annibaldi

Received: 17 February 2022
Accepted: 3 March 2022
Published: 8 March 2022

1. Introduction

Polycyclic aromatic hydrocarbons (PAHs) are a large group of toxic [1,2] organic compounds, made up of a variable number of fused aromatic rings of carbon and hydrogen [3,4].

Depending on their origin, PAHs can be classified as petrogenic and pyrolytic (or pyrogenic). Pyrolytic PAHs result from the incomplete combustion of organic matter, such as the combustion of wood, oil, vehicular and industrial emissions and forest fires; petrogenic PAHs, instead, derive from fossil fuels, and can occur as a result of oil spills and petroleum production [5,6].

As regards chemical structure, low molecular weight (LMW) PAHs comprise two to three aromatic rings, while PAHs comprising four aromatic rings are defined as middle molecular weight (MMW) PAHs, and those made of five or more rings are referred to as high molecular weight (HMW) PAHs [3]. In the marine environment, MMW- and HMW PAHs are mostly pyrolytic in their origin, while LMW PAHs are from petrogenic sources [7,8].

From a chemical point of view, LMW- and HMW PAHs behave differently, with MMW PAHs showing intermediate behavior. In particular, HMW PAHs are much less insoluble in water, and so they tend to bind organic particulate matter, being less bioavailable for their uptake from the water. Despite this, they can still be absorbed by aquatic organisms from sediments and particulate matter present in the water column [3,9].

For non-smokers, the major source of exposure to PAHs is diet; seafood, along with cereals, is a major contributor to the dietary intake of these compounds in Europe [10]. In 2011, the EU set the maximum allowable levels of benzo(a)pyrene (BaP) and of the sum of four HMW PAHs (PAH4; i.e., benzo(a)anthracene BaA, chrysene Chr, benzo(b)fluoranthene BbF, and BaP) in several fishery products (Table 1) [11].

Table 1. Maximum levels for PAHs in seafood, set by EU in 2011 [11]. Concentrations are expressed in mg/kg.

Foodstuffs	Maximum Levels for BaP (mg/kg)	Maximum Levels for PAH4 [1] (mg/kg)
Muscle meat of smoked fish and smoked fishery products	0.002	0.012
Bivalve mollusks (fresh, chilled or frozen)	0.005	0.030
Bivalve mollusks (smoked)	0.006	0.035

[1] *PAH4* sum of BaA, Chr, BbF and BaP.

Several factors could affect PAHs concentrations in seafood. As already mentioned, physico-chemical properties of PAHs influence their bioavailability, HMW PAHs being less available for uptake from the water, compared to LMW and MMW PAHs. Moreover, in fish, HMW PAHs are readily excreted, thanks to their fast metabolism, whose rate is higher compared to that of lighter PAHs [9,12]. PAHs levels in seafood also depend on the nutritional condition of the organism [13] and on the season [14], likely as an effect of changes in the pollutant environmental inputs [12], seasonal variation of the pollutant elimination rate [15], hydrodynamic processes [16], and/or because of factors pertaining to the reproductive cycle of the species [17]. Even the age of the fish is related to PAH accumulation, with earlier stages of life being more prone to accumulating higher amounts of contaminants because of the immaturity of detoxification pathways [15]. Finally, the level of PAHs varies with the species taken into account, and this is also because of differences in the PAH-metabolizing capability [18]. Most notably, filter feeding organisms show the slowest PAHs elimination rates [14]. Accordingly, metabolic efficiency is greater in fish, intermediate in crustaceans and lowest in mollusks [19].

Over the years, several studies have investigated the levels of PAHs in seafood caught in the Adriatic [12,20–35], Ionian [9,20,34,36–39], and Tyrrhenian Seas [7,14,16,17,19,30,40–50], with the aims of monitoring the environmental status of particularly impacted areas [39,48], improving the knowledge of factors influencing PAHs levels in marine species [12,35], and, ultimately, informing decisionmakers when it comes to fixing the limits to PAH levels in fishery products [9].

The aim of this work was to collect and harmonize the results of those studies, in order to obtain a database of georeferenced observations on PAHs in seafood caught in the Western and Central Mediterranean Sea. For each observation, information on the taxonomy and the ecology of the investigated species are reported, as well as any details on the hydrocarbon, temporal and spatial information on sampling. Additionally, two health risk indexes were calculated for each record and included in the database. Finally, we conducted a meta-analysis of the data on Mediterranean mussel, Manila clam, red mullet and common sole, identifying trends that could be related to the biology of the investigated organisms, as well as to the physico-chemical properties of each hydrocarbon and to the water circulation in this part of the Mediterranean.

2. Materials and Methods

The literature search was carried out using Scopus and PubMed, aiming for the retrieval of publications reporting PAHs levels in seafood, i.e., in marine species included in D.M. 19105, 22 September 2017, Annex 1, which details commercially relevant fish species in Italy, or having "commercial", "minor commercial", "subsistence fisheries", or "of potential interest" in the "Human uses" section on FishBase (https://www.fishbase.se/search.php, accessed on 15 January 2022), or whose specimens were recovered from the fish market. Moreover, publications had to report PAH concentrations measured in marine species caught in the Adriatic (FAO geographical subareas 17 and 18), Ionian (FAO geographical subareas 13–16, 19–21) or Tyrrhenian Seas (FAO geographical subareas 8–10, 11.2, 12).

From each publication, we extrapolated several data, including the sample location and date, species and biological tissue investigated, sample length (cm) and weight (g), sample size, and PAH concentrations detected (mg/kg) (Table S1 for the full list and description of variables included in the database). Moreover, the database comprises a column specifying whether the record comes from wild, farmed or transplanted animals.

Each record was georeferenced, providing latitude and longitude in decimal degrees of the sampling location. Following [51], a geographic precision code was assigned to each record (Table S2), based on whether the study provided the exact coordinates or a more or less precise description of the sample location, from which geographical coordinates were inferred.

The trophic level of each species was obtained from online resources (Table S3) and included in the database.

Abbreviations for the biological tissue where PAHs concentration was measured, as well as for each PAH and molecular weight class (i.e., LMW, MMW or HMW PAHs), are reported in the Supplementary Materials (Tables S4–S6, respectively).

A column specifying if the reported PAH concentration is expressed in fresh (FW), wet (WW) or dry weight (DW) was added to the database (Table S7). Moreover, whenever a study provided concentration in DW, we converted that measure in WW, reporting both the original and the inferred value in separate columns. In the analyses, FW measurements were considered WW. For statistical analyses, when the measured concentration was below the limit of quantification (LOQ) or limit of detection (LOD), and the authors specified those limits, we assigned to that record a value equal to half the LOQ or LOD.

For conversion from DW to WW, we used the following formula:

$$C\ (\text{mg/kg w.w.}) = ((100 - \%\ \text{of water}) \div 100) \times C\ (\text{mg/kg d.w.}) \quad (1)$$

where C is the PAH concentration, and % of water is the percentage of water in the analyzed tissue. The percentage of water in Mediterranean mussel (*Mytilus galloprovincialis*) and Manila clam (*Ruditapes philippinarum*) was assumed to be equal to 85%, based on personal data not shown. The percentage of water in common sole (*Solea solea*) liver, gills and muscle was assumed to be equal to 72.75%, 70% and 74.4%, respectively, based on [33]. Finally, the percentage of water in red mullet (*Mullus barbatus*) fillet was assumed to be equal to 80%, following [51].

Two health risk indexes (Table S8), excess lifetime cancer risk (ELCR) and target hazard quotient (THQ), were calculated for each record and included in the database. Specifically, ELCR was calculated according to the following equation:

$$ELCR = EF\ (\text{day/yr}) \times ED\ (\text{yr}) \times IR\ (\text{kg/day}) \times CSF\ (\text{mg kg}-1\ \text{day}) \div BW\ (\text{kg}) \times AT \quad (2)$$

where EF is the exposure frequency (365 days/year); ED is the exposure duration, which was assumed to be equal to the Italian mean life expectancy (83,226 yr) (ISTAT, 2019); IR is the ingestion rate, which is equal to the PAH concentration times the mean ingestion rate of the species (FAOSTAT, 2018); CSF is the cancer slope factor for each of the analyzed PAH (OEHHA, 2009); BW is the body weight, which was assumed to be equal to the mean body weight of the Italian population (67 Kg) (ANSA, 2013); and AT is the averaging time, which is equal to EF×ED. An ELCR above 10^{-5}, which is the acceptable lifetime risk (ALR) [52], indicates a probability greater than 1 chance over 100,000 of developing cancer [46].

THQ, which indicates the ratio between exposure and the reference dose, was calculated according to the following equation:

$$THQ = EF\ (day/yr) \times ED\ (yr) \times IR\ (kg/day) \times C\ (mg/kg) \div RFD\ (mg\ kg-1\ day) \times BW\ (kg) \times AT \quad (3)$$

where C is the PAH concentration, and RFD the oral reference dose for PAH. When THQ risk is above 1, it means that THQ is higher than the reference dose, and systemic effects may occur [46].

Graphs and statistics were produced using RStudio version3.6.1. All the analyses were carried out after grouping records according to PAHs molecular weight, i.e., analyses were conducted separately on LMW, MMW and HMW PAHs in each selected species.

After checking the assumption of normality of residual distributions through the Shapiro–Wilk test (R function: shapiro_test), we used the Wilcoxon rank sum test (R function: wilcox.test) with data on Mediterranean mussel, Manila clam, red mullet and common sole to calculate the statistical significance of differences between the mean concentrations of LMW, MMW and HMW PAHs in those species.

To look for seasonal trends, we used the Wilcoxon rank sum test to calculate the statistical significance of differences between the mean concentrations of each class of PAHs in cold (October–March) and warm (April–September) months. Additionally, to control for potential confounding factors, we repeated the above analysis using non-parametric ANCOVA (R function: ancova.np), with a sampling depth and sampling year as covariates. In the present work, we used the same month clustering as in [12,53], which is based on sea water temperature: January–March (winter), April–June (spring), July–September (summer) and October–December (autumn).

Finally, to see if there is a relationship between the latitude, sampling depth and sampling year and the PAH concentrations in Mediterranean mussel caught along the Italian coast of the Adriatic and the Tyrrhenian Seas, we used Kendall's rank correlation (R function: pcor; method = "kendall").

3. Results

3.1. Database Description

Of the 10,704 records included in the database, 5790 were extracted from a database on contaminants in Mediterranean biota available at https://www.emodnet-chemistry.eu/data (accessed on 1 December 2021), while the other 4914 are from 38 scientific publications in peer-reviewed journals. The database was compiled along the lines of the work of Cinnirella and colleagues (https://doi.org/10.1594/PANGAEA.899723) on mercury concentration in Mediterranean biota [51].

The geographical distribution of sampling sites is shown in Figure 1.

In total, 1025 records are from the Ionian Sea, 4469 from the Adriatic Sea, and 5210 from the Tyrrhenian Sea (Figure 2).

Mediterranean mussel (*Mytilus galloprovincialis*, code: Mytgal) is the most studied species, with 7710 records from 22 sources, followed by Manila clam (*Ruditapes philippinarum*, code: Rudphi), with 982 records from two sources, and common sole (*Solea solea*, code: Solsol), with 344 records from four sources. All the other species are present in this database with fewer than 250 observations (Figure 3).

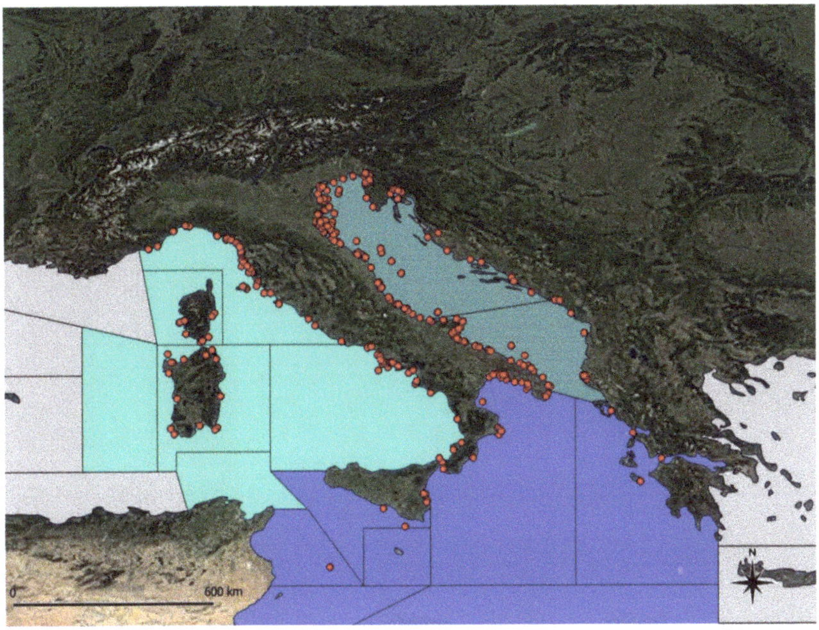

Figure 1. Geographical distribution of sampling sites (red dots). FAO geographical subareas are reported and colored based whether they belong to Adriatic (cadet blue), Ionian (royal blue) or Tyrrhenian Seas (turquoise).

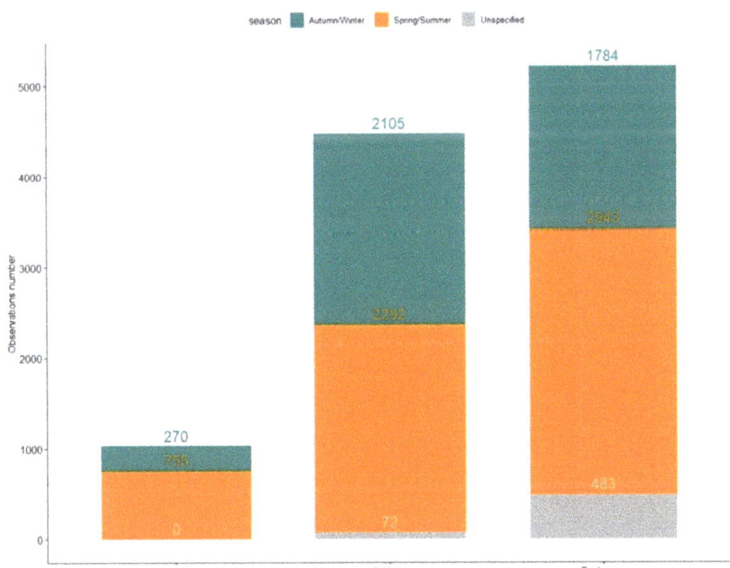

Figure 2. Bar plot showing the number of observations (x axis) by sea (i.e., Ionian, Adriatic, and Tyrrhenian Seas) in this database (y axis), with colors indicating the season in which sampling took place.

Year of publication goes from 1990 to 2021, while sampling year goes from 1981 to 2019.

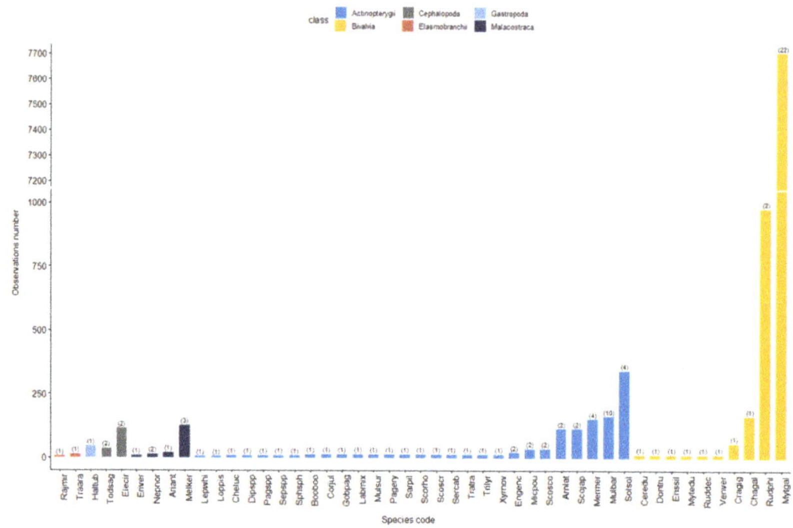

Figure 3. Bar plot showing the number of observations (x axis) for each marine species present in this database (y axis). The bars are arranged in ascending order of number of observations, by taxonomic class. On top of each bar, in brackets, is the number of sources from which the above observations were obtained. Each color corresponds to the taxonomic class of the species. Species codes are on the X-axis and are made of the first three letter of genus and species. The graph was realized using ggbreak library (version 0.0.7) in R [54].

3.2. PAHs Concentration by Molecular Weight and by Season

We compared the concentrations of LMW, MMW and HMW PAHs in Mediterranean mussel, Manila clam, common sole and red mullet, being that these species are the ones on which more records are available in this database. Summary statistics for each class of PAHs in each species are reported in Table 2. Statistics and *p*-values for each test are reported in Supplementary Materials (Tables S9–S12).

Table 2. Summary statistics on LMW (low molecular weight), MMW (middle molecular weight) and HMW (high molecular weight) PAHs in Mediterranean mussel, Manila clam, common sole and red mullet caught in the Adriatic, Ionian and Tyrrhenian Seas. Concentrations are in mg/kg wet weight.

Species	PAHs Class [1]	C_{mean} [2]	C_{min} [3]	C_{max} [4]
Mytilus galloprovincialis	LMW	0.00942	0.00000	3.96000
	MMW	0.01068	0.00000	1.05960
	HMW	0.00570	0.00000	0.34500
Ruditapes philippinarum	LMW	0.00137	0.00000	0.03770
	MMW	0.00197	0.00041	0.00810
	HMW	0.00058	0.00000	0.00690
Solea solea	LMW	0.00079	0.00006	0.01000
	MMW	0.00189	0.00009	0.01037
	HMW	0.00914	0.00000	0.73500
Mullus barbatus	LMW	0.01030	0.00011	0.09385
	MMW	0.00299	0.00042	0.01050
	HMW	0.00090	0.00000	0.00539

[1] *PAHs class*—class of PAHs based on molecular weight, [2] C_{mean}—mean concentration, [3] C_{min}—minimum concentration, [4] C_{max}—maximum concentration.

In all four species, MMW PAHs show the highest mean concentration, followed by LMW PAHs and, finally, HMW PAHs, which show the lowest mean concentration (Figure 4). In all cases, the difference between the mean concentration of each PAHs class was statistically significant ($p < 0.05$), except for the difference between MMW and LMW PAHs in red mullet.

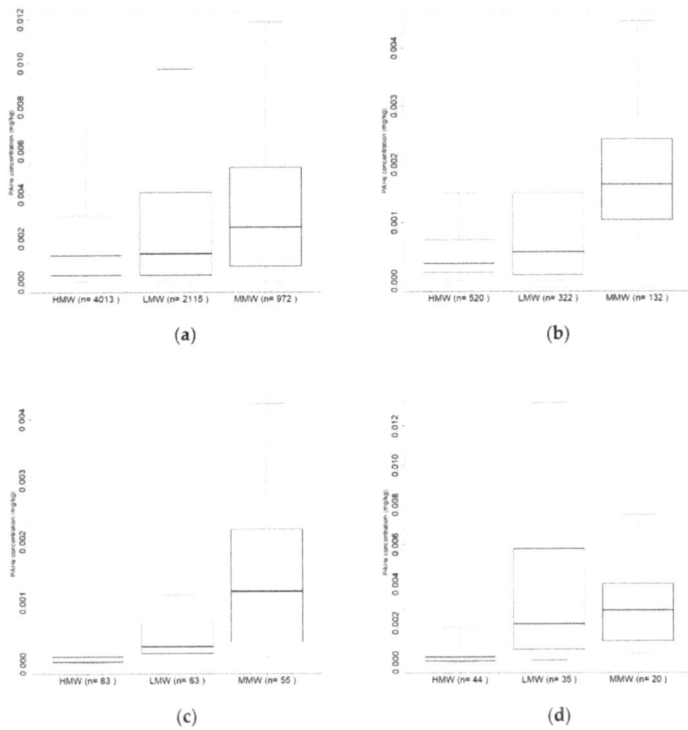

Figure 4. PAHs concentration (mg/kg wet weight) by molecular weight in (**a**) Mediterranean mussel, (**b**) Manila clam, (**c**) common sole, and (**d**) red mullet. Sample sizes are in brackets; outliers not shown.

For each class of PAHs, we compared mean concentrations measured in cold and warm months in Mediterranean mussel, Manila clam, and red mullet, controlling for the effect of sampling depth and sampling year (Figure 5). We excluded common sole from the analysis because all samples for which sampling date was specified in the original sources were collected in autumn (October–December). Summary statistics on each variable included in the present analysis are reported in Table 3.

In Mediterranean mussel, the Wilcoxon rank sum test reveals that both MMW- and HMW PAHs are present at significantly ($p < 0.05$) higher concentrations in cold months (Table S13). Moreover, ANCOVA shows that the same trend is statistically significant, even after controlling for the sampling depth and sampling year (Table S14).

In Manila clam, the Wilcoxon rank sum test reveals that all PAHs classes (i.e., LMW-, MMW- and HMW PAHs) are present at significantly higher concentrations in cold months (Table S15). However, as shown by ANCOVA, the above trend remains statistically significant after controlling for sampling year only in LMW- and MMW PAHs (Table S16). Data on sampling depth for Manila clam were not sufficient to use this variable as covariate in ANCOVA.

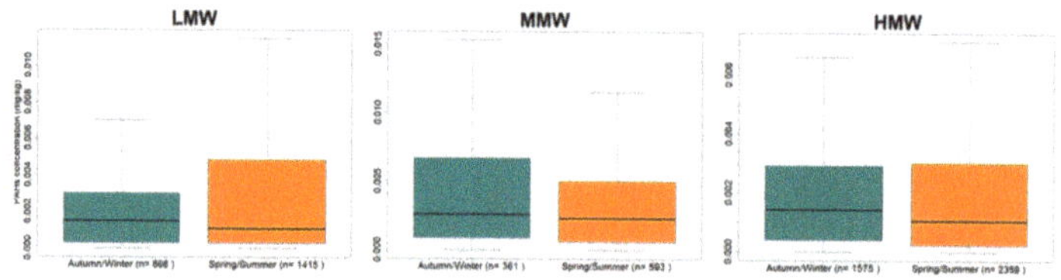

(**a**)

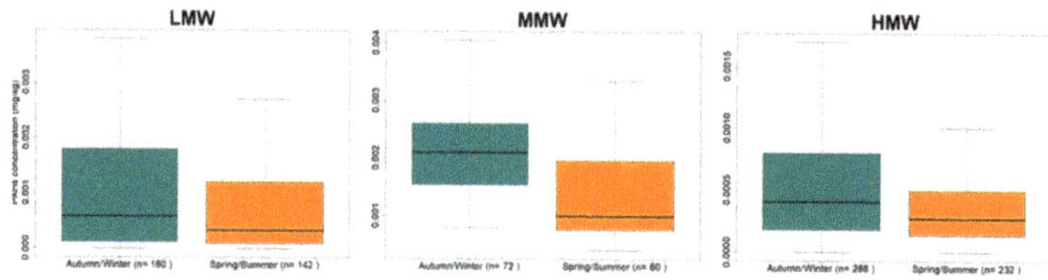

(**b**)

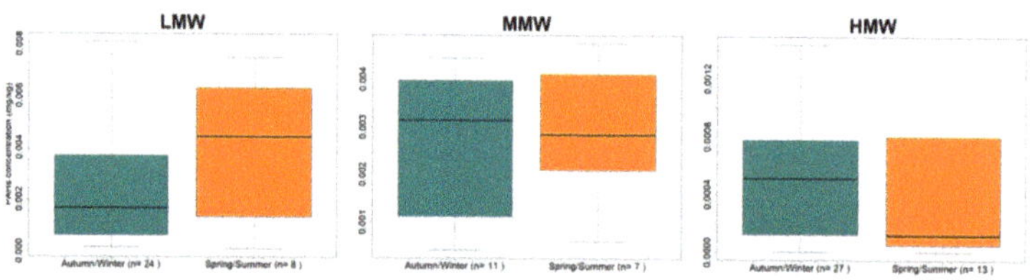

(**c**)

Figure 5. Comparison of PAHs concentrations (mg/kg wet weight) in cold and warm months in (**a**) Mediterranean mussel, (**b**) Manila clam, and (**c**) red mullet; sample sizes are in brackets; outliers not shown.

Table 3. Summary statistics on LMW (low molecular weight), MMW (middle molecular weight) and HMW (high molecular weight) PAHs in Mediterranean mussel, Manila clam and red mullet caught in different seasons in Adriatic, Ionian and Tyrrhenian Seas. Concentrations are in mg/kg wet weight; sampling depth is in meters.

Species	Period	PAHs Class [1]	C_{mean} [2]	C_{min} [3]	C_{max} [4]	$Depth_{min}$ [5]	$Dept_{max}$ [6]	$Year_{min}$ [7]	$Year_{max}$ [8]
Mytilus gallo-provincialis	Autumn/Winter	LMW	0.00434	0.00000	0.18090	0.00	77.60	1995	2018
		MMW	0.01292	0.00000	0.96000	0.00	118.80	1995	2018
		HMW	0.00512	0.00000	0.34500	0.00	118.80	1995	2018
	Spring/Summer	LMW	0.01155	0.00000	3.96000	0.00	108.00	1981	2017
		MMW	0.00908	0.00000	1.05960	0.00	108.00	1981	2017
		HMW	0.00601	0.00000	0.18040	0.00	108.00	1981	2017
Ruditapes philip-pinarum	Autumn/Winter	LMW	0.00177	0.00000	0.03770	-	-	2005	2013
		MMW	0.00228	0.00080	0.00770	-	-	2005	2013
		HMW	0.00066	0.00000	0.00690	-	-	2005	2013
	Spring/Summer	LMW	0.00087	0.00000	0.00890	-	-	2001	2014
		MMW	0.00160	0.00041	0.00810	-	-	2001	2014
		HMW	0.00047	0.00000	0.00600	-	-	2001	2014
Mullus barbatus	Autumn/Winter	LMW	0.00649	0.00016	0.09265	70	70	2004	2019
		MMW	0.00315	0.00042	0.01050	70	70	2004	2019
		HMW	0.00088	0.00000	0.00539	70	70	2004	2019
	Spring/Summer	LMW	0.01477	0.00011	0.09385	70	70	2004	2019
		MMW	0.00334	0.00060	0.00749	70	70	2004	2019
		HMW	0.00118	0.00000	0.00536	70	70	2004	2019

[1] *PAHs class*—class of PAHs based on molecular weight, [2] C_{mean}—mean concentration, [3] C_{min}—minimum concentration, [4] C_{max}—maximum concentration, [5] $Depth_{min}$—minimum sampling depth, [6] $Depth_{max}$—maximum sampling depth, [7] $Year_{min}$—year of the first sampling campaign, [8] $Year_{max}$—year of the last sampling campaign.

Finally, in red mullet, both the Wilcoxon rank sum test and ANCOVA show that for any of the PAH classes, concentrations in warm months are not significantly different from those in cold months (Tables S17 and S18). Additionally, in the case of red mullet, ANCOVA was carried out with only sampling year as a covariate, as data on sampling depth were not sufficient.

3.3. Latitude, Depth and Sampling Year Effect in Mediterranean Mussel

We tested the presence of a relationship between the latitude, sampling depth and sampling year and the PAH concentrations in Mediterranean mussel caught along the Italian coast of the Adriatic and the Tyrrhenian Seas. Kendall's rank correlation coefficients (τb) and *p*-values are reported in the Supplementary Materials (Tables S19–S30), along with maps showing the geographical distributions of sampling sites from where data used in each analysis derive (Figures S1–S4).

3.3.1. Adriatic

Summary statistics on each variable included in the correlation analysis on data from the Adriatic Sea are reported in Table 4.

Table 4. Summary statistics on LMW (low molecular weight), MMW (middle molecular weight) and HMW (high molecular weight) PAHs in Mediterranean mussels caught in different seasons along the Italian coast of the Adriatic Sea. Concentrations are in mg/kg wet weight; sampling depth is in meters; latitude are in decimal degrees. Reported statistics were calculated after missing-data removal from concentration, sampling depth, sampling year and latitude columns.

Period	PAHs Class [1]	C_{mean} [2]	C_{min} [3]	C_{max} [4]	$Depth_{min}$ [5]	$Depth_{max}$ [6]	$Year_{min}$ [7]	$Year_{max}$ [8]	Lat_{min} [9]	Lat_{max} [10]
Autumn/Winter	LMW	0.00463	0.00002	0.18090	1.20	24.80	2006	2009	41.60	45.76
	MMW	0.00591	0.00015	0.11010	1.20	24.80	2005	2009	41.60	45.76
	HMW	0.00382	0.00002	0.34500	1.20	24.80	2005	2009	41.60	45.76
Spring/Summer	LMW	0.01093	0.00000	1.15500	0.01	35.00	2006	2011	40.20	45.77
	MMW	0.00396	0.00003	0.03780	0.01	30.60	2005	2017	40.20	45.77
	HMW	0.00209	0.00000	0.03150	0.01	30.60	2005	2017	40.20	45.77

[1] *PAHs class*—class of PAHs based on molecular weight, [2] C_{mean}—mean concentration, [3] C_{min}—minimum concentration, [4] C_{max}—maximum concentration, [5] $Depth_{min}$—minimum sampling depth, [6] $Depth_{max}$—maximum sampling depth, [7] $Year_{min}$—year of the first sampling campaign, [8] $Year_{max}$—year of the last sampling campaign, [9] Lat_{min}—minimum latitude of sampling sites, [10] Lat_{max}—maximum of sampling sites.

As Kendall's rank correlation revealed, after removing the effect of the sampling depth and sampling year, concentrations of LMW-, MMW- and HMW PAHs in Mediterranean mussel caught along the Italian coast of the Adriatic Sea turned out to be negatively correlated with latitude in warm months, while the correlation becomes positive in cold months. Results are all statistically significant ($p < 0.05$), except for MMW PAHs in both periods of the year.

In Mediterranean mussel caught along the Italian coast of the Adriatic Sea, after removing the effect of the latitude and sampling year, concentrations of LMW-, MMW- and HMW PAHs are always negatively correlated with the sampling depth. Results are all statistically significant ($p < 0.05$), except in the case of MMW PAHs in warm months.

Finally, concentrations of all three classes of PAHs in warm months increase over the years, while concentrations of all three classes of PAHs in cold months decrease over the years, after removing the effect of latitude and sampling depth. Results are all statistically significant ($p < 0.05$), except for those on MMW PAHs in cold months.

3.3.2. Tyrrhenian

Summary statistics on each variable included in the correlation analysis on data from the Tyrrhenian Sea are reported in Table 5.

As Kendall's rank correlation revealed, after removing the effect of the sampling depth and sampling year, concentrations of LMW-, MMW- and HMW PAHs in Mediterranean mussel caught in the Tyrrhenian Sea increase with latitude in both cold and warm periods, and this increase is statistically significant ($p < 0.05$) for every PAH class and period of the year, excluding MMW PAHs.

Table 5. Summary statistics on LMW (low molecular weight), MMW (middle molecular weight) and HMW (high molecular weight) PAHs in Mediterranean mussels caught in different seasons in the Tyrrhenian Sea. Concentrations are in mg/kg wet weight; sampling depth is in meters; latitude are in decimal degrees. Reported statistics were calculated after missing data removal from concentration, sampling depth, sampling year and latitude columns.

Period	PAHs Class [1]	C_{mean} [2]	C_{min} [3]	C_{max} [4]	$Depth_{min}$ [5]	$Depth_{max}$ [6]	$Year_{min}$ [7]	$Year_{max}$ [8]	Lat_{min} [9]	Lat_{max} [10]
Autumn/Winter	LMW	0.00434	0.00000	0.12000	0.10	49.02	1999	2017	38.54	43.50
	MMW	0.02649	0.00000	0.96000	0.10	49.02	1999	2017	38.54	43.50
	HMW	0.00953	0.00000	0.20237	0.10	49.02	1999	2017	38.54	43.50
Spring/Summer	LMW	0.00685	0.00000	0.20000	0.19	108.00	1981	2017	38.54	44.42
	MMW	0.01234	0.00000	1.05960	0.19	108.00	1981	2017	38.97	44.42
	HMW	0.00766	0.00000	0.18040	0.19	108.00	1981	2017	38.54	44.42

[1] *PAHs class*—class of PAHs based on molecular weight, [2] C_{mean}—mean concentration, [3] C_{min}—minimum concentration, [4] C_{max}—maximum concentration, [5] $Depth_{min}$—minimum sampling depth, [6] $Depth_{max}$—maximum sampling depth, [7] $Year_{min}$—year of the first sampling campaign, [8] $Year_{max}$—year of the last sampling campaign, [9] Lat_{min}—minimum latitude of sampling sites, [10] Lat_{max}—maximum of sampling sites.

In Mediterranean mussel caught in the Tyrrhenian Sea, after removing the effect of the latitude and sampling year, concentrations of LMW in warm months and of MMW in both periods of the year are negatively correlated with the sampling depth. These results are always statistically significant ($p < 0.05$), apart from those on MMW in cold months. On the contrary, in all other cases (i.e., LMW PAHs in cold months, and HMW PAHs in both periods of the year), the PAH concentrations and sampling depth turned out to be positively correlated, always reaching statistical significance, except for HMW PAHs in warm months.

Finally, concentrations of LMW-, MMW- and HMW PAHs in Mediterranean mussels caught in the Tyrrhenian Sea, in both cold and warm months, increase over the years, and this trend is statistically significant ($p < 0.05$) for LMW- and MMW PAHs in warm months, and for HMW PAHs in both periods of the year.

3.4. Human Health Risks Assessment

In total, 92 out of 651 (~14%) records of BaP in bivalves exceed the limit of 0.005 mg/kg set by the EU [11] (Table 1), with most of these records (57) being on Mediterranean mussels caught in the Tyrrhenian Sea. Conversely, none of the 26 records of PAH4 in bivalves exceed the limit of 0.03 mg/kg set by the EU [11] (Table 1).

In this database, assuming a consumption rate equal to the average per capita consumption in Italy (FAOSTAT, 2018), ELCR values range from a minimum of 1.36×10^{-10} to a maximum of 4.52×10^{-4}, which is reached by a record of dibenzo(a,i)pyrene (DaiP) in Mediterranean mussel. A total of 324 out of 5275 records (~6%) exceeds the threshold value of 10^{-5} for ELCR. Records exceeding ELCR threshold value are mostly on Mediterranean mussel (225), while dibenzo(a,h)anthracene (DahA) is the compound most frequently associated with such high ELCR values.

THQ values range from a minimum of 2.2×10^{-9} to a maximum of ~0.13, which is reached by a record of BaP in Mediterranean mussel. Therefore, none of the records in this database exceeds the threshold value of 1 for THQ. THQ distributions for some of the most consumed species in Italy [55] are reported in Figure 6. As can be seen, among those included in the graph, European anchovy (*Engraulis encrasicolus*), blue mussel (*Mytilus edulis*) and European hake (*Merluccius merluccius*) are the first three species showing the highest median THQ.

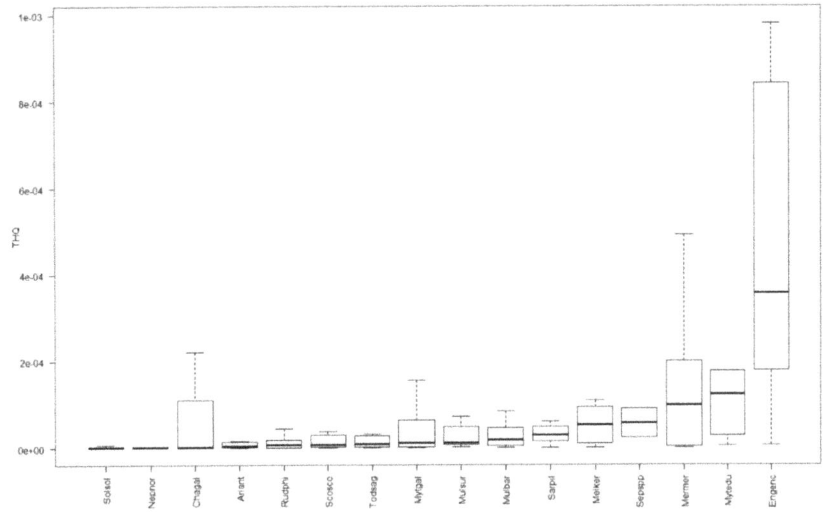

Figure 6. Boxplots showing the THQs distribution in several marine species included in this database. Species codes are on the X-axis and are made of the first three letter of genus and species. Boxplots are in ascending order from left to right, based on median values; outliers not shown.

4. Discussion

The development of this database on PAHs in seafood caught in the Western and Central Mediterranean Sea showed that most of the studies carried out from 1990 to 2021 focused on the Tyrrhenian and Adriatic Seas, with fewer records coming from the Ionian Sea. Moreover, Mediterranean mussel was by far the most studied species, accounting for more than half of records in the database. This is not surprising, given that, along with red mullet, Mediterranean mussel is considered one of the most suitable organisms to be used in biomonitoring studies, because of the widespread distribution, the ability to accumulate contaminants to a degree proportional to their bioavailability, as well as the ease of sampling [41,56].

In this database, MMW PAHs and LMW PAHs showed higher concentrations than HMW PAHs in Mediterranean mussel, Manila clam, common sole and red mullet. This is in line with the expectations, given the greater solubility and bioavailability of lighter PAHs and the faster metabolism of the heavier ones [12].

In a recent investigation carried out using large-scale monitoring data on PAHs in sediments of the Mediterranean Sea [57], the authors found that the two most prevalent PAHs in the Western Mediterranean basin were fluoranthene (Flu) and phenanthrene (Phe), with the former being the most abundant also in the Adriatic Sea and in the Central Mediterranean basin. In view of this, it is interesting to note that, in our meta-analysis, Flu and Phe are also the most abundant PAHs in both Mediterranean mussel and Manila clam (Figures S5 and S6), which are benthic filter-feeding bivalves, and so may be particularly prone to absorbing PAHs accumulated in bottom sediments, after their remobilization.

The analysis on seasonality shows that PAH concentrations tend to be higher in specimens sampled in cold months (October–March), and this trend is statistically significant in both Mediterranean mussel and Manila clam, confirming what several studies, including those in this database, found independently [8,16,19,29,32,41].

As suggested by previous investigators, the reason for such a seasonal pattern may lie both in the biology of these species and in changes in the emission and mobilization of PAHs in the environment. PAHs are lipophilic compounds [49], and as such, they accumulate preferentially in lipid-rich tissues [50]. Given that the lipid content of tissues of marine species can vary under the effect of, for example, nutritional [13] and reproductive status [35,58], one might speculate that seasonal fluctuations in such parameters may be reflected in PAH concentration changes. However, this should not be the case with Mediterranean mussel, since several investigations carried out in the Mediterranean Sea found higher lipid content in mussel sampled during summer [59,60], and this is likely due to the depletion of lipids that takes place after spawning [58,60]. Another possibility is that higher PAHs concentrations in winter are the result of increased PAH emission through, for example, domestic heat [12,17,61,62] and industrial activities [63]. Moreover, in summer, the degradation of PAHs, especially of the lighter ones (i.e., LMW PAHs) [63,64], is enhanced by UV radiation, high temperatures, and ozone [65]. On the other hand, low seawater temperature in winter can inhibit the microbial degradation of PAHs [66,67]. Additionally, the resuspension of particulate matter that takes place during winter via sea storms [19] and water mixing [12] could foster PAHs availability and accumulation in filter feeding organisms, such as mussels, and this is particularly true for heavier and more recalcitrant PAHs [12,68].

In red mullet, lighter (i.e., LMW) and heavier (i.e., MMW and HMW) PAHs show opposite trends, with the former being present at greater concentrations during warm months, and the latter during cold months. Frapiccini and colleagues [35] suggested that higher LMW PAHs in summer months might be related to an increase in maritime traffic during this period of the year, while the increase in heavier PAHs concentrations observed in winter might be caused by a reduced expression in detoxification enzymes in red mullet [69], along with the environmental factors that are hypothesized to determine the same pattern in Mediterranean mussel. Nevertheless, in our meta-analysis, the difference between PAHs concentrations in warm and cold months was not statistically significant. Moreover, it is important to note that, analyzing the liver of red mullets caught in the Northern Adriatic Sea, Guerranti and colleagues [30] found higher levels of PAHs in autumn than in spring, but they found the opposite trend in the liver of red mullets caught in the Southern Adriatic and Tyrrhenian Seas.

In this database, concentrations of PAHs in Mediterranean mussel caught along the Italian coast of the Adriatic Sea are negatively correlated with latitude in warm months, (average $\tau b \cong -0.2$). On the contrary, concentrations of PAHs and latitude show a positive correlation (average $\tau b \cong 0.2$) in cold months. We tentatively attribute the above results to the water masses circulation that characterizes the Adriatic Sea, and/or to the seasonal changes in the river discharge from the Italian coast. The major riverine freshwater input in the Adriatic Sea [70], the Po River, flows into the northern part of the basin, with greater discharge in spring and autumn, and lower discharge in summer [53]. In winter, the Po plume flows mostly southward [53] along the Italian coast, forming the Western Adriatic Current, a buoyant fresher water layer more than 50 m deep [71]. Moreover, in this period of the year, the cold Bora wind that affects the region causes the cooling of the sea surface layer, resulting in the complete vertical mixing of the sea water. This in turn leads to the resuspension of the bottom sediments and, therefore, to the remobilization of contaminants accumulated therein [53]. On the other hand, during spring and summer, the strong thermal stratification of the water results in the offshore propagation of the Po plume, which reaches the center of the basin [53].

Considering all the above, we hypothesize that the positive correlation between the PAH concentration and latitude observed in cold months may be due to, or at least favored by, the increased input of pollutants from the Po River, coupled with the resuspension of contaminants determined by the cooling action of the Bora wind in the northern Adriatic Sea. This latitudinal trend may be disrupted in warm months because of a greater contribution of pollutants from local rivers in the middle Adriatic, given the weaker influence of the Po plume on the regions south of the Po delta.

Our results seem to agree with a previous study analyzing nutrients transport along the Western Adriatic coast [70]. Indeed, what that study found is that, in winter, nutrient concentrations in water decreased southward from the Po River, while in spring, concentrations off Pescara were higher than concentrations in the Po area.

Although above explanation might sound plausible, our results should be interpreted with caution, considering the high data heterogeneity and the different geographical distribution of observations during warm and cold months, the latter being more discontinuous (Figure S2).

The negative correlation (average $\tau b \cong -0.2$) observed between the sampling depth and PAH concentrations in Mediterranean mussel caught along the Italian coast of the Adriatic Sea is in accordance with several past investigations. A study carried out in the Western Mediterranean Sea revealed a higher PAH content in suspended particulate matter collected near the sea surface compared to that sampled from deep water [72], while another study found similar results in water samples collected in the Baltic Sea [66]. Moreover, in [7], concentrations of PAHs in Mediterranean mussels collected in the Gulf of Naples were negatively correlated with sampling depth. We hypothesize that the observed bathymetric gradient may be due, at least in the Adriatic Sea, to a greater impact of river discharge at a low depth.

Concerning the discrepancy between results of the correlation analyses on the PAH concentration and sampling year, with concentrations increasing over the years in warm months and decreasing over the years in cold months, we were not able to find a likely explanation based on information retrieved from the literature. Overall, we cannot exclude that our results are an artefact stemming from the high degree of heterogeneity of the data collected in the database. Moreover, it is notable that data on PAH concentrations in Mediterranean mussel caught along the Western Adriatic coast in cold months are from the periods 2006–2009 and 2005–2009, for LMW PAHs and for MMW- and HMW PAHs, respectively, while data on mussels caught in warm months cover a longer period of time (2006–2011 and 2005–2017 for LMW PAHs and for MMW- and HMW PAHs, respectively) (Table 4); therefore, we cannot rule out the possibility of an actual declining trend in PAH concentrations in the period covered by data on cold months, whose signal is lost in the broader period that the data on warm months span. In this regard, it is interesting to note that an oscillating temporal trend of PAHs concentrations was observed in Adriatic sediments by Rizzi and colleagues [57], with several PAHs showing a decrease in concentrations until the years between 2005 and 2010, when a new increase took place (Figure S2 of [57]).

Turning to mussels from the Tyrrhenian Sea, we observed a latitudinal trend, consistent across the three PAHs classes, in both periods of the year, that is, an increase in PAH concentrations from the southern to the northern part of the basin (average $\tau b \cong 0.1$). Interpreting the above results is challenging, given the low number of sampling sites, especially in cold months, and their uneven geographical distribution along the Italian coast. Therefore, we limit ourselves to observe that, in the Tyrrhenian Sea, the salinity of the shallower water mass (0–150 m) follows the same latitudinal trend that we observed for the PAH concentration in mussels, passing from 36.2 psu in the southern region to 38.4 in the northern [73]; salinity superior or equal to 37 psu has been shown to slow down the degradation of PAH molecules [64]. Moreover, water circulation in the Tyrrhenian Sea is dominated by a wide cyclonic path that enters the basin through the Sardinia Channel and flows along the Sicilian and Italian coasts [73]. Thereby, it could be that mussels at more northerly latitudes are affected not only by local input of contaminants, but also by the substances that currents catch along their path. Finally, it is worth noting that, as can be seen from the marine traffic density map [74] of the European Atlas of the Seas [75], the Tyrrhenian coast of Calabria, in the southern part of the basin, is the least affected by vessel traffic.

The inconsistency between the results of the correlation analysis on PAH concentrations and sampling depth in the Tyrrhenian Sea is not easy to interpret, and again, it may represent a simple artefact of data heterogeneity. The statistically significant negative correlation between sampling depth and LMW- and MMW PAHs concentrations in warm months is in line with the results on LMW- and HMW PAHs in the Adriatic Sea and may be attributable to a greater impact of river discharge at low depth. On the contrary, we found that LMW- and HMW PAHs measured in the Tyrrhenian Sea in cold months significantly increase with sampling depth. Although we cannot find a convincing explanation to such a pattern, it is noteworthy that the Tyrrhenian Sea is rich in volcanic submarine structures [76,77], which can be important contributors of pyrolytic PAHs in the environment [78,79]. As such, we hypothesize that Tyrrhenian mussels that inhabit greater depths may be more susceptible to contamination driven by submarine volcanism, being at the same time less affected by river discharge.

The increasing trend of PAH concentrations in Mediterranean mussels from Tyrrhenian Sea is in line with recent findings suggesting that, in recent years, concentrations of these compounds in Mediterranean Sea sediments have increased, especially in the western part of the basin, probably as an effect of a parallel rise in PAH emissions from forest fires [57].

Both human and animal studies point to PAH exposure as being detrimental for the health, due to, for example, their carcinogenicity, teratogenicity and endocrine-disrupting effects [2]. Accordingly, the EU set a maximum level of several PAHs in fresh and smoked seafood to be sold [11] (Table 1). None of the records included in this database exceed those limits. Moreover, based on FAOSTAT data on per capita seafood consumption in Italy (FAOSTAT, 2018), none of the records in this database exceed the threshold value of 1 for THQ, pointing to a minimum risk of running into systemic effects at a consumption rate like the one typical of the Italian population. Conversely, in around 6% of cases, samples exceed the threshold value of 10^{-5} for ELCR, pointing to a probability of greater than 1 chance over 100,000 of developing cancer [46]. Ultimately, at a consumption rate like the one typical of the Italian population, seafood caught from the area considered in the present work seems to pose a minimal risk to health. However, it should be considered that the seafood ingestion rate is variable among the population [80], and that an individual is simultaneously exposed to several PAH sources, from both ingestion and other routes. Furthermore, it is vital to note the emerging role of genetics in shaping individual susceptibility to PAHs [81].

5. Conclusions

Gathering the results of several investigations, we produced a database on PAHs in seafood from the Western and Central Mediterranean Sea. A clear imbalance in favor of studies addressing PAHs in bivalve mollusks emerged.

The meta-analysis carried out on the database led us to obtain potential hints on factors (e.g., reproductive status, water masses circulation, and river discharge seasonal variability) that could determine differences in the PAH contamination of marine species.

The assessment of human health risks posed by PAH seafood contamination showed that, at a consumption rate like the one typical of the Italian population, seafood caught from the study areas seems to pose a minimal risk to health. Despite this, concerns may arise considering the individual susceptibility to PAHs exposure as well as the apparent increasing trend of PAHs levels observed in both environmental matrices and sea animals.

Supplementary Materials: The following are available online at https://www.mdpi.com/article/10.3390/app12062776/s1. Table S1: Columns of the database Table S2: Geographic precision code used in the database. Table S3: Sources of the trophic levels of the organisms included in the database. Table S4: Tissues codes used in the database. Table S5: PAHs included in the database. Table S6: PAHs molecular weight abbreviations used in the database. Table S7: Water content abbreviations used in the database. Table S8: Indices and formulas included in the database. Table S9: Pairwise comparisons between the three PAH classes in Mediterranean mussel, using Wilcoxon rank sum test. *P*-value adjustment method: Benjamini–Hochberg. Table S10: Pairwise comparisons between the three PAHs classes in Manila clam, using Wilcoxon rank sum test. *P*-value adjustment method: Benjamini–Hochberg. Table S11: pairwise comparisons between the three PAHs classes in common sole, using Wilcoxon rank sum test. *P*-value adjustment method: Benjamini–Hochberg. Table S12: Pairwise comparisons between the three PAHs classes in red mullet, using Wilcoxon rank sum test. *P*-value adjustment method: Benjamini–Hochberg. Table S13: Comparison between mean concentrations measured in cold and warm months in Mediterranean mussel, using Wilcoxon rank sum test with continuity correction. Table S14: Comparison between mean concentrations measured in cold and warm months in Mediterranean mussel, controlling for the effect of sampling depth and sampling year with ANCOVA. Table S15: Comparison between mean concentrations measured in cold and warm months in Manila clam, using Wilcoxon rank sum test with continuity correction. Table S16: Comparison between mean concentrations measured in cold and warm months in Manila clam, controlling for the effect of sampling year with ANCOVA. Table S17: Comparison between mean concentrations measured in cold and warm months in red mullet, using Wilcoxon rank sum test with continuity correction. Table S18: Comparison between mean concentrations measured in cold and warm months in red mullet, controlling for the effect of sampling year with ANCOVA. Table S19: Pairwise correlation coefficients with p-values (among brackets) between latitude, sampling depth and sampling year and LMW PAHs concentrations in Mediterranean mussel caught along the Italian coast of the Adriatic Sea in warm months, Table S20: Pairwise correlation coefficients with p-values (among brackets) between latitude, sampling depth and sampling year and LMW PAHs concentrations in Mediterranean mussel caught along the Italian coast of the Adriatic Sea in cold months. Table S21: Pairwise correlation coefficients with p-values (among brackets) between latitude, sampling depth and sampling year and MMW PAHs concentrations in Mediterranean mussel caught along the Italian coast of the Adriatic Sea in warm months. Table S22: Pairwise correlation coefficients with p-values (among brackets) between latitude, sampling depth and sampling year and MMW PAHs concentrations in Mediterranean mussel caught along the Italian coast of the Adriatic Sea in cold months. Table S23: Pairwise correlation coefficients with p-values (among brackets) between latitude, sampling depth and sampling year and HMW PAHs concentrations in Mediterranean mussel caught along the Italian coast of the Adriatic Sea in warm months. Table S24: Pairwise correlation coefficients with p-values (among brackets) between latitude, sampling depth and sampling year and HMW PAHs concentrations in Mediterranean mussel caught along the Italian coast of the Adriatic Sea in cold months. Table S25: Pairwise correlation coefficients with p-values (among brackets) between latitude, sampling depth and sampling year and LMW PAHs concentrations in Mediterranean mussel caught in the Tyrrhenian Sea in warm months. Table S26: Pairwise correlation coefficients with p-values (among brackets) between latitude, sampling depth and sampling year and LMW PAHs concentrations in Mediterranean mussel caught in the Tyrrhenian Sea in cold months. Table S27: Pairwise correlation coefficients with p-values (among brackets) between latitude, sampling depth and sampling year and MMW PAHs concentrations in Mediterranean mussel caught in the Tyrrhenian Sea in warm months. Table S28: Pairwise correlation coefficients with p-values (among brackets) between latitude, sampling depth and sampling year and MMW PAHs concentrations in Mediterranean mussel caught in the Tyrrhenian Sea in cold months. Table S29: Pairwise correlation coefficients with p-values (among brackets) between latitude, sampling depth and sampling year and HMW PAHs concentrations in Mediterranean mussel caught in the Tyrrhenian Sea in warm months. Table S30: Pairwise correlation coefficients with p-values (among brackets) between latitude, sampling depth and sampling year and HMW PAHs concentrations in Mediterranean mussel caught in the Tyrrhenian Sea in cold months. Figure S1: Geographical distributions of sampling sites of Mediterranean mussel caught along the Italian coast of the Adriatic Sea in warm months. Figure S2: Geographical distributions of sampling sites of Mediterranean mussel caught along the Italian coast of the Adriatic Sea in cold months. Figure S3: Geographical distributions of sampling sites of Mediterranean mussel caught in the Tyrrhenian Sea in warm months. Figure S4. Geographical distributions of sampling sites of Mediterranean mussels caught in the Tyrrhenian Sea in cold months. Figure S5: Boxplots of the concentration of single PAHs in Mediterranean mussel. The color of the boxes indicates the molecular weight of the given PAH. Boxplots are arranged in ascending order by median. Figure S6: Boxplots of the concentration of single PAHs in Manila clam. The color of the boxes indicates the molecular weight of the given PAH. Boxplots are arranged in ascending order by median.

Author Contributions: Conceptualization, A.D.G. and M.M.; methodology, A.D.G. and M.M.; formal analysis, A.D.G., P.A.; data curation, A.D.G.; writing—original draft preparation, A.D.G.; writing—review and editing, P.A., E.F., D.L., M.M.; supervision, E.F., D.L., M.M.; project administration, D.L., M.M. All authors have read and agreed to the published version of the manuscript.

Funding: This research received no external funding.

Institutional Review Board Statement: Not applicable.

Informed Consent Statement: Not applicable.

Data Availability Statement: The comma-separated values file of the database presented in this study is in the Supplementary Materials.

Acknowledgments: The research leading to these results was conceived under the collaboration between the University of Bologna and the National Research Council for the implementation of the International PhD Program "Innovative Technologies and Sustainable Use of Mediterranean Sea Fishery and Biological Resources" (www.FishMed-PhD.org, accessed on 15 January 2022).

Conflicts of Interest: The authors declare no conflict of interest.

References

1. Pavanello, S.; Campisi, M.; Mastrangelo, G.; Hoxha, M.; Bollati, V. The effects of everyday-life exposure to polycyclic aromatic hydrocarbons on biological age indicators. *Environ. Health* **2020**, *19*, 128. [CrossRef] [PubMed]
2. Sun, K.; Song, Y.; He, F.; Jing, M.; Tang, J.; Liu, R. A review of human and animals exposure to polycyclic aromatic hydrocarbons: Health risk and adverse effects, photo-induced toxicity and regulating effect of microplastics. *Sci. Total Environ.* **2021**, *773*, 145403. [CrossRef] [PubMed]
3. Capuano, E.; van Ruth, S.M. Infrared Spectroscopy: Applications. In *Encyclopedia of Food and Health*, 1st ed.; Caballero, B., Finglas, P.M., Toldrá, F., Eds.; Academic Press: Oxford, UK, 2016; pp. 424–431, ISBN 978-0-12-384953-3.
4. Sakshi; Haritash, A.K. A comprehensive review of metabolic and genomic aspects of PAH-degradation. *Arch. Microbiol.* **2020**, *202*, 2033–2058. [CrossRef] [PubMed]
5. Ifegwu, O.C.; Anyakora, C. Polycyclic Aromatic Hydrocarbons. In *Advances in Clinical Chemistry*; Elsevier: Amsterdam, The Netherlands, 2015; Volume 72, pp. 277–304. [CrossRef]
6. Lucas, J.; Percelay, I.; Larcher, T.; Lefrançois, C. Effects of pyrolytic and petrogenic polycyclic aromatic hydrocarbons on swimming and metabolic performance of zebrafish contaminated by ingestion. *Ecotoxicol. Environ. Saf.* **2016**, *132*, 145–152. [CrossRef]
7. Mercogliano, R.; Santonicola, S.; De Felice, A.; Anastasio, A.; Murru, N.; Ferrante, M.C.; Cortesi, M.L. Occurrence and distribution of polycyclic aromatic hydrocarbons in mussels from the gulf of Naples, Tyrrhenian Sea, Italy. *Mar. Pollut. Bull.* **2016**, *104*, 386–390. [CrossRef]
8. Grigoriou, C.; Costopoulou, D.; Vassiliadou, I.; Chrysafidis, D.; Tzamtzis, V.; Bakeas, E.; Leondiadis, L. Monitoring of Polycyclic Aromatic Hydrocarbon Levels in Mussels (*Mytilus galloprovincialis*) from Aquaculture Farms in Central Macedonia Region, Greece, Using Gas Chromatography—Tandem Mass Spectrometry Method. *Molecules* **2021**, *26*, 5953. [CrossRef]
9. Bua, R.O.; Contino, A.; Giuffrida, A. Polycyclic aromatic hydrocarbons in Mullus surmuletus from the Catania Gulf (Sicily, Italy): Distribution and potential health risks. *Environ. Sci. Pollut. Res.* **2020**, *28*, 7756–7765. [CrossRef]
10. SCF. Opinion of the Scientific Committee on Food on the risks to human health of Polycyclic Aromatic Hydrocarbons in food. *EFSA J.* **2008**, *724*, 1–114. [CrossRef]
11. Commission Regulation (EU) No 835/2011 of 19 August 2011 Amending Regulation (EC) No 1881/2006 as Regards Maxi-mum Levels for Polycyclic Aromatic Hydrocarbons in FoodstuffsText with EEA Relevance. 5. Available online: https://eur-lex.europa.eu/LexUriServ/LexUriServ.do?uri=OJ:L:2011:215:0004:0008:En:PDF (accessed on 15 January 2022).
12. Frapiccini, E.; Cocci, P.; Annibaldi, A.; Panfili, M.; Santojanni, A.; Grilli, F.; Marini, M.; Palermo, F.A. Assessment of seasonal relationship between polycyclic aromatic hydrocarbon accumulation and expression patterns of oxidative stress-related genes in muscle tissues of red mullet (*M. barbatus*) from the Northern Adriatic Sea. *Environ. Toxicol. Pharmacol.* **2021**, *88*, 103752. [CrossRef]
13. González-Fernández, C.; Albentosa, M.; Campillo, J.A.; Viñas, L.; Romero, D.; Franco, A.; Bellas, J. Effect of nutritive status on Mytilus galloprovincialis pollution biomarkers: Implications for large-scale monitoring programs. *Aquat. Toxicol.* **2015**, *167*, 90–105. [CrossRef]
14. Conte, F.; Copat, C.; Longo, S.; Conti, G.O.; Grasso, A.; Arena, G.; Dimartino, A.; Brundo, M.V.; Ferrante, M. Polycyclic aromatic hydrocarbons in *Haliotis tuberculata* (*Linnaeus*, 1758) (Mollusca, Gastropoda): Considerations on food safety and source investigation. *Food Chem. Toxicol.* **2016**, *94*, 57–63. [CrossRef] [PubMed]
15. Recabarren-Villalón, T.; Ronda, A.C.; Oliva, A.L.; Cazorla, A.L.; Marcovecchio, J.E.; Arias, A.H. Seasonal distribution pattern and bioaccumulation of Polycyclic aromatic hydrocarbons (PAHs) in four bioindicator coastal fishes of Argentina. *Environ. Pollut.* **2021**, *291*, 118125. [CrossRef] [PubMed]
16. Esposito, M.; Canzanella, S.; Lambiase, S.; Scaramuzzo, A.; La Nucara, R.; Bruno, T.; Picazio, G.; Colarusso, G.; Brunetti, R.; Gallo, P. Organic pollutants (PCBs, PCDD/Fs, PAHs) and toxic metals in farmed mussels from the Gulf of Naples (Italy): Monitoring and human exposure. *Reg. Stud. Mar. Sci.* **2020**, *40*, 101497. [CrossRef]
17. Fiorito, F.; Amoroso, M.G.; Lambiase, S.; Serpe, F.P.; Bruno, T.; Scaramuzzo, A.; Maglio, P.; Fusco, G.; Esposito, M. A relationship between environmental pollutants and enteric viruses in mussels (*Mytilus galloprovincialis*). *Environ. Res.* **2018**, *169*, 156–162. [CrossRef] [PubMed]
18. Baali, A.; Yahyaoui, A. Polycyclic Aromatic Hydrocarbons (PAHs) and Their Influence to Some Aquatic Species. In *Biochemical Toxicology-Heavy Metals and Nanomaterials*; Ince, M., Kaplan Ince, O., Ondrasek, G., Eds.; IntechOpen: London, UK, 2020; ISBN 978-1-78984-696-6.
19. Perugini, M.; Visciano, P.; Manera, M.; Turno, G.; Lucisano, A.A.; Amorena, M. Polycyclic Aromatic Hydrocarbons in Marine Organisms from the Gulf of Naples, Tyrrhenian Sea. *J. Agric. Food Chem.* **2007**, *55*, 2049–2054. [CrossRef]

20. Corsi, I.; Mariottini, M.; Menchi, V.; Sensini, C.; Balocchi, C.; Focardi, S. Monitoring a Marine Coastal Area: Use of Mytilus galloprovincialis and Mullus barbatus as Bioindicators. *Mar. Ecol.* **2002**, *23*, 138–153. [CrossRef]
21. Vassura, I.; Foschini, F.; Baravelli, V.; Fabbri, D. Distribution of alternant and non-alternant polycyclic aromatic hydrocarbons in sediments and clams of the Pialassa Baiona Lagoon (Ravenna, Italy). *Chem. Ecol.* **2005**, *21*, 415–424. [CrossRef]
22. Fabbri, D.; Baravelli, V.; Giannotti, K.; Donnini, F.; Fabbri, E. Bioaccumulation of cyclopenta[cd]pyrene and benzo[ghi]fluoranthene by mussels transplanted in a coastal lagoon. *Chemosphere* **2006**, *64*, 1083–1092. [CrossRef]
23. Perugini, M.; Visciano, P.; Giammarino, A.; Manera, M.; Di Nardo, W.; Amorena, M. Polycyclic aromatic hydrocarbons in marine organisms from the Adriatic Sea, Italy. *Chemosphere* **2007**, *66*, 1904–1910. [CrossRef]
24. Bihari, N.; Fafanđel, M.; Piškur, V. Polycyclic Aromatic Hydrocarbons and Ecotoxicological Characterization of Seawater, Sediment, and Mussel *Mytilus galloprovincialis* from the Gulf of Rijeka, the Adriatic Sea, Croatia. *Arch. Environ. Contam. Toxicol.* **2007**, *52*, 379–387. [CrossRef]
25. Della Torre, C.; Corsi, I.; Nardi, F.; Perra, G.; Tomasino, M.P.; Focardi, S. Transcriptional and post-transcriptional response of drug-metabolizing enzymes to PAHs contamination in red mullet (Mullus barbatus, Linnaeus, 1758): A field study. *Mar. Environ. Res.* **2010**, *70*, 95–101. [CrossRef] [PubMed]
26. Corsi, I.; Tabaku, A.; Nuro, A.; Beqiraj, S.; Marku, E.; Perra, G.; Tafaj, L.; Baroni, D.; Bocari, D.; Guerranti, C.; et al. Ecotoxicological Assessment of Vlora Bay (Albania) by a Biomonitoring Study Using an Integrated Approach of Sublethal Toxicological Effects and Contaminant Levels in Bioindicator Species. *J. Coast. Res.* **2011**, *270*, 116–120. [CrossRef]
27. Trisciani, A.; Corsi, I.; Della Torre, C.; Perra, G.; Focardi, S. Hepatic biotransformation genes and enzymes and PAH metabolites in bile of common sole (Solea solea, Linnaeus, 1758) from an oil-contaminated site in the Mediterranean Sea: A field study. *Mar. Pollut. Bull.* **2011**, *62*, 806–814. [CrossRef] [PubMed]
28. Storelli, M.M.; Barone, G.; Perrone, V.G.; Storelli, A. Risk characterization for polycyclic aromatic hydrocarbons and toxic metals associated with fish consumption. *J. Food Compos. Anal.* **2013**, *31*, 115–119. [CrossRef]
29. Gomiero, A.; Volpato, E.; Nasci, C.; Perra, G.; Viarengo, A.; Dagnino, A.; Spagnolo, A.; Fabi, G. Use of multiple cell and tissue-level biomarkers in mussels collected along two gas fields in the northern Adriatic Sea as a tool for long term environmental monitoring. *Mar. Pollut. Bull.* **2015**, *93*, 228–244. [CrossRef] [PubMed]
30. Guerranti, C.; Grazioli, E.; Focardi, S.; Renzi, M.; Perra, G. Levels of chemicals in two fish species from four Italian fishing areas. *Mar. Pollut. Bull.* **2016**, *111*, 449–452. [CrossRef]
31. Glad, M.; Bihari, N.; Jaksic, Z.; Fafanđel, M. Comparison between resident and caged mussels: Polycyclic aromatic hydrocarbon accumulation and biological response. *Mar. Environ. Res.* **2017**, *129*, 195–206. [CrossRef]
32. Cacciatore, F.; Bernarello, V.; Brusà, R.B.; Sesta, G.; Franceschini, G.; Maggi, C.; Gabellini, M.; Lamberti, C.V. PAH (Polycyclic Aromatic Hydrocarbon) bioaccumulation and PAHs/shell weight index in *Ruditapes philippinarum* (Adams & Reeve, 1850) from the Vallona lagoon (northern Adriatic Sea, NE Italy). *Ecotoxicol. Environ. Saf.* **2018**, *148*, 787–798. [CrossRef]
33. Frapiccini, E.; Annibaldi, A.; Betti, M.; Polidori, P.; Truzzi, C.; Marini, M. Polycyclic aromatic hydrocarbon (PAH) accumulation in different common sole (Solea solea) tissues from the North Adriatic Sea peculiar impacted area. *Mar. Pollut. Bull.* **2018**, *137*, 61–68. [CrossRef]
34. Bajt, O.; Ramšak, A.; Milun, V.; Andral, B.; Romanelli, G.; Scarpato, A.; Mitrić, M.; Kupusović, T.; Kljajić, Z.; Angelidis, M.; et al. Assessing chemical contamination in the coastal waters of the Adriatic Sea using active mussel biomonitoring with *Mytilus galloprovincialis*. *Mar. Pollut. Bull.* **2019**, *141*, 283–298. [CrossRef]
35. Frapiccini, E.; Panfili, M.; Guicciardi, S.; Santojanni, A.; Marini, M.; Truzzi, C.; Annibaldi, A. Effects of biological factors and seasonality on the level of polycyclic aromatic hydrocarbons in red mullet (*Mullus barbatus*). *Environ. Pollut.* **2019**, *258*, 113742. [CrossRef] [PubMed]
36. Storelli, M.M.; Marcotrigiano, G.O. Polycyclic Aromatic Hydrocarbons in Mussels (*Mytilus galloprovincialis*) from the Ionian Sea, Italy. *J. Food Prot.* **2001**, *64*, 405–409. [CrossRef] [PubMed]
37. Conti, G.O.; Copat, C.; Ledda, C.; Fiore, M.; Fallico, R.; Sciacca, S.; Ferrante, M. Evaluation of Heavy Metals and Polycyclic Aromatic Hydrocarbons (PAHs) in Mullus barbatus from Sicily Channel and Risk-Based Consumption Limits. *Bull. Environ. Contam. Toxicol.* **2012**, *88*, 946–950. [CrossRef]
38. Marrone, R.; Smaldone, G.; Pepe, T.; Mercogliano, R.; De Felice, A.; Anastasio, A. Polycyclic Aromatic Hydrocarbons (Pahs) in Seafoods Caught in Corigliano Calabro Gulf (Cs, Italy). *Ital. J. Food Saf.* **2012**, *1*, 41–46. [CrossRef]
39. Traina, A.; Ausili, A.; Bonsignore, M.; Fattorini, D.; Gherardi, S.; Gorbi, S.; Quinci, E.; Romano, E.; Manta, D.S.; Tranchida, G.; et al. Organochlorines and Polycyclic Aromatic Hydrocarbons as fingerprint of exposure pathways from marine sediments to biota. *Mar. Pollut. Bull.* **2021**, *170*, 112676. [CrossRef] [PubMed]
40. Cocchieri, R.A.; Arnese, A.; Minicucci, A.M. Polycyclic aromatic hydrocarbons in marine organisms from Italian central Mediterranean coasts. *Mar. Pollut. Bull.* **1990**, *21*, 15–18. [CrossRef]
41. Piccardo, M.; Coradeghini, R.; Valerio, F. Polycyclic Aromatic Hydrocarbon Pollution in Native and Caged Mussels. *Mar. Pollut. Bull.* **2001**, *42*, 951–956. [CrossRef]
42. Amoroso, S.; Arnese, A.; Cirillo, T.; Montuori, P.; Triassi, M.; Amodio-Cocchieri, R. Pollution by Mercury, Arsenic, Lead, Chromium, Cadmium, and Polycyclic Aromatic Hydrocarbons of Fish and Mussels from the Gulf of Naples, Italy. *Bull. Environ. Contam. Toxicol.* **2003**, *71*, 551–560. [CrossRef]

43. Serpe, F.P.; Esposito, M.; Gallo, P.; Salini, M.; Maglio, P.; Hauber, T.; Serpe, L. Determination of heavy metals, polycyclic aromatic hydrocarbons and polychlorinated biphenyls in *Mytilus galloprovincialis* from Campania coasts, Italy. *Fresenius Environ. Bull.* **2010**, *19*, 2292–2296.
44. Serpe, F.P.; Esposito, M.; Gallo, P.; Serpe, L. Optimisation and validation of an HPLC method for determination of polycyclic aromatic hydrocarbons (PAHs) in mussels. *Food Chem.* **2010**, *122*, 920–925. [CrossRef]
45. Marrone, R.; Mercogliano, R.; Palma, G.; Chirollo, C.; Smaldone, G.; Anastasio, A. Polycyclic aromatic hydrocarbons (PAHS) in seafood caught off in napoli gulf (Italy). *Ital. J. Food Saf.* **2011**, *1*, 61–65. [CrossRef]
46. Ferrante, M.; Zanghì, G.; Cristaldi, A.; Copat, C.; Grasso, A.; Fiore, M.; Signorelli, S.S.; Zuccarello, P.; Conti, G.O. PAHs in seafood from the Mediterranean Sea: An exposure risk assessment. *Food Chem. Toxicol.* **2018**, *115*, 385–390. [CrossRef] [PubMed]
47. Fasano, E.; Arnese, A.; Esposito, F.; Albano, L.; Masucci, A.; Capelli, C.; Cirillo, T.; Nardone, A. Evaluation of the impact of anthropogenic activities on arsenic, cadmium, chromium, mercury, lead, and polycyclic aromatic hydrocarbon levels in seafood from the Gulf of Naples, Italy. *J. Environ. Sci. Health Part A* **2018**, *53*, 786–792. [CrossRef] [PubMed]
48. Arienzo, M.; Toscanesi, M.; Trifuoggi, M.; Ferrara, L.; Stanislao, C.; Donadio, C.; Grazia, V.; Gionata, D.V.; Carella, F. Contaminants bioaccumulation and pathological assessment in Mytilus galloprovincialis in coastal waters facing the brownfield site of Bagnoli, Italy. *Mar. Pollut. Bull.* **2019**, *140*, 341–352. [CrossRef] [PubMed]
49. Lambiase, S.; Ariano, A.; Serpe, F.P.; Scivicco, M.; Velotto, S.; Esposito, M.; Severino, L. Polycyclic aromatic hydrocarbons (PAHs), arsenic, chromium and lead in warty crab (*Eriphia verrucosa*): Occurrence and risk assessment. *Environ. Sci. Pollut. Res.* **2021**, *28*, 35305–35315. [CrossRef]
50. Baumard, P.; Budzinski, H.; Garrigues, P. Polycyclic aromatic hydrocarbons in sediments and mussels of the western Mediterranean sea. *Environ. Toxicol. Chem.* **1998**, *17*, 765–776. [CrossRef]
51. Cinnirella, S.; Bruno, D.E.; Pirrone, N.; Horvat, M.; Živković, I.; Evers, D.C.; Johnson, S.; Sunderland, E.M. Mercury concentrations in biota in the Mediterranean Sea, a compilation of 40 years of surveys. *Sci. Data* **2019**, *6*, 205–211. [CrossRef]
52. US-EPA. *Guidance for Assessing Chemical Contamination Data for Use in Fish Advisories, Volume 2. Risk Assessment and Fish Consumption Limits*; EPA/823-B94-004; Office of Science and Technology Office of Water U.S. Environmental Protection Agency: Washington, DC, USA, 2000.
53. Grilli, F.; Accoroni, S.; Acri, F.; Aubry, F.B.; Bergami, C.; Cabrini, M.; Campanelli, A.; Giani, M.; Guicciardi, S.; Marini, M.; et al. Seasonal and Interannual Trends of Oceanographic Parameters over 40 Years in the Northern Adriatic Sea in Relation to Nutrient Loadings Using the EMODnet Chemistry Data Portal. *Water* **2020**, *12*, 2280. [CrossRef]
54. Xu, S.; Chen, M.; Feng, T.; Zhan, L.; Zhou, L.; Yu, G. Use ggbreak to Effectively Utilize Plotting Space to Deal With Large Datasets and Outliers. *Front. Genet.* **2021**, *12*, 774846. [CrossRef]
55. Navarra, S.; Carbonari, F.; Bambi, C. Ismea il Pesce a Tavola: Percezioni e Stili di Consumo degli Italiani, 2011. Ismea Mercati web site. Available online: https://www.ismea.it/flex/cm/pages/ServeAttachment.php/L/IT/D/f%252Fb%252F5%252FD.914a4f340b52602fcc38/P/BLOB%3AID%3D6191/E/pdf (accessed on 15 January 2022).
56. Galgani, F.; Martínez-Gómez, C.; Giovanardi, F.; Romanelli, G.; Caixach, J.; Cento, A.; Scarpato, A.; Benbrahim, S.; Messaoudi, S.; Deudero, S.; et al. Assessment of polycyclic aromatic hydrocarbon concentrations in mussels (*Mytilus galloprovincialis*) from the Western basin of the Mediterranean Sea. *Environ. Monit. Assess.* **2010**, *172*, 301–317. [CrossRef]
57. Rizzi, C.; Villa, S.; Chimera, C.; Finizio, A.; Monti, G. Spatial and temporal trends in the ecological risk posed by polycyclic aromatic hydrocarbons in Mediterranean Sea sediments using large-scale monitoring data. *Ecol. Indic.* **2021**, *129*, 107923. [CrossRef]
58. Çelik, M.Y.; Karayücel, S.; Karayücel, I.; Öztürk, R.; Eyüboğlu, B. Meat Yield, Condition Index, and Biochemical Composition of Mussels (*Mytilus galloprovincialis* Lamarck, 1819) in Sinop, South of the Black Sea. *J. Aquat. Food Prod. Technol.* **2012**, *21*, 198–205. [CrossRef]
59. Prato, E.; Danieli, A.; Maffia, M.; Biandolino, F. Lipid and Fatty Acid Compositions of *Mytilus galloprovincialis* Cultured in the Mar Grande of Taranto (Southern Italy): Feeding Strategies and Trophic Relationships. *Zool. Stud.* **2010**, *49*, 211–219.
60. Bongiorno, T.; Iacumin, L.; Tubaro, F.; Marcuzzo, E.; Sensidoni, A.; Tulli, F. Seasonal changes in technological and nutritional quality of *Mytilus galloprovincialis* from suspended culture in the Gulf of Trieste (North Adriatic Sea). *Food Chem.* **2015**, *173*, 355–362. [CrossRef] [PubMed]
61. Miura, K.; Shimada, K.; Sugiyama, T.; Sato, K.; Takami, A.; Chan, C.K.; Kim, I.S.; Kim, Y.P.; Lin, N.-H.; Hatakeyama, S. Seasonal and annual changes in PAH concentrations in a remote site in the Pacific Ocean. *Sci. Rep.* **2019**, *9*, 12591. [CrossRef] [PubMed]
62. Yu, Q.; Ding, X.; He, Q.; Yang, W.; Zhu, M.; Li, S.; Zhang, R.; Shen, R.; Zhang, Y.; Bi, X.; et al. Nationwide increase of polycyclic aromatic hydrocarbons in ultrafine particles during winter over China revealed by size-segregated measurements. *Atmos. Chem. Phys.* **2020**, *20*, 14581–14595. [CrossRef]
63. Guigue, C.; Tedetti, M.; Ferretto, N.; Garcia, N.; Méjanelle, L.; Goutx, M. Spatial and seasonal variabilities of dissolved hydrocarbons in surface waters from the Northwestern Mediterranean Sea: Results from one year intensive sampling. *Sci. Total Environ.* **2014**, *466-467*, 650–662. [CrossRef]
64. Marini, M.; Frapiccini, E. Persistence of polycyclic aromatic hydrocarbons in sediments in the deeper area of the Northern Adriatic Sea (Mediterranean Sea). *Chemosphere* **2012**, *90*, 1839–1846. [CrossRef]

65. Kodnik, D.; Carniel, F.C.; Licen, S.; Tolloi, A.; Barbieri, P.; Tretiach, M. Seasonal variations of PAHs content and distribution patterns in a mixed land use area: A case study in NE Italy with the transplanted lichen *Pseudevernia furfuracea*. *Atmos. Environ.* **2015**, *113*, 255–263. [CrossRef]
66. Witt, G. Polycyclic aromatic hydrocarbons in water and sediment of the Baltic Sea. *Mar. Pollut. Bull.* **1995**, *31*, 237–248. [CrossRef]
67. Frapiccini, E.; Marini, M. Polycyclic Aromatic Hydrocarbon Degradation and Sorption Parameters in Coastal and Open-Sea Sediment. *Water Air Soil Pollut.* **2015**, *226*, 246. [CrossRef]
68. Guigue, C.; Tedetti, M.; Dang, D.H.; Mullot, J.-U.; Garnier, C.; Goutx, M. Remobilization of polycyclic aromatic hydrocarbons and organic matter in seawater during sediment resuspension experiments from a polluted coastal environment: Insights from Toulon Bay (France). *Environ. Pollut.* **2017**, *229*, 627–638. [CrossRef]
69. Mathieu, A.; Lemaire, P.; Carriere, S.; Drai, P.; Giudicelli, J.; Lafaurie, M. Seasonal and Sex-Linked Variations in Hepatic and Extrahepatic Biotransformation Activities in Striped Mullet (*Mullus barbatus*). *Ecotoxicol. Environ. Saf.* **1991**, *22*, 45–57. [CrossRef]
70. Marini, M.; Jones, B.H.; Campanelli, A.; Grilli, F.; Lee, C.M. Seasonal variability and Po River plume influence on biochemical properties along western Adriatic coast. *J. Geophys. Res. Earth Surf.* **2008**, *113*, C05S90. [CrossRef]
71. Marini, M.; Campanelli, A.; Sanxhaku, M.; Kljajić, Z.; Betti, M.; Grilli, F. Late Spring Characterization of Different Coastal Areas of the Adriatic Sea. *Acta Adriat.* **2015**, *56*, 27–46.
72. Dachs, J.; Bayona, J.M.; Raoux, C.; Albaigés, J. Spatial, Vertical Distribution and Budget of Polycyclic Aromatic Hydrocarbons in the Western Mediterranean Seawater. *Environ. Sci. Technol.* **1997**, *31*, 682–688. [CrossRef]
73. Iacono, R.; Napolitano, E.; Palma, M.; Sannino, G. The Tyrrhenian Sea Circulation: A Review of Recent Work. *Sustainability* **2021**, *13*, 6371. [CrossRef]
74. Marine Traffic Density Map. Available online: https://ec.europa.eu/maritimeaffairs/atlas/maritime_atlas/#lang=EN;p=w;bkgd=1;theme=2:0.75,5007:0.75;c=1417747.2062308616,4941995.124365145;z=6;e=t (accessed on 10 February 2022).
75. Barale, V.; Dusart, J.; Assouline, M.; Niceta, F. European Atlas of the Seas: "A picture is worth a thousand words". *J. Coast. Conserv.* **2017**, *22*, 105–113. [CrossRef]
76. Pensa, A.; Pinton, A.; Vita, L.; Bonamico, A.; De Benedetti, A.A.; Giordano, G. Atlas of Italian Submarine Volcanic Structures. In *Memorie Descrittive della Carta Geologica d'Italia*; ISPRA—Servizio Geologico d'Italia: Rome, Italy, 2019; Volume 104.
77. Saroni, A.; Sciarra, A.; Grassa, F.; Eich, A.; Weber, M.; Lott, C.; Ferretti, G.; Ivaldi, R.; Coltorti, M. Shallow submarine mud volcano in the northern Tyrrhenian sea, Italy. *Appl. Geochem.* **2020**, *122*, 104722. [CrossRef]
78. Kozak, K.; Ruman, M.; Kosek, K.; Karasiński, G.; Stachnik, Ł.; Polkowska, Ż. Impact of Volcanic Eruptions on the Occurrence of PAHs Compounds in the Aquatic Ecosystem of the Southern Part of West Spitsbergen (Hornsund Fjord, Svalbard). *Water* **2017**, *9*, 42. [CrossRef]
79. Remizovschi, A.; Carpa, R.; Forray, F.L.; Chiriac, C.; Roba, C.-A.; Beldean-Galea, S.; Andrei, A.-S.; Szekeres, E.; Baricz, A.; Lupan, I.; et al. Mud volcanoes and the presence of PAHs. *Sci. Rep.* **2020**, *10*, 1253. [CrossRef] [PubMed]
80. Domingo, J.L. Nutrients and Chemical Pollutants in Fish and Shellfish. Balancing Health Benefits and Risks of Regular Fish Consumption. *Crit. Rev. Food Sci. Nutr.* **2014**, *56*, 979–988. [CrossRef] [PubMed]
81. De Giovanni, A.; Giuliani, C.; Marini, M.; Luiselli, D. Methylmercury and Polycyclic Aromatic Hydrocarbons in Mediterranean Seafood: A Molecular Anthropological Perspective. *Appl. Sci.* **2021**, *11*, 11179. [CrossRef]

applied sciences

MDPI

Article

Polycyclic Aromatic Hydrocarbons and Polychlorinated Biphenyls in Seawater, Sediment and Biota of Neritic Ecosystems: Occurrence and Partition Study in Southern Ligurian Sea

Luca Rivoira [1,*], Michele Castiglioni [1,*], Nicola Nurra [2,3], Marco Battuello [2,3], Rocco Mussat Sartor [2,3], Livio Favaro [2] and Maria Concetta Bruzzoniti [1,*]

1 Department of Chemistry, Università degli Studi di Torino, Via Pietro Giuria 7, 10125 Torino, Italy
2 Department of Life Sciences and Systems Biology, Università degli Studi di Torino, Via Accademia Albertina 13, 10123 Torino, Italy; nicola.nurra@unito.it (N.N.); marco.battuello@unito.it (M.B.); rocco.mussat@unito.it (R.M.S.); livio.favaro@unito.it (L.F.)
3 Pelagosphera Marine Environmental Services Cooperative, Via U. Cosmo 17bis, 10131 Torino, Italy
* Correspondence: luca.rivoira@unito.it (L.R.); michele.castiglioni@unito.it (M.C.); mariaconcetta.bruzzoniti@unito.it (M.C.B.); Tel.: +39-011-670-5245 (L.R.); Fax: +39-011-230-5245 (L.R.)

Abstract: The Mediterranean Sea is subjected to a high anthropic pressure, which determines direct or indirect discharges of persistent organic pollutants deriving from intensive industrial activities. These compounds could easily enter and contaminate the whole marine compartment, with possible transfers (and contamination) among water, sediment and biota. Based on the above-mentioned assumptions, in this work we studied the presence of 16 polycyclic aromatic hydrocarbons (PAHs) and 14 dioxin and non-dioxin-like polychlorinated biphenyls (PCBs) in the neritic protected marine area of the Southern Ligurian Sea, affected by the impact of human activities. The study was focused on the possible partition of micropollutants within seawater, sediment and zooplankton. Results showed that both seasonal and anthropic causes strongly affect contaminant transfer behaviors, with summertime periods more impacted by PAH and PCB contamination. Regarding the PAH contamination, low molecular weight congeners were mainly detected in the target matrices, revealing concentrations up to 1 µg/L in seawater (anthracene), 250 µg/Kg in sediments (benzo[b]fluoranthene) and 2.3 mg/Kg in carnivorous copepods. Concerning PCBs, only few congeners were detected in the matrices studied. To better understand the occurrence of preferential bioaccumulation pathways in zooplankton, partition studies were also performed in several taxa (hyperbenthic Isopoda, holoplanktonic crustacean copepods and ichthyoplankton) through the calculation of BAF values, observing that both living and feeding habits could influence the bioaccumulation process.

Keywords: PAH; PCB; neritic environment; seawater; sediment; biota; partition; contamination

Citation: Rivoira, L.; Castiglioni, M.; Nurra, N.; Battuello, M.; Sartor, R.M.; Favaro, L.; Bruzzoniti, M.C. Polycyclic Aromatic Hydrocarbons and Polychlorinated Biphenyls in Seawater, Sediment and Biota of Neritic Ecosystems: Occurrence and Partition Study in Southern Ligurian Sea. *Appl. Sci.* **2022**, *12*, 2564. https://doi.org/10.3390/app12052564

Academic Editors: Mauro Marini and Anna Annibaldi

Received: 30 December 2021
Accepted: 25 February 2022
Published: 1 March 2022

1. Introduction

Neritic environments are peculiar marine areas acting as interface between the atmosphere, the sea and the continental masses [1]. Due to their proximity to land and to sunlight infiltration, they are rich in nutrients and, consequently, in biologic activities, showing a remarkable biodiversity and biomass of algae, seagrasses and animal organisms inhabiting the coastal ecosystems.

Zooplankton, along with phytoplankton, representing the most abundant form of life in terms of biomass and biodiversity in neritic ecosystems, is comprised of heterotrophic microscopic, unicellular or multicellular organisms with size classes ranging from a few microns (picozooplankton) to a millimeter or more (mesozooplankton), such as the gelatinous zooplankton [2]. The marine zooplankton can be divided into two ecological categories: holoplankton, spending their entire lifecycle in the water column (e.g., crustaceans such as copepods and krill) and meroplankton, spending only the larval stages of their lifecycle as zooplankton, such as decapod crustaceans, echinoderms and fish larvae that, once they mature, adopt a benthic or nektonic lifestyle.

Taxa can be further distinguished depending on their typical living habits (permanently in the water column or diel vertical migration from benthos to sea surface such as hyperbenthic organisms e.g., Isopoda) and on their feeding habits and behavior (herbivorous, carnivorous and omnivorous).

Zooplankton plays a critical ecological role in marine food webs, both in the neritic and pelagic ecosystems, since it is involved in the conservation of energy from primary producers (phytoplankton) to higher trophic levels [3], in biogeochemistry cycles and in supporting the ocean's biological pump of carbon export. Hence, an alteration of this category, i.e., through chemical contaminations, can lead to disruptions up to the highest trophic levels.

Due to their peculiar conformation, neritic areas are strongly vulnerable to harmful effects of several anthropogenic pollutants derived from human activities, even when located far away from the pollution sources [4]. Such compounds could easily enter the whole marine ecosystem, with possible transfer (and contamination) along its main compartments, namely the water column, the sediment and the biota [5–7]. Transfer and contamination pathways mainly depend on the physicochemical properties of the pollutant molecules [8] or on the living habits of biota. Moreover, it is worth noting that the fate of pollutants is strongly influenced by the characteristics of the sea, such as water exchange capacity, as the enclosed or semi-enclosed basins (i.e., the Mediterranean Sea) are more susceptible to the accumulation of pollutants in marine sediments than open basins [9]. Among the organic micropollutants frequently detected in marine waters, are polycyclic aromatic hydrocarbons (PAHs) and polychlorinated biphenyls (PCBs) [10,11]. PAHs represent pollutants of natural and anthropogenic origins (e.g., volcanic eruptions, combustion) while PCBs mainly derive from use in industrial activities (e.g., dielectric fluids, motor oils).

Seawater contamination from PAHs and PCBs derives from the not negligible impact of ports, industries and touristic activities located on coastal areas [12,13], as well as from wastewater treatment plants [14] and river effluents [15]. Atmospheric deposition presents an additional source of marine water contamination [16]. Since both PAHs and PCBs are mutagenic/carcinogenic recalcitrant pollutants [17], there are serious concerns regarding their presence in the environment [18–20], particularly in neritic compartments, and also with consideration of their tendency to partition between water and sediments and from water and sediments to biota, due to their high octanol/water partition coefficients ($logK_{OW}$ ranging from 2.96 to 5.60 for PAHs and from 5.41 to 7.83 for PCBs) [21].

In Europe, the environmental status of water compartments (including neritic areas) is regulated by the 2013/39/EU directive [22], adopted in Italy by the Legislative Decree 172/2015 [23], in which a list of priority substances (including selected PAHs and PCBs compounds) to be monitored in water, sediments and biota is reported. For individual congeners of PAHs, Environmental Quality Standards (EQS) are fixed for water, sediments and biota, defining annual average values (AA-EQS) and maximum allowable concentrations (MAC-EQS). Conversely, AA-EQS for PCBs in biota refer only to dioxin-like congeners, while with regards to sediments the same dioxin-like compounds, together with other eight congeners, are considered. It should be highlighted that, to date, the limits for PCBs in marine waters remain unregulated.

Due to the strategic importance of coastal areas in the marine ecosystems, as previously presented, several studies investigated the contamination from PAHs and PCBs in neritic environments, as well as their partition between water columns, sediments and biota [24–26]. Within this literature, an extensive bibliography is dedicated to the Mediterranean Sea which, due to its morphological peculiarities, is strongly subjected to anthropic impact. In particular, studies on PAH and PCB contamination of the Venice Lagoon [27], Tyrrhenian Sea [28], and Ionian Sea [29] are reported. On the other hand, few studies were focused on Ligurian Sea [30–32], even though its pollution causes serious concern, due to intense anthropic activities (e.g., ports, industries, tourism, etc.) [32]. Furthermore, studies investigating the possible distribution pathways of PAHs and PCBs within neritic compartments of Ligurian Sea are, as far as we know, absent thus far, despite the natural (marine protected area) and economic (shipping, industries, harbors) strategical importance of this area.

Based on the above considerations, the aim of this work is the study of the contamination of PAHs and PCBs (both dioxin and non-dioxin-like congeners) in the neritic environment of the Southern Ligurian Sea, which hosts a protected marine area called "Cetacean Sanctuary". The study focuses comprehensively on the possible partition of micropollutants among the environmental compartments of the marine ecosystem (seawater, sediment and zooplankton), and on the factors that could influences these transfer behaviors (i.e., seasonality). With this aim, analytical protocols for the extraction and quantitation of PAHs and PCBs from seawater, sediment and zooplankton were developed and validated. Furthermore, to better understand the presence of preferential pathways of bioaccumulation in zooplankton, partition studies were also carried out in four taxa, chosen on the basis of dietary habits (herbivorous or carnivorous) and living environment: hyperbenthic Isopoda, herbivores and carnivorous holoplanktonic crustacean copepods and fish larvae (ichthyoplankton). To the best of our knowledge, this study is the first to describe the partition of PAHs and PCBs in this protected area of the Ligurian Sea, with unique natural characteristics, providing insights on the contamination of the whole bioma.

2. Materials and Methods

2.1. Sampling Area and Sample Pre-Treatments

Waters, sediments and biota were sampled in the Ligurian Sea, in an offshore area (12.5 nautical miles off the Italian coast), above the continental shelf, at the border with Northern Tyrrhenian Sea, the same study area of previous research [5] and included the following stations: (i) 43°29′40″ N–10°01′45″ E, (ii) 43°28′10″ N, 10°01′55″ E and (iii) 43°27′10″ N, 10°03′00″ E (see Figure S1 of Supplementary Material). The area was considered of great interest for several reasons. In fact, it is characterized by one of the highest levels of shipping in the Mediterranean basin; moreover, a strong anthropogenic influence and impact are present due to several commercial, industrial and harbor activities. The investigated area hosts the "Cetacean Sanctuary", an extensive marine protected area where the number of cetaceans is at least double that of any other part of the Mediterranean. Finally, the area is subjected to a high dynamic water current equilibrium, since both perpetual and seasonal currents influence the possible dispersion and partition of micropollutants originating from the highly developed coastline of Italy.

Two samplings were carried out for both water column and marine biota during summer 2017 and winter 2018, in order to assess the possible contribution of seasonality. For the sediment, a single sample was collected in summer 2017, due to the low effect of the seasonality on the circalittoral mud sediments.

Sampling methods were performed as follows: (i) water column, 5 L collection for each sampling using a PVC Niskin bottle water sampler (collected at 5 m depth). Subsequently, the samples were filtered via qualitative paper and refrigerated until analysis; (ii) sediments, sampled using a 18 L Van Veen grab sampler on the seabed at 110 m depth. Moisture evaporation (at 60 °C for 12 h) was performed prior to micropollutant extraction and analysis; (iii) marine biota (mesozooplankton and ichthyoplankton) was collected in a surface haul, using a WP-2 standard net (200 μm mesh size and diameter 57 cm). The horizontal sampling time was approximately 15 min (2 knots vessel cruising speed). The net was fitted with a flow-meter (KC Denmark model 23.090) to measure volume of water filtered. After collection, each zooplankton sample was washed, on board first, with filtered seawater from the sampling site, in order to remove terrigenous or inorganic particles, and hence washed with distilled water. After the washing procedure, samples were immediately fixed in 70% ethanol and seawater and stored in the dark.

In the laboratory, a qualitative–quantitative analysis of the marine biota was performed with a Leica stereomicroscope and microscope, which allowed the identification and regrouping of the following four taxa: hyperbenthic Isopoda, holoplanktonic crustaceans' copepods both herbivores and carnivorous and fish larvae (ichthyoplankton). Samples were dried at 80 °C for 12 h before extraction and analysis.

2.2. Reagents

Acetone ≥ 99.8%, dichloromethane ≥ 99.9%, 2-propanol ≥ 99.8%, methanol ≥ 99.9%, sodium hydroxide ≥ 98.0%, sulfuric acid 96–97%, magnesium sulphate anhydrous ≥ 99.5% and NaCl ≥ 99.5% were obtained from Honeywell Riedel-de-Haën, Fisher Scientific Italia, Rodano, MI (Italy). Cyclohexane 99.5% and dichloromethane were obtained from VWR International (Radnor, PA, USA).

High-purity water (18.2 MΩ cm resistivity at 25 °C), produced by an Elix-Milli Q Academic system (Millipore-Billerica, MA, USA) was used.

The d-SPE sorbent used was Primary Secondary Amine (PSA) from Agilent Technologies (Santa Clara, CA, USA).

The 16 PAHs studied, i.e., naphthalene (NaPh), acenaphthylene (AcPY), acenaphthene (AcPh), fluorene (Flu), phenanthrene (Phe), anthracene (Ant), fluoranthene (Flth), pyrene (Pyr), benzo[a]anthracene (BaA), chrysene (Chr), benzo[b]fluoranthene (BbFl), benzo[k]fluoranthene (BkFl), benzo[a]pyrene (BaP), indeno[1,2,3-cd]pyrene (Ind), dibenzo[a,h]anthracene (DBA) and benzo[ghi]perylene (BP), were the compounds listed by the United States Environmental Protection Agency (US-EPA) and were purchased from Wellington Laboratories (Guelph, ON, Canada).

The 14 PCBs studied were purchased from Chemical Research 2000 (Rome, Italy). They were chosen according to the results of the main monitoring campaigns and included 3,3′-dichlorobiphenyl (PCB 11), 4,4′-dichlorobiphenyl (PCB 15), 2,4,4′-trichlorobiphenyl (PCB 28), 2,2′,5,5′-tetrachlorobiphenyl (PCB 52), 2,2′,4,5,5′-pentachlorobiphenyl (PCB 101), 2,2′,3,4,4′,5-hexachlorobiphenyl (PCB 138), 2,2′,4, 4′,5,5′-hexachlorobiphenyl (PCB 153) and 2,2′,3,4,4′,5,5′-heptachlorobiphenyl (PCB 180). Furthermore, the following dioxin-like PCBs were included 3,4,4′,5-tetrachlorobiphenyl (PCB 81), 2,3′,4,4′,5-pentachlorobiphenyl (PCB 118), 2′,3,4,4′, 5-penta-chlorobiphenyl (PCB 123), and 2,3′,4,4′,5, 5′-hexachlorobiphenyl (PCB 167), 3,3′,4,4′,5,5′-hexachlorobiphenyl (PCB 169) and 2,3,3′,4,4′,5,5′-heptachlorobiphenyl (PCB 189)

Isotope labelled compounds for PAHs (5 mg/L) and for PCBs (2 mg/L), both purchased from Wellington Laboratories, were employed as internal standards and surrogates to obtain calibration curves and to calculate extraction recoveries. The deuterated PAH surrogate solution included the following compounds: benzo[a]anthracene-d12 (BaA-d12), chrisene-d12 (Chr-d12), benzo[b]fluoranthene-d12 (BbFl-d12), benzo[k]fluoranthene-d12 (BkFl-d12), benzo[a]pyrene-d12 (BaP-d12), indeno[1,2,3-cd]pyrene-d12 (Ind-d12), dibenzoanthracene-d14 (DBA-d14), benzoperylene-d12 (BP-d12). The ^{13}C-PCB surrogate solution included the following congeners: $^{13}C_{12}$-PCB28, $^{13}C_{12}$-PCB52, $^{13}C_{12}$-PCB118, $^{13}C_{12}$-PCB153, and $^{13}C_{12}$-PCB180. ^{2}H-anthracene and $^{13}C_{12}$-PCB70 were used as internal standards.

2.3. Chromatographic Analysis

PAHs and PCBs, extracted from neritic matrices as detailed in the following paragraphs, were analyzed by gas chromatography coupled with mass spectrometry (GC–MS) using an Agilent 6980 series gas chromatograph and an Agilent 5973 Network MS detector controlled by Agilent ChemStation software. The gas chromatograph was provided with an autosampler of the Agilent 7683 Series.

The GC column was a (5%-Phenyl)-methylpolysiloxane column (DB-5 ms, 30 m × 0.25 mm × 25 μm; Agilent). Helium was employed as gas carrier (1 mL/min). MS detection was performed in Single Ion Monitoring (SIM) mode at proper m/z ratio (m/z ratio available upon request). Injections (2 μL) were performed by the pulsed splitless mode (pressure at 40 psi for 2.5 min, injector temperature 280 °C). The oven ramp for CH_3CN and CH_2Cl_2 was set as follows: starting temperature: 80 °C, hold for 2 min; ramp to 176 °C, 12 °C/min rate; ramp to 196 °C, 5 °C/min rate, hold for 3 min; ramp to 224 °C, 12 °C/min rate; ramp to 244 °C, 12 °C/min rate, hold for 3 min; ramp to 270 °C, 7 °C/min rate, hold for 3 min; final ramp to 300 °C, 5 °C/min, hold for 10 min to completely clean and restore the GC column. The complete separation of the 13 PAHs and 14 PCBs was obtained within 49 min.

2.4. Extraction of PAHs and PCBs from Neritic Matrices

All the extraction procedures used for water, sediment and biota were specifically optimized for this study and are hereafter detailed. Aliquots of all the matrixes were extracted in triplicate. Blank samples were run in parallel to evaluate possible ambient contamination.

2.4.1. Water Column

Extraction was performed via solid-phase extraction (SPE) using an SPE Vacuum manifold and a polymeric reversed-phase (RP) cartridge (STRATA-XL-100 µm; Phenomenex, Torrance, CA, USA), The cartridge was conditioned (20 psi) with 5 mL CH_2Cl_2, 5 mL 2-propanol, and 5 mL H_2O. Then, 200 mL water sample added with 20 mL 2-propanol was loaded (50 psi) on the SPE cartridge. The water sample container was subsequently washed with 20 mL 2-propanol–water solution (10 + 90, *v*/*v*). After loading, the cartridge was washed (20 psi) with 5 mL H_2O, and 5 mL 2-propanol–water solution (85 + 15, *v*/*v*). The cartridge was dried for 10 min, and the analytes were finally eluted with two aliquots of 1.0 mL CH_2Cl_2. The eluted extract was spiked with the internal standard solution of PAHs and PCBs to achieve a final concentration of 5 µg/L and injected for GC-MS analysis. To evaluate the extraction recoveries of PAHs and PCBs, before extraction, the water samples were spiked with surrogates (as detailed in "Reagents" section) to achieve a final concentration of 2 µg/L.

2.4.2. Sediment

Sediments were extracted through a QuEChERS procedure. In detail, 5 g sediment (previously spiked with surrogates to achieve a final concentration of 2 µg/L in the extract solvent) were extracted in 10 mL CH_2Cl_2 with the addition of 1 g NaCl and 0.4 g $MgSO_4$. The extraction mixture was shaken for 5 min and subsequently centrifuged for 10 min (1534× *g*). For the d-SPE purification step, 6 mL of the supernatant was transferred into a 15 mL tube containing 50 mg of PSA and 150 mg $MgSO_4$. The mixture was again shaken for 5 min and centrifuged for 10 min (7871× *g*). A 2 mL aliquot of the supernatant was spiked with internal standards to achieve a final concentration of 5 µg/L and injected for GCMS analysis.

2.4.3. Biota

Each taxon was separately extracted to study possible partition pathways. A total of 100 mg of each taxon were weighted inside a 50 mL tube and spiked with surrogates to achieve a final concentration of 2 µg/L in the extract solvent. Subsequently, 16 mL 4 M NaOH, 4 mL methanol and 10 mL CH_2Cl_2 were added. The extraction mixture was shaken for 1 min, sonicated for 15 min and refrigerated at 4 °C for 2 h. Finally, the tube was centrifuged for 10 min (7871× *g*) and a 2 mL aliquot of the supernatant was spiked with internal standards to achieve a final concentration of 5 µg/L and injected for GCMS analysis.

2.5. Recovery Evaluation

To evaluate the apparent recoveries of each extraction method developed, each sample was spiked, before extraction, with surrogate solutions of PAHs and PCBs to achieve a final concentration of 2 µg/L (C_S). After extraction, the concentrations were calculated by using an external standard calibration curve prepared in dichloromethane and the extraction yield ($E\%$) was calculated according to the following equation:

$$E\% = \frac{C_e}{C_s} * 100$$

where C_e is the calculated concentration of the surrogate after extraction expressed as μg/L.

2.6. Validation of Analytical Protocols

The protocols developed for the extraction and quantitation of PAHs and PCBs from water, sediment and biota were validated through the evaluation of linearity, limits of detection (MDL) and quantitation (MQL). Linearity was evaluated over ten concentration levels, within a concentration range included between 0.87 μg/L and 14 μg/L for PAHs and between 0.42 μg/L and 6.75 μg/L for PCBs in CH_2Cl_2. The values of MDL and MQL for the 30 target compounds were calculated by means of the response error and the slope of the calibration curve, using the expression MDL = 3.3 Sy/m, and MQL = 10 Sy/m, where Sy = response error; m = slope of the calibration.

2.7. Bioaccumulation Factor

Bioaccumulation factor (BAF) is defined as the ratio of the concentration of a micropollutant in an organism to the concentration of the same compound in water [33]. Logarithmic BAFs were individually estimated for all the analytes detected both in biota matrices, as follows:

$$log\ BAF = \log \frac{C_{biota}}{C_{water}} \tag{1}$$

Finally, for both herbivorous and carnivorous zooplankton, the average of *logBAF* values was calculated and results were compared and discussed to understand possible effects of feeding habits in the bioaccumulation process.

3. Results and Discussion

3.1. Optimization of Analytical Protocols

To verify the effectiveness of the analytical methods developed for the analysis of PAH and PCB occurrence and partition in water, sediment and biota, extraction yields, methods detection (MDL) and quantitation limits (MQL) were assessed.

For each matrix, the extraction recoveries were determined using surrogates and were identified in the following ranges, water: from 52% (Ind-d12) to 88% (BaA-d12) for PAHs and from 54% ($^{13}C_{12}$-PCB118) to 77% ($^{13}C_{12}$-PCB28) for PCBs, with RSD% lower than 10% for all the analytes; sediment: from 54% (Ind-d12) to 96% (BaA-d12) for PAHs and from 56% (13C12-PCB52) to 86% (13C12-PCB28) for PCBs, with RSD% lower than 7% for all the analytes; biota: from 80% (DBA-d14) to 93% (BaA-d12) for PAHs and from 79% ($^{13}C_{12}$-PCB28) to 93% ($^{13}C_{12}$-PCB118) for PCBs, with RSD% lower than 8% for all the analytes. Extraction yields are reported in Figures S2–S4 of Supplementary Materials for both PAHs and PCBs in the three matrices.

The results here obtained for all the matrices are in the same range (or higher) than other studies devoted to the analysis of PAHs and PCBs in marine compartments [25,34,35]. Differently from the procedures typically applied to sediments, the proposed QuEChERS protocol is green and sustainable since it does not require high volumes of organic solvents. Furthermore, the method does not require a final evaporation step (which increases the risk of losing target analytes) [25].

MDL and MQL for water, sediment and biota are reported in Tables S1 and S2, together with regulation limits (if present). MQL were in the following ranges, water: from 4.6 ng/L (NaPh) to 27 ng/L (DBA) for PAHs and from 6 ng/L (PCB11) to 32 ng/L (PCB123) for PCBs; sediment: from 3 μg/Kg (DBA) to 20 μg/Kg (AcPY) for PAHs and from 1 μg/Kg (PCB153) to 3 μg/Kg (PCB123) for PCBs; biota: from 10 μg/Kg (BaP and Ind) to 21 μg/Kg

(BP) for PAHs and from 15 μg/Kg (PCB138) to 35 μg/Kg (PCB189) for PCBs. These values fully satisfy the limits fixed by the Italian and European regulations (previously described).

3.2. Chemical Characterization of Seawater

PAHs detected in seawater (both in summer and winter samples) are summarized in Figure 1.

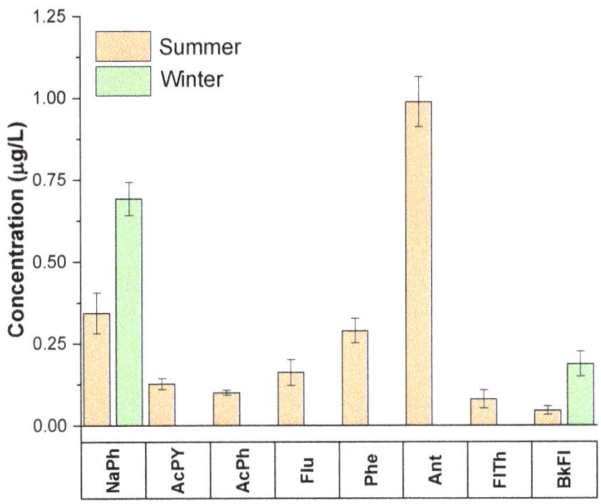

Figure 1. Concentration of PAHs detected in seawater samples, both in summer (orange) and winter (green) sampling. Analytical conditions are detailed in the Experimental section.

It is interesting to observe that PAHs with a higher number of aromatic rings (i.e., BbFl, BaP, DBA, BP and Ind) were not detected in seawater, regardless of the season, in accordance with their higher hydrophobicity behavior (and hence lower solubility in water).

In addition, an inter-seasonal variability of PAH concentrations (among summer and winter sampling campaigns) was also clearly present for all the detected analytes. Indeed, a higher number of congeners was detected in summer (8), rather than in winter (2), with a total PAH concentration about three times higher in summer than winter (2.2 μg/L and 0.8 μg/L, respectively).

To explain these differences, both climatological and anthropogenic effects should be considered. To elucidate, one of the main aspects to consider is the seasonal variation of thermocline. Thermocline refers to the water layer below the surface, characterized by a massive temperature gradient which decreases up to 4 °C (or lower for deep waters). During summertime, surface waters increase in temperature while the deeper waters remain cold. Stratification causes rapid vertical changes in the density of water, which largely prevents the mixing of the water column [36]. This phenomenon could affect the sinking and the accumulation of micropollutants [37]. Conversely, during winter, temperatures of surface water can undergo those of deeper layers, which tends to rise due to its lower density (homothermic conditions). Consequently, the movement of water masses promotes a mixing and, therefore, a diffusion of dissolved compounds, such as pollutants, thus justifying their lower concentrations.

In addition to the thermocline effect, the higher concentration of PAHs in the summer season could be addressed to the effects of the Northern current, the northern part of the cyclonic surface circulation of the North-Western Mediterranean Sea. Indeed, the Northern current exhibits its weakest intensities during summertime [38], thus promoting a possible accumulation of land and air derived pollutants, while during winter its higher intensities boosts their dispersion into the sea waters.

Finally, the intense touristic activities that characterize the Ligurian Sea during summertime (with private boat traffic, cruises, etc.) with respect to winter, could act as a possible source of PAHs, which can be released by thermic engines. Evidence of the effects of maritime traffic on the increment of PAH contamination were also investigated elsewhere [39]. In addition, the origin of PAHs (evaluated by studying the ration between congeners [40]) confirmed this hypothesis. In fact, both the ratio between the concentration of low (NaPh, AcPh, AcPY, Flu) and high molecular weight (Phe, Ant, FlTh, BkFl) PAHs (0.92) and the ratio between phenanthrene and anthracene (0.30) are lower than 1, thus suggesting a dominant pyrogenic origin of the detected PAHs (i.e., transport activities, use of petroleum derived products, accumulation of small emissions of gasoline, engine oils, incomplete combustion of organic substances) [41,42].

It should be noted that PAH concentrations detected in sea water within this study were in the same level of other Mediterranean areas [43], with a pollution classification from moderate to high, according to the indications proposed by Baumard and colleagues [44]. Concerning regulation limits, benzo[k]fluoranthene was the only congener detected among the compounds included in the regulated list, with an average concentration of 45 ng/L, which exceeds the maximum allowable concentrations (17 ng/L) [22].

Concerning PCBs, only congeners 15 and 138 were detected, with concentrations below the quantitation limits.

3.3. Chemical Characterization of Sediment

As discussed in the previous section, the presence of PAHs and PCBs was observed in seawater. Hence, the analysis of sediment belonging to the same sea area is justified to assess whether a partition took place from the waters. Results are reported in Figure 2.

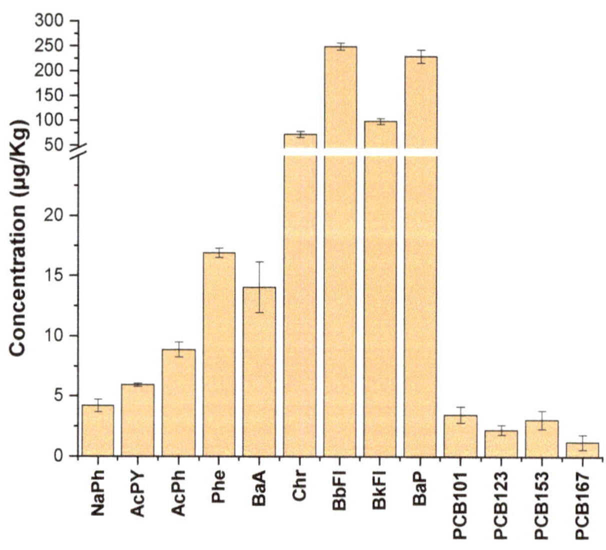

Figure 2. Concentration of PAHs and PCBs detected in sediment. Analytical conditions are detailed in the Experimental section.

Results showed that, as for seawater, PAHs containing the highest number of aromatic rings (i.e., DBA, BP and Ind) were not present (Figure 2), except for BbFl and BaP that were not previously detected in the water column.

The total amount of PAHs in sediment was assessed to reach about 700 µg/Kg, with most of the contribution ascribed to higher molecular weight compounds (i.e., BaA (14 µg/Kg), Chr (73 µg/kg), BbFl (250 µg/Kg), BkFl (99 µg/Kg) and BaP (230 µg/Kg)). The concentration of these compounds accounted for almost 95% of the total PAH concentration. Upon comparing these latter results (Figure 2) with those previously presented for the water column (Figure 1), it was observed that the contribution of the compounds with higher molecular weight to the total PAH content measured higher in sediment rather than in water. Such differences remain in good agreement with the physicochemical properties of PAHs and may be ascribed to the different partitioning attitudes of the individual compounds. Indeed, the higher the molecular weight, the higher the octanol–water partition coefficients, thus enhancing the sorption on the organic matter present in the sediment [45]. In addition, it should be mentioned that neritic areas are highly subjected to sediment resuspension [46] which causes organic matter to rise temporarily in water and, hence, favoring the sorption of PAHs dissolved in water.

As for seawater, the ratios between PAH congeners (see "Chemical characterization of seawater" section) confirm a pyrogenic origin of the PAHs detected in sediment (ΣLowPAHs/ΣHighPAHs = 0.04 and BaP/(BaP + Chr) = 0.76). Hence, detected contamination should be tentatively ascribed to industrial activities, as well as to maritime transport.

Concerning PCBs, detected concentrations were about two orders of magnitude lower in comparison to PAHs (Figure 2), ranging from 1.2 μg/Kg (PCB167) to 3.3 μg/Kg (PCB101). Even if no congeners were detected in seawater, their presence in sediment at quantifiable concentrations could be ascribed to the sorption properties of sediment towards microorganic pollutants, as described for PAHs [47]. A similar behavior was also confirmed by studies performed in other marine areas, where PCBs were detected at fractions of ng/L in water and at μg/Kg in sediments [48].

It should be mentioned that PAH concentrations detected in sediment exceeded the limit imposed by Italian and European regulations for BbFl (limit: 40 μg/Kg), BkFl (limit: 40 μg/Kg) and BaP (limit: 40 μg/Kg). As for the sum of PCB congeners (10.5 μg/Kg), this value partially exceeded the limit of 8 μg/Kg. However, similar values are not uncommonly found in the Mediterranean sea, due to the high impact of anthropic activities, as shown by other studies [44,49].

3.4. *Chemical Characterization of Biota and Partition Studies*

PAH and PCB contamination was measured in four zooplankton taxa, namely *hyperbenthic Isopoda*, herbivores and carnivorous *holoplanktonic crustaceans copepods* and fish larvae-*ichthyoplankton*, previously sorted after summer and winter sampling campaigns. The results obtained are hereafter discussed.

3.4.1. Copepods

Regarding *crustaceans* copepods, results are summarized in Figure 3A (herbivorous) and Figure 3B (carnivorous).

Data showed that PAHs were detected in both taxa at higher concentrations than PCBs, with a total concentration from 270 μg/Kg (winter) to 1.3 mg/Kg (summer) for PAHs witn respect to 43 μg/Kg for PCBs (detected only in summer campaign) for herbivorous copepods. For carnivorous copepods, PAHs concentrations from 230 μg/Kg (winter) to 4.3 mg/Kg (summer) were measured, with no detection of PCBs. The measured concentrations are in the same range of those obtained by Cailleaud and colleagues in estuarine water (in which seasonal variability was also assessed) [50] or by Tiano and colleagues in Marseille Bay (Mediterranean Sea) [51].

Since copepods tend to spend their entire lives mostly in the water column [52], comparison with PAH and PCB contamination in this matrix seems to be fully justified. Concerning PAHs, the same congeners (NaPh, AcPY, Flu, Phe, Ant, FlTh and BkFl) detected in seawater were detected in both herbivorous and carnivorous copepods (Figure 1 vs. Figure 3A,B). Furthermore, the same seasonal trend observed for water (i.e., higher number of congeners and concentrations in summer sampling, rather than in winter) was observed for copepods. This behavior suggests that a possible transfer of micropollutants from water to copepods had occurred. To support this hypothesis, it is important to note that Phe and Ant, which were found to be almost the most abundant PAH compounds in water, confirmed this trend also in herbivorous copepods.

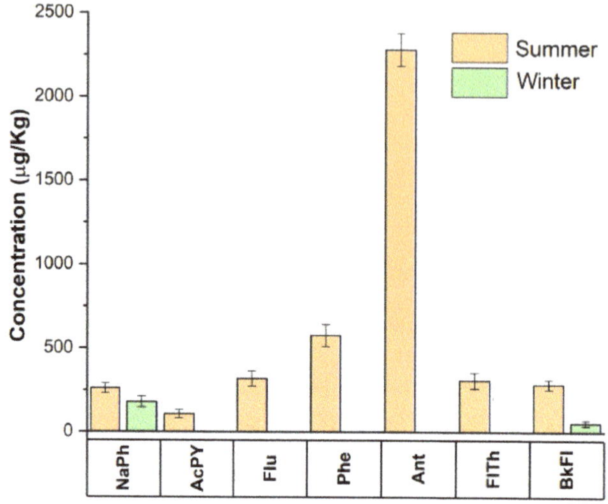

(A)

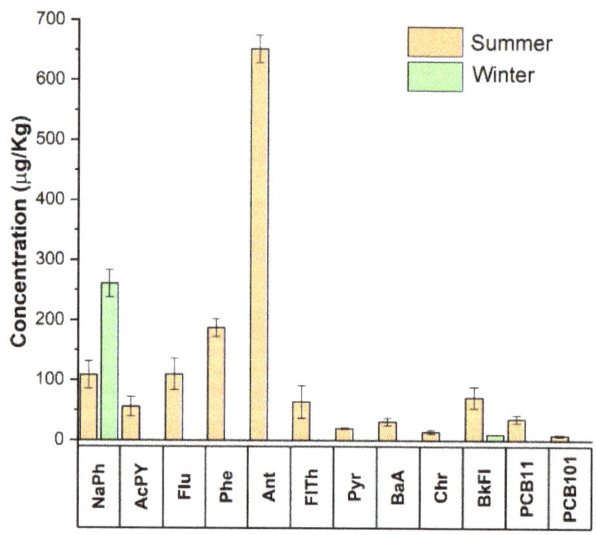

(B)

Figure 3. Concentration of PAHs and PCBs detected in herbivorous (**A**) and carnivorous (**B**) copepods. Both summer (orange) and winter (green) sampling are reported. Analytical conditions are detailed in the Experimental section.

In contrast, Pyr, BaA and Chr, which were detected in the herbivorous copepods, were not present in carnivorous copepods and in seawater. The presence of these congeners could be explained taking into account feeding habits, since it was recently demonstrated that herbivorous copepods could be deceived in their feeding by microplastics and nanoplastics, upon which algae and phytoplankton grow [53]. Such micropolymers have been shown to adsorb organic and inorganic pollutants (such as PAHs and PCBs) [54], thus suggesting that they could be an additional source of PAH congeners once ingested by organisms [55].

Moreover, the same hypothesis could be formulated to explain the detection of PCB11 and 101 in herbivorous copepods (Figure 3A), which were not previously detected in the water column and that were below detection limits in carnivorous organisms. The observed PCB contamination does not exceed limits imposed by European and Italian regulation, since both congeners are not dioxin-like and limits on biota are referred only to dioxin-like compounds.

3.4.2. Hyperbenthic Isopoda

PAH and PCB contamination detected in *hyperbenthic Isopoda* was observed to be almost absent, apart from NaPh, detected at 280 ± 20 μg/Kg, and PCB101, detected at 140 μg/Kg (summer-winter average concentrations). RSD% between summer and winter measured below 8% for both analytes, indicating that, for this taxon, a seasonal variance is not present.

Moreover, It is interesting to note that PCB101, the most abundant pollutant detected in sediment (see "Chemical characterization of sediment" paragraph), is the only congener detected in *hyperbenthic Isopoda*. Since *hyperbenthic Isopoda* was assessed to live half of its life in contact with sediment [56], a possible transfer could not be excluded.

No comparison of contamination level with previous studies could be performed (as for copepods), since, to the best of our knowledge, this is the first study investigating PAH and PCB occurrence in *Hyperbenthic Isopoda*.

3.4.3. Ichthyoplankton

Finally, the PAH and PCB occurrence was also determined for the fourth taxon, fish larvae (*ichthyoplankton*). Results revealed, regarding PAHs, only NaPh and AcPy were detected, with concentrations below quantitation limits, while PCB congeners were not detected.

Comparing to previous taxa, *ichthyoplankton* showed the lowest contamination; such behavior could be ascribed to fish larvae feeding habits. Indeed, *ichthyoplankton* feeding is typically based on copepods larvae that are characterized by a short lifetime and, hence, a short exposure to pollution agents dissolved in water. The transfer of pollutants is, therefore, further hindered.

Finally, it should be mentioned that PAH concentrations detected in the four target taxa did not exceed regulation limits (5 μg/Kg of BaP).

3.5. Bioaccumulation Factors

The BAF (Bioaccumulation Factor) represents one of the most used models to predict the partitioning between an exposure medium (such as seawater) and biota [57]. Two possible mechanisms could influence the bioaccumulation of copepods (and hence the BAF values), namely the ingestion of contaminated dietary sources and the diffusive sorption of pollutants from water [58].

In the present study, BAF values were calculated for the PAH congeners detected both in seawater and in copepods (NaPh, AcPY, Flu, Phe, Ant, FlTh and BkFl). To evaluate a possible effect of feeding habits in the bioaccumulation, BAF values were calculated for herbivorous and carnivorous copepods. PCBs were not included since they were seldom detected in all the sea compartments investigated. The logarithms of BAF values are summarized in Table 1.

Table 1. Logarithmic Bioaccumulation Factors (BAF) from seawater calculated for both herbivorous and carnivorous copepods.

	$LogBAF_{Herb}$ [L/Kg]	$LogBAF_{Carn}$ [L/Kg]
NaPh	2.50	2.88
AcPY	2.64	2.92
Flu	2.83	3.30
Phe	2.81	3.30
Ant	2.82	3.36
FlTh	2.90	3.59
BkFl	3.19	3.79
Taxon average	2.81	3.31

The logBAFs obtained revealed positive values for both herbivorous and carnivorous copepods, thus highlighting that, for both taxa, a bioaccumulation from seawater occurred. Average values, 2.8 and 3.4 for herbivorous and carnivorous copepods, respectively, are in good agreement with those discovered by Arias et al. in Pseudodiaptomus marinus copepods [59].

LogBAF average values of herbivores and carnivores were shown to be statistically different (t-test, N = 7, $p = 0.001$), meaning that higher values observed for carnivorous copepods corresponded to a higher bioaccumulation. Even if it is not possible to univocally explain this behavior, this higher value could represent a first biomagnification effect, since carnivorous copepods feed mainly on herbivorous ones. Hence, an increase of pollutant concentrations along the trophic chain is not unexpected.

The plotting of logBAFs against logKow (Figure 4) exhibited a sound correlation for both taxa ($R^2 = 0.92$ for herbivorous and $R^2 = 0.81$ for carnivorous), supporting the hypothesis that the lipophilicity of the molecule provides the main factor driving the type of PAH congeners in copepods.

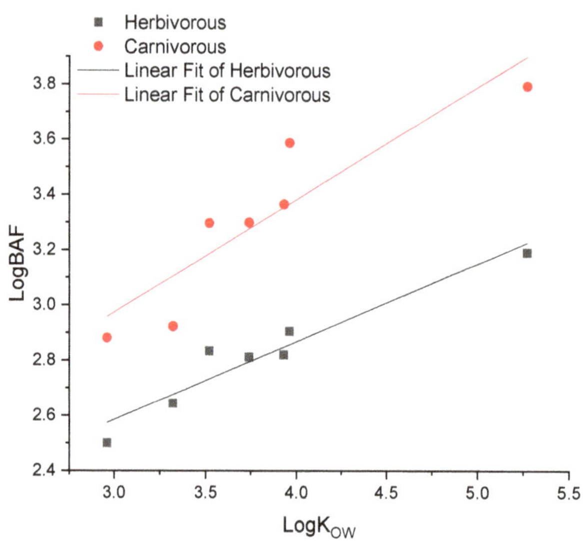

Figure 4. Linear modelling of LogBAFs against LogKow of PAHs for both herbivorous (black squared symbols) and carnivorous (red circled symbols) copepods.

4. Conclusions

This work represents the first study devoted to the comprehensive investigation of 16 PAHs and 14 PCBs (including dioxin-like congeners) in the seawater, sediment and biota of the Ligurian Sea.

For both PAHs and PCBs, the overall results indicate that the contamination levels are equivalent to those typically found in other areas of the Mediterranean Sea characterized by moderate to high pollution. Moreover, this study reveals that PAHs origin is mainly ascribed to pyrogenic activities (combustion, transport, etc). A seasonal trend was observed for the presence of PAHs and PCBs in seawater, since micropollutant concentrations measured in summer were higher than those measured in winter, in agreement with increased touristic and harbor activities and with the higher stratification of water layers.

Micropollutant partition between water and sediment was confirmed to be influenced by the hydrophobicity of the pollutant molecules. To elucidate, higher molecular weight pollutants were preferentially detected in sediment, with organic matter acting as adsorbent. Bioaccumulation on biota was also assessed for copepods.

In addition, partition pathways were investigated within four different taxa (hyperbenthic Isopoda, herbivores and carnivorous holoplanktonic crustaceans copepods and fish larvae-ichthyoplankton), by the calculation of BAF values. The effects of living habits were shown to strongly influence bioaccumulation, as the congeners detected in sediment were preferentially discovered in taxa that spent most of their lives there. Moreover, feeding habits were found to influence the bioaccumulation of PAHs and PCBs, and a positive correlation between BAF values and hydrophobicity was confirmed. This study performed in the Southern Ligurian Sea, provides a novel insight into PAH and PCB distribution in marine zooplankton, which represents the main food constituent of most fish of the coastal waters, thus playing an important role in the transfer of pollutants through the food web. The experimental approach here followed can be transposed to other geographical areas of strategic importance to understand peculiar distributions patterns.

Supplementary Materials: The following supporting information can be downloaded at: https://www.mdpi.com/article/10.3390/app12052564/s1, Figure S1. Ligurian Sea (Western Mediterranean): study area and location of sampling stations; Figure S2: average (n = 3) PAH and PCB recoveries from seawater samples. Recoveries were calculated using surrogates as described in the Materials and Methods Section; Figure S3: average (n = 3) PAH and PCB recoveries from sediment samples. Recoveries were calculated using surrogates as described in the Materials and Methods Section; Figure S4: average (n = 3) PAH and PCB recoveries from biota samples. Recoveries were calculated using surrogates as described in the Materials and Methods Section; Table S1: MDL and MQL of PAHs in seawater, sediment and biota, using optimized analytical protocols, compared with their regulation limits reported by 2013/39/EU normative and Italian Legislative Decree 172/2015; Table S2: MDL and MQL of PCBs in seawater, sediment and biota, using optimized analytical protocols, compared with their regulation limits reported by 2013/39/EU normative and Italian Legislative Decree 172/2015.

Author Contributions: Conceptualization, L.R., M.C., N.N. and M.C.B.; Formal analysis, M.C.; Funding acquisition, M.C.B.; Investigation, L.R., N.N., M.B., R.M.S. and M.C.B.; Methodology, L.R., M.C., N.N. and M.C.B.; Project administration, M.C.B.; Supervision, M.C.B.; Validation, L.R. and M.C.; Visualization, M.B., L.F. and M.C.B.; Writing—original draft, L.R.; Writing—review and editing, L.R., N.N., R.M.S. and M.C.B. All authors have read and agreed to the published version of the manuscript.

Funding: This research was funded by Ministero dell'Università e della Ricerca, Ex60%.

Institutional Review Board Statement: Not applicable.

Informed Consent Statement: Not applicable.

Data Availability Statement: Not applicable.

Acknowledgments: The authors would like to thank Andrea Dalla Libera and Lamberto Giusti for their kind assistance in laboratory experiments. Financial support from Ministero della Ricerca e dell'Università (Ex-60%) is gratefully acknowledged.

Conflicts of Interest: The authors declare no conflict of interest.

References

1. Kumar, S. *Modern Treatment Strategies for Marine Pollution*; Elsevier: Amsterdam, The Netherlands, 2020.
2. Simeonov, L.I.; Macaev, F.Z.; Simeonova, B.G. *Environmental Security Assessment and Management of Obsolete Pesticides in Southeast Europe*; Springer: Berlin/Heidelberg, Germany, 2014.
3. Steinberg, D.K.; Landry, M.R. Zooplankton and the ocean carbon cycle. *Annu. Rev. Mar. Sci.* **2017**, *9*, 413–444. [CrossRef] [PubMed]
4. Vikas, M.; Dwarakish, G. Coastal pollution: A review. *Aquat. Procedia* **2015**, *4*, 381–388. [CrossRef]
5. Battuello, M.; Brizio, P.; Sartor, R.M.; Nurra, N.; Pessani, D.; Abete, M.; Squadrone, S. Zooplankton from a North Western Mediterranean area as a model of metal transfer in a marine environment. *Ecol. Indic.* **2016**, *66*, 440–451. [CrossRef]
6. Vernberg, J.F. *Physiological Responses of Marine Biota to Pollutants*; Elsevier: Amsterdam, The Netherlands, 2012.
7. Wakeham, S.G.; McNichol, A.P. Transfer of organic carbon through marine water columns to sediments–insights from stable and radiocarbon isotopes of lipid biomarkers. *Biogeosciences* **2014**, *11*, 6895–6914. [CrossRef]
8. Nematollahi, M.J.; Keshavarzi, B.; Moore, F.; Vogt, R.D.; Nasrollahzadeh Saravi, H. Trace elements in the shoreline and seabed sediments of the southern Caspian Sea: Investigation of contamination level, distribution, ecological and human health risks, and elemental partition coefficient. *Environ. Sci. Pollut. Res.* **2021**, *28*, 1–24. [CrossRef]
9. Cheng, J.; Han, J.; Zheng, B.; Wang, X.; Yang, Z.; Zhang, X. Exploring the influence of water exchange on the distribution of polycyclic aromatic hydrocarbons in marine sediments by numerical calculation model. *J. Hydrol.* **2021**, *603*, 126874. [CrossRef]
10. Alshemmari, H. An overview of persistent organic pollutants along the coastal environment of Kuwait. *Open Chem.* **2021**, *19*, 149–156. [CrossRef]
11. Gioia, R.; Dachs, J.; Nizzetto, L.; Berrojalbiz, N.; Galbán, C.; Del Vento, S.; Méjanelle, L.; Jones, K.C. Sources, transport and fate of organic pollutants in the oceanic environment. In *Persistent Pollution—Past, Present and Future*; Springer: Berlin/Heidelberg, Germany, 2011; pp. 111–139.
12. Lawson, M.C.; Cullen, J.A.; Nunnally, C.C.; Rowe, G.T.; Hala, D.N. PAH and PCB body-burdens in epibenthic deep-sea invertebrates from the northern Gulf of Mexico. *Mar. Pollut. Bull.* **2021**, *162*, 111825. [CrossRef]
13. Sprovieri, M.; Feo, M.L.; Prevedello, L.; Manta, D.S.; Sammartino, S.; Tamburrino, S.; Marsella, E. Heavy metals, polycyclic aromatic hydrocarbons and polychlorinated biphenyls in surface sediments of the Naples harbour (southern Italy). *Chemosphere* **2007**, *67*, 998–1009. [CrossRef]
14. Moon, H.-B.; Yoon, S.-P.; Jung, R.-H.; Choi, M. Wastewater treatment plants (WWTPs) as a source of sediment contamination by toxic organic pollutants and fecal sterols in a semi-enclosed bay in Korea. *Chemosphere* **2008**, *73*, 880–889. [CrossRef]
15. Foster, G.D.; Cui, V. PAHs and PCBs deposited in surficial sediments along a rural to urban transect in a Mid-Atlantic coastal river basin (USA). *J. Environ. Sci. Health Part A* **2008**, *43*, 1333–1345. [CrossRef] [PubMed]
16. Golomb, D.; Barry, E.; Fisher, G.; Varanusupakul, P.; Koleda, M.; Rooney, T. Atmospheric deposition of polycyclic aromatic hydrocarbons near New England coastal waters. *Atmos. Environ.* **2001**, *35*, 6245–6258. [CrossRef]
17. Nakata, H.; Sakai, Y.; Miyawaki, T.; Takemura, A. Bioaccumulation and toxic potencies of polychlorinated biphenyls and polycyclic aromatic hydrocarbons in tidal flat and coastal ecosystems of the Ariake Sea, Japan. *Environ. Sci. Technol.* **2003**, *37*, 3513–3521. [CrossRef] [PubMed]
18. Bruzzoniti, M.C.; Rivoira, L.; Castiglioni, M.; El Ghadraoui, A.; Ahmali, A.; El Mansour, T.E.H.; Mandi, L.; Ouazzani, N.; Del Bubba, M. Extraction of polycyclic aromatic hydrocarbons and polychlorinated biphenyls from urban and olive mill wastewaters intended for reuse in agricultural irrigation. *J. AOAC Int.* **2020**, *103*, 382–391. [CrossRef] [PubMed]
19. Rivoira, L.; Castiglioni, M.; Kettab, A.; Ouazzani, N.; Al-Karablieh, E.; Boujelben, N.; Fibbi, D.; Coppini, E.; Giordani, E.; Del Bubba, M. Impact of effluents from wastewater treatments reused for irrigation: Strawberry as case study. *Environ. Eng. Manag. J.* **2019**, *18*, 2133–2143.
20. Cardellicchio, N.; Buccolieri, A.; Giandomenico, S.; Lerario, V.L.; Lopez, L.; Pizzulli, F. Distribution and occurence of polycyclic aromatic hydrocarbons (PAHs) in sediments from the Mar Grande and Gulf of Taranto (Ionian Sea, Southern Italy). *Ann. Chim. J. Anal. Environ. Cult. Herit. Chem.* **2006**, *96*, 51–64. [CrossRef]
21. ChemAxon. Chemicalize. Available online: https://chemicalize.com/ (accessed on 29 December 2021).
22. European Parliament and Council. *Directive 2013/39/EU as Regards Priority Substances in the Field of Water Policy*; European Parliament and Council: Strasbourg, France, 2013.
23. Repubblica Italiana. *DECRETO LEGISLATIVO 13 Ottobre 2015, n. 172 Sulle Sostanze Prioritarie Nel Settore della Politica delle Acque*; Gazzetta Ufficiale: Rome, Italy, 2015.
24. Ma, M.; Feng, Z.; Guan, C.; Ma, Y.; Xu, H.; Li, H. DDT, PAH and PCB in sediments from the intertidal zone of the Bohai Sea and the Yellow Sea. *Mar. Pollut. Bull.* **2001**, *42*, 132–136. [CrossRef]
25. Sinaei, M.; Mashinchian, A. Polycyclic aromatic hydrocarbons in the coastal sea water, the surface sediment and Mudskipper Boleophthalmus dussumieri from coastal areas of the Persian Gulf: Source investigation, composition pattern and spatial distribution. *J. Environ. Health Sci. Eng.* **2014**, *12*, 1–11. [CrossRef]
26. Sobek, A.; Reigstad, M.; Gustafsson, Ö. Partitioning of polychlorinated biphenyls between Arctic seawater and size-fractionated zooplankton. *Environ. Toxicol. Chem. Int. J.* **2006**, *25*, 1720–1728. [CrossRef]

27. Pizzini, S.; Morabito, E.; Gregoris, E.; Vecchiato, M.; Corami, F.; Piazza, R.; Gambaro, A. Occurrence and source apportionment of organic pollutants in deep sediment cores of the Venice Lagoon. *Mar. Pollut. Bull.* **2021**, *164*, 112053. [CrossRef]
28. Pitacco, V.; Mistri, M.; Granata, T.; Moruzzi, L.; Maria, L.M.; Massara, F.; Sfriso, A.; Sfriso, A.A.; Munari, C. Sediment Contamination by Heavy Metals and PAH in the Piombino Channel (Tyrrhenian Sea). *Water* **2021**, *13*, 1487.
29. Cardellicchio, N.; Annicchiarico, C.; Di Leo, A.; Giandomenico, S.; Spada, L. The Mar Piccolo of Taranto: An interesting marine ecosystem for the environmental problems studies. *Environ. Sci. Pollut. Res.* **2016**, *23*, 12495–12501. [CrossRef] [PubMed]
30. Bertolotto, R.; Ghioni, F.; Frignani, M.; Alvarado-Aguilar, D.; Bellucci, L.; Cuneo, C.; Picca, M.; Gollo, E. Polycyclic aromatic hydrocarbons in surficial coastal sediments of the Ligurian Sea. *Mar. Pollut. Bull.* **2003**, *46*, 907–913. [CrossRef]
31. Deyme, R.; Bouloubassi, I.; Taphanel-Valt, M.-H.; Miquel, J.-C.; Lorre, A.; Marty, J.-C.; Mejanelle, L. Vertical fluxes of organic contaminants in the Ligurian Sea. In Proceedings of the EGU General Assembly Conference Abstracts, Vienna, Austria, 2–7 May 2010; p. 9955.
32. Pane, L.; Boccardo, S.; Bonfiglioli, F.; Mariottini, G.L.; Priano, F.; Conio, O. Polycyclic aromatic hydrocarbons in water, seston and copepods in a harbour area in the Western Mediterranean (Ligurian Sea). *Mar. Ecol.* **2005**, *26*, 89–99. [CrossRef]
33. Gobas, F.; Morrison, H.A. Bioconcentration and biomagnification in the aquatic environment. In *Handbook of Property Estimation Methods for Chemicals*; CRC Press: Boca Raton, FL, USA, 2000; pp. 189–231.
34. Albarano, L.; Zupo, V.; Caramiello, D.; Toscanesi, M.; Trifuoggi, M.; Guida, M.; Libralato, G.; Costantini, M. Sub-Chronic Effects of Slight PAH-and PCB-Contaminated Mesocosms in Paracentrotus lividus Lmk: A Multi-Endpoint Approach and De Novo Transcriptomic. *Int. J. Mol. Sci.* **2021**, *22*, 6674. [CrossRef]
35. Li, Q.; Xu, X.; Sen-Chun, L.F.; Wang, X. Determination of trace PAHs in seawater and sediment pore-water by solid-phase microextraction (SPME) coupled with GC/MS. *Sci. China Ser. B Chem.* **2006**, *49*, 481–491. [CrossRef]
36. Leppäranta, M.; Myrberg, K. *Physical Oceanography of the Baltic Sea*; Springer Science & Business Media: Berlin/Heidelberg, Germany, 2009.
37. Fabbri, E.; Franzellitti, S. Human pharmaceuticals in the marine environment: Focus on exposure and biological effects in animal species. *Environ. Toxicol. Chem.* **2016**, *35*, 799–812. [CrossRef]
38. Bourg, N.; Molcard, A. Northern boundary current variability and mesoscale dynamics: A long-term HF RADAR monitoring in the North-Western Mediterranean Sea. *Ocean. Dyn.* **2021**, *71*, 851–870. [CrossRef]
39. Rice, S.D.; Holland, L.; Moles, A. Seasonal increases in polycyclic aromatic hydrocarbons related to two-stroke engine use in a small Alaskan lake. *Lake Reserv. Manag.* **2008**, *24*, 10–17. [CrossRef]
40. Ali, H.R.; Nour, S.; El-Ezbewy, S.; El-gemaee, G.H.; Moustafa, Y.; Roushdy, M. Assessment of polycyclic aromatic hydrocarbons contamination in water, sediment and fish of Temsah Lake, Suez Canal, Egypt. *Curr. World Environ.* **2006**, *1*, 11–22. [CrossRef]
41. Magi, E.; Bianco, R.; Ianni, C.; Di Carro, M. Distribution of polycyclic aromatic hydrocarbons in the sediments of the Adriatic Sea. *Environ. Pollut.* **2002**, *119*, 91–98. [CrossRef]
42. Tolosa, I.; de Mora, S.; Sheikholeslami, M.R.; Villeneuve, J.-P.; Bartocci, J.; Cattini, C. Aliphatic and aromatic hydrocarbons in coastal Caspian Sea sediments. *Mar. Pollut. Bull.* **2004**, *48*, 44–60. [CrossRef]
43. Odabasi, M.; Dumanoglu, Y.; Kara, M.; Altiok, H.; Elbir, T.; Bayram, A. Spatial variation of PAHs and PCBs in coastal air, seawater, and sediments in a heavily industrialized region. *Environ. Sci. Pollut. Res.* **2017**, *24*, 13749–13759. [CrossRef] [PubMed]
44. Baumard, P.; Budzinski, H.; Garrigues, P. Polycyclic aromatic hydrocarbons in sediments and mussels of the western Mediterranean Sea. *Environ. Toxicol. Chem. Int. J.* **1998**, *17*, 765–776. [CrossRef]
45. Guigue, C.; Tedetti, M.; Dang, D.H.; Mullot, J.-U.; Garnier, C.; Goutx, M. Remobilization of polycyclic aromatic hydrocarbons and organic matter in seawater during sediment resuspension experiments from a polluted coastal environment: Insights from Toulon Bay (France). *Environ. Pollut.* **2017**, *229*, 627–638. [CrossRef]
46. Komada, T.; Schofield, O.M.; Reimers, C.E. Fluorescence characteristics of organic matter released from coastal sediments during resuspension. *Mar. Chem.* **2002**, *79*, 81–97. [CrossRef]
47. Iwata, H.; Tanabe, S.; Aramoto, M.; Sakai, N.; Tatsukawa, R. Persistent organochlorine residues in sediments from the Chukchi Sea, Bering Sea and Gulf of Alaska. *Mar. Pollut. Bull.* **1994**, *28*, 746–753. [CrossRef]
48. Jafarabadi, A.R.; Bakhtiari, A.R.; Mitra, S.; Maisano, M.; Cappello, T.; Jadot, C. First polychlorinated biphenyls (PCBs) monitoring in seawater, surface sediments and marine fish communities of the Persian Gulf: Distribution, levels, congener profile and health risk assessment. *Environ. Pollut.* **2019**, *253*, 78–88. [CrossRef]
49. Barakat, A.O.; Mostafa, A.; Wade, T.L.; Sweet, S.T.; El Sayed, N.B. Distribution and characteristics of PAHs in sediments from the Mediterranean coastal environment of Egypt. *Mar. Pollut. Bull.* **2011**, *62*, 1969–1978. [CrossRef]
50. Cailleaud, K.; Forget-Leray, J.; Souissi, S.; Hilde, D.; LeMenach, K.; Budzinski, H. Seasonal variations of hydrophobic organic contaminant concentrations in the water-column of the Seine Estuary and their transfer to a planktonic species Eurytemora affinis (Calanoida, copepoda). Part 1: PCBs and PAHs. *Chemosphere* **2007**, *70*, 270–280. [CrossRef]
51. Tiano, M.; Tronczyński, J.; Harmelin-Vivien, M.; Tixier, C.; Carlotti, F. PCB concentrations in plankton size classes, a temporal study in Marseille Bay, Western Mediterranean Sea. *Mar. Pollut. Bull.* **2014**, *89*, 331–339. [CrossRef] [PubMed]
52. Kiørboe, T. What makes pelagic copepods so successful? *J. Plankton Res.* **2011**, *33*, 677–685. [CrossRef]
53. Cole, M.; Lindeque, P.; Fileman, E.; Halsband, C.; Goodhead, R.; Moger, J.; Galloway, T.S. Microplastic ingestion by zooplankton. *Environ. Sci. Technol.* **2013**, *47*, 6646–6655. [CrossRef] [PubMed]

54. Johnson, W.S.; Allen, D.M. *Zooplankton of the Atlantic and Gulf Coasts: A Guide to Their Identification and Ecology*; JHU Press: Baltimore, MA, USA, 2012.
55. Rivoira, L.; Castiglioni, M.; Rodrigues, S.; Freitas, V.; Bruzzoniti, M.; Ramos, S.; Almeida, C. Microplastic in marine environment: Reworking and optimisation of two analytical protocols for the extraction of microplastics from sediments and oysters. *MethodsX* **2020**, *7*, 101116. [CrossRef]
56. Koulouri, P. Preliminary study of hyperbenthos in Heraklion Bay (Cretan Sea). *Biomare Newsl.* **2002**, *3*, 12.
57. Newman, M.C. *Quantitative Methods in Aquatic Ecotoxicology*; CRC Press: Boca Raton, FL, USA, 1994.
58. Berrojalbiz, N.; Lacorte, S.; Calbet, A.; Saiz, E.; Barata, C.; Dachs, J. Accumulation and cycling of polycyclic aromatic hydrocarbons in zooplankton. *Environ. Sci. Technol.* **2009**, *43*, 2295–2301. [CrossRef]
59. Arias, A.H.; Souissi, A.; Roussin, M.; Ouddane, B.; Souissi, S. Bioaccumulation of PAHs in marine zooplankton: An experimental study in the copepod Pseudodiaptomus marinus. *Environ. Earth Sci.* **2016**, *75*, 1–9. [CrossRef]

applied sciences

MDPI

Article

Natural Versus Anthropic Influence on North Adriatic Coast Detected by Geochemical Analyses

Eliana Barra [1,2], Francesco Riminucci [1,3,*], Enrico Dinelli [2], Sonia Albertazzi [1], Patrizia Giordano [4], Mariangela Ravaioli [1] and Lucilla Capotondi [1,*]

1. Istituto di Scienze Marine-Consiglio Nazionale delle Ricerche ISMAR-CNR, Via P. Gobetti 101, 40129 Bologna, Italy; eliana.barra@studio.unibo.it (E.B.); sonia.albertazzi@bo.ismar.cnr.it (S.A.); mariangela.ravaioli@bo.ismar.cnr.it (M.R.)
2. BiGeA-Dipartimento di Scienze Biologiche, Geologiche e Ambientali, Università di Bologna, Via Zamboni 33, 40126 Bologna, Italy; enrico.dinelli@unibo.it
3. Consorzio Proambiente S.c.r.l., Tecnopolo Bologna CNR, Via Gobetti 101, 40129 Bologna, Italy
4. Istituto di Scienze Polari-Consiglio Nazionale delle Ricerche ISP-CNR, Via P. Gobetti 101, 40129 Bologna, Italy; patrizia.giordano@cnr.it

* Correspondence: francesco.riminucci@bo.ismar.cnr.it (F.R.); lucilla.capotondi@bo.ismar.cnr.it (L.C.)

Received: 28 July 2020; Accepted: 15 September 2020; Published: 21 September 2020

Abstract: This study focused on the geochemical and sedimentological characterization of recent sediments from two marine sites (S1 and E1) located in the North Adriatic Sea, between the Po River prodelta and the Rimini coast. Major and trace metal concentrations reflect the drainage area of the Po River and its tributaries, considered one of the most polluted areas in Europe. Sediment geochemistry of the two investigated sites denote distinct catchment areas. High values of Cr, Ni, Pb and Zn detected in sediments collected in the Po River prodelta (S1 site) suggest the Po River supply, while lower levels of these elements characterize sediments collected in front of the Rimini coast (E1 site), an indication of Northern Apennines provenance. Historical trends of Pb and Zn reconstructed from the sedimentary record around the E1 site document several changes that can be correlated with the industrialization subsequent to World War II, the implementation of the environmental policy in 1976 and the effects of the Comacchio dumping at the end of 1980. At the S1 site, the down core distributions of trace elements indicate a reduction of contaminants due to the introduction of the Italian Law 319/76 and the implementation of anti-pollution policies on automotive Pb (unleaded fuels) in the second half of the 1980s.

Keywords: northern Adriatic Sea; anthropic signal; trace metals concentration; marine sediments

1. Introduction

Sedimentary deposits represent an important natural archive of environmental change. Their study allows the evaluation of the extent and the effect of anthropogenic inputs [1]. Indeed, marine sediments act as sinks and sources of contaminants delivered to the sea from inland activities. Therefore, the knowledge of the type and concentration of contaminants, along with their spatial distribution, pathways and historical trends are essential for supporting measures to mitigate future impacts on marine ecosystems. Hence, an exhaustive assessment of the natural geochemical background is mandatory, when dealing with inorganic contaminants [2].

The Po River plain is one of the most industrialized and agriculturally developed areas in Europe, generating nearly 40% of the Italian Gross Domestic Product. The high concentration of industrial, zootechnical and agricultural activities critically intensifies the risk of heavy-metal contamination of the prodelta area, coastal lagoons [3–6] and in the neighboring marine environments [7,8].

To evaluate the anthropic effects on coastal environments, we report integrated geochemical and sedimentological analyses performed on marine sediment cores collected in the North Adriatic at the Po River delta and the Rimini coast (Figure 1). In this study, we aimed: (1) to provide additional information about the geochemical composition and distribution of the surface marine sediments, including major and trace element concentrations; and (2) to investigate the recent (the last 100 years) anthropic influence on the *Emilia-Romagna* coast.

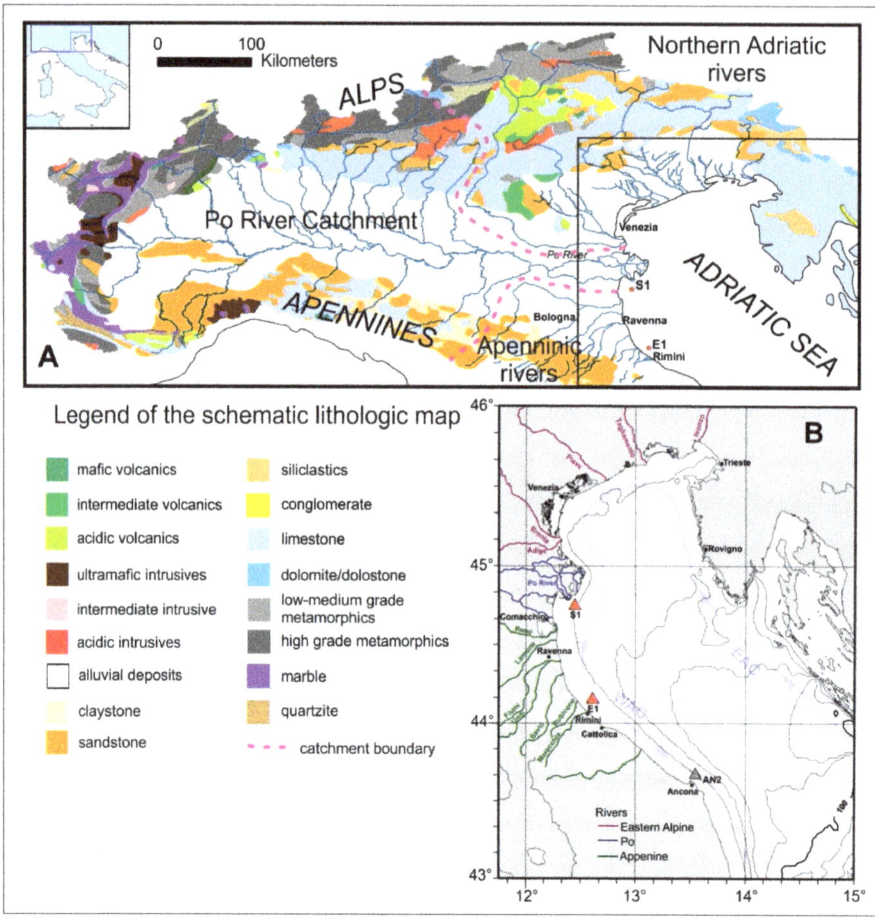

Figure 1. (**A**) Lithologic map of the Po River drainage basin (data from 1:1 Million Surface Geology—European Geological Data Infrastructure: https://egdi-v6.geology.cz/?ak=detailall&uuid=5729ffdf-2558-48fc-a5d2-645a0a010855). Major Po contributor rivers are also indicated. (**B**) Schematic surface circulation pattern of the Northern Adriatic Sea. EAC: Eastern Adriatic Current; WAC: Western Adriatic Current [9]. Triangles represent S1, E1 (this work, red triangles) and AN2 (grey triangle) [10] coring sites.

Study Area

The North Adriatic Sea (NAS) is an epicontinental basin characterized by a global cyclonic circulation with seawater inflow to the northwest along the Croatian coast (EAC) and outflow to the southeast along the Italian coast (WAC) [9] (Figure 1). This general pattern is greatly influenced by winds, mainly Bora and Scirocco [11,12], and by lateral freshwater inputs [13]. Freshwater runoff enters the NAS from eastern Alpine and northern Apennine rivers located mainly along the

Italian coastline. The largest source of freshwater and sediments is represented by the Po River (mean annual discharge of 1500 $m^3 s^{-1}$) [14–16]. The Po River catchment extends from the Central and Western Alps up to the northwestern Apennines and includes more than 140 Alpine and Apennine tributaries that discharge water and sediment into the Po River [17] (Figure 1). The hydrologic behavior of the Po River is characterized by the mixed Alpine/Apennine discharge regime resulting in two flooding periods in late fall and spring, due to intense precipitation and snow melting, respectively [18].

Sediment accumulation rates range 1.2-30 mm/year with the highest value recorded in the study area, and then accumulation rates decrease southward and toward the open sea [15]. Several studies [18–21] pointed out a decrease of the sediment supply in terms of both suspended and bed loads, from the Po and the Northern Apennines rivers, after 1945. An additional decrease was observed in the 1980s when the *Emilia-Romagna* region underwent heavy urbanization of both coastal and inland areas, as well as other parts of the Po River plain [18,19,22,23]. The Apennine rivers—which flow into the NAS encompassing the area between Ravenna and Rimini (Reno, Lamone, Bevano, Fiumi Uniti, Savio, Rubicone, Uso and Marecchia)—are characterized by smaller catchments areas. Their courses are almost straight, and their flow rate is subject to the rainfall regime resulting in very low discharges in summer, when hydric demand and evapotranspiration are the highest [20,24,25].

Geochemical characterization of Po Plain late Quaternary deposits [17,26–31] document that they preserve fingerprints of the related catchment basins. In detail, nickel and chromium have been used as tracers of sediment provenance since Cr-rich (and Ni-rich) sediments have been found to be representative of the Po River drainage basin, whereas sediments with lower Cr and Ni values have been related to the Apennines supply [26,28]. Their concentration is due to the contribution of ultramafic rocks in the Po River catchment (Figure 1). Although Cr and Ni could also be affected by anthropogenic sources, they seem to be minor, since comparable results were reported for present day Po River sediments and Holocene marine deposits in the area [27].

2. Materials and Methods

2.1. Sampling and Analyses

We investigated the marine area around two fixed observatories (the E1 meteo-oceanographic buoy and the S1-GB dynamic pylon [32,33]), part of the Italian LTER network (long-term ecological research network, https://deims.org/6869436a-80f4-4c6d-954b-a730b348d7ce) deployed to study the ecosystems and the long-term response to environmental, societal and economic drivers [34]. Six sediment cores were collected with a box-corer in 2014, 2016 and 2017 during the oceanographic cruises EL-14, LTER-ANOC16 and INTERNOS17 onboard the R/V Dallaporta (Table 1).

Table 1. Details of sampling stations.

Cores	Latitude	Longitude	Site	Water Depth (m)	Core Length (cm)
S1-EL14	44.737° N	12.447° E	S1	21	17
E1-EL14	44.141° N	12.573° E	E1	10.5	15
S1-ANOC16	44.742° N	12.452° E	S1	21	17.5
E1-ANOC16	44.148° N	12.577° E	E1	10.5	16
S1-INT17	44.742° N	12.459° E	S1	21	18
E1-INT17	44.143° N	12.572° E	E1	10.5	10.5

Core locations and study areas, namely *S1 (Delta del Po)* and *E1 (Costa Romagnola)* (summarized above as S1 and E1, respectively), are shown in Figure 1. Each box-core was subsampled with smaller PVC cores (diameter 8 cm) and stored at 4 °C [35,36] directly on the vessel. At the ISMAR-CNR laboratories, each core was described and subsampled at 1 or 2 cm of thickness. The samples were dried at 60 °C and the porosity was calculated assuming a sediment density of 2.55 g cm^{-3} [37]. Dried sediment levels from all six cores were slightly disaggregated and analyzed for the radiotracer ^{137}Cs, by non-destructive gamma-ray spectrometry using two intrinsic Ge detectors.

Sediment samples were put into cylindrical transparent plastic jars, pressed and counted for 24 h. Activities were measured following Albertazzi et al. [38] and Frignani et al. [39]. Both detectors were connected to an integrated gamma-spectrometry system, the *DSPEC-Ortec*. Output spectra were processed and calibrated by using the *Gammavision-Ortec* software. All concentrations and activities are relative to dry weight and are expressed in Bq kg^{-1}. Detector efficiency was estimated using samples of sediment spiked with multipeak standard solutions (QCY58) counted in standard geometry jars (5–10–15–20 ml). Accuracy was checked periodically by using certified reference sediments (IAEA-TEL-2012-03).

To assess grain-size distributions, the organic content was firstly removed by treating the samples with H_2O_2; subsequently, sediments were wet sieved with a 63-µm-net size to separate coarse- and fine-grained fractions. Finally, the coarse fraction was analyzed using a stack of ASTM sieves, while the finer sediments by an X-ray sedigraph *Micromeritics 5120*, in order to quantify the silt and clay fractions. Results are relative to dry weight and are expressed as percentages of grain-size fractions. Sediments were also analyzed for geochemical composition at the BiGeA laboratories, by means of X-ray fluorescence spectrometry. At least 3 g of each dried sediment layer were ground in an agate mortar and then pressed with 7 g of H_3BO_3 to obtain powder pellets. Geochemical analyses were carried out by using a *Panalytical Axios 4000* X-ray spectrometer, to quantify major (Si, Ti, Al, Fe, Mn Mn, Mg, Na, K and P) and trace element (Sc, V, Cr, Co, Ni, Cu, Zn, Ga, As, Rb, Sr, Y, Zr, Nb, Ba, La, Ce, Pb, Th, S, Ba and Mo) contents, expressed as dry weight percentages (dw %) and in mg kg^{-1}, respectively (Appendix A). In this work, the trace metal concentrations are normalized to Al_2O_3, one of the most commonly used normalizers in sediment studies [40]. For a more detailed description of the historical trend of anthropic influence at the S1 site, the results for Zn and Pb were integrated with the literature data (see Section 4.3).

The Loss On Ignition (LOI), which includes contribution from organic matter, crystalline water and carbonates, was estimated in terms of weight loss after heating at 950 °C to constant weight.

2.2. Background Levels

To evaluate the anthropogenic signals in S1 and E1 sites, we compared the data with the background trace metal values. As work strategy, for the S1 site, we considered as background values from adjacent areas available from the literature [10,41,42] (Table 2), and, for the E1 site, we chose the mean values detected in the 1900–1920 interval (Table 2 and interval 12–15 cm in Figure 5) as representative of the pre-industrial values.

Table 2. Mean and standard deviation of trace metals in S1 and E1 sediment cores (ANOC-2016) and background (Bkg) levels.

	Zn	Pb	Cu	Cr	Ni
S1-Mean	129 ± 8	36 ± 3	34 ± 3	183 ± 4	128 ± 3
S1-Bkg *	84	14	30	152	61
E1-Mean	79 ± 7	22 ± 4	18 ± 3	139 ± 4	80 ± 7
E1-Bkg	80	20	16	142	91

Metal concentrations are reported in mg kg^{-1}. * Data from literature [10].

3. Results

3.1. Porosity

Porosity vertical profiles of S1 sediment cores do not show evident variations (values around 0.65–0.69, Figure 2). A slight decrease of porosity is detected only in the bottom layer documenting an increase of sediment compaction. On the other hand, the E1 sedimentary interval displays a larger variability of porosity (ranging from 0.50 to 0.69, Figure 2). The higher values are recorded in the first 4 cm with a clear decreasing trend and with the downward porosity stabilizing toward the lower values (Figure 2).

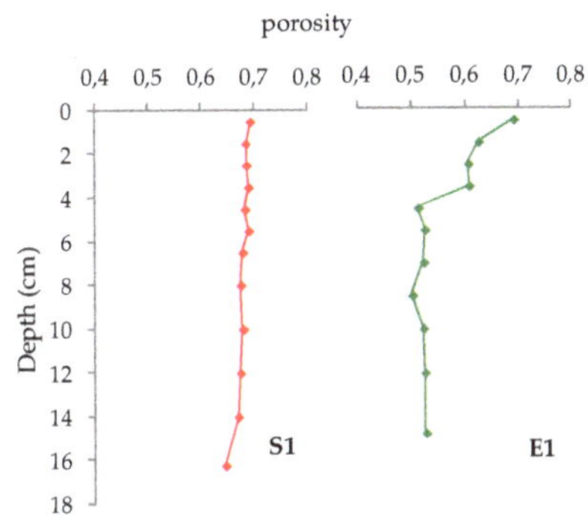

Figure 2. Porosity vs. depth profiles for S1 and E1 cores (cruise ANOC-16).

3.2. ^{137}Cs Activity

Sedimentation rates were estimated using the vertical distribution of ^{137}Cs activity [43]. In the S1 investigated cores, results point out a ^{137}Cs activity range of 10.22 ± 1.55 to 29.01 ± 4.15 Bq kg^{-1} with a median value of 17.74 Bq kg^{-1} (Figure 3a–c). The trends are very similar, and the highest value are detected in the S1-INT17 core at 2 cm of depth (Figure 3c). Cores from E1 (Figure 3d–f) display very low ^{137}Cs activity, between 3.28 ± 2.1 and 6.45 ± 2.78 Bq kg^{-1} with a median value of 5.39 Bq kg^{-1}.

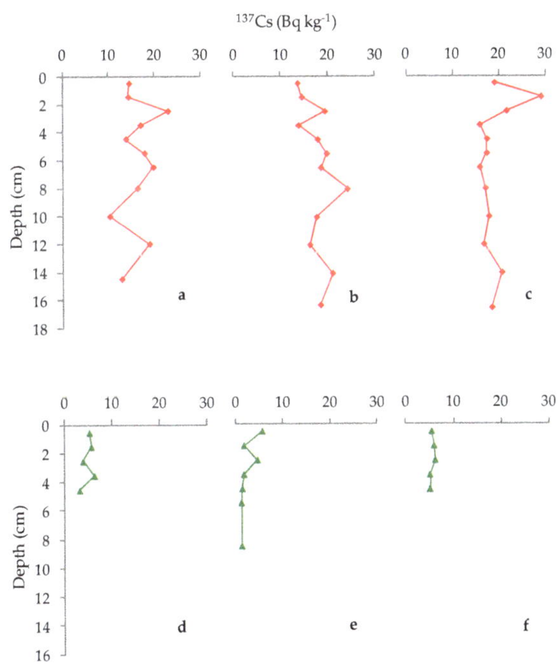

Figure 3. Depth profiles of ^{137}Cs for: (**a**) S1-EL14; (**b**) S1-ANOC16; (**c**) S1-INT17; (**d**) E1-EL14; (**e**) E1-ANOC16; and (**f**) E1-INT17.

3.3. *Grain Size*

Grain size results are reported in Figure 4. The S1 sediments are composed prevalently by clay and silt fractions (around the 90%) in almost all levels, in line with the literature data [44,45]. The E1 sediments are characterized by a more heterogeneous composition; they consist of silty clays with high sand percentages (40%) in the surficial levels (Figure 4).

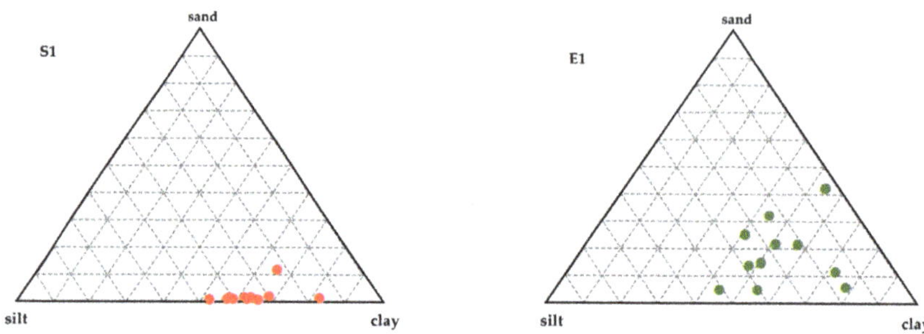

Figure 4. Shepard diagram of the sediment composition in S1 and E1 (cruise ANOC-16) sediment cores.

3.4. *Geochemistry*

The chemical composition of sediments collected at the S1 and E1 sites is reported in Table 3 (complete geochemical dataset in Appendix A). High MgO content characterizes sediments from the S1 site, whereas higher CaO content is observed in sediments from E1. Regarding trace elements, sediments from the S1 area have higher concentrations of Cr, Cu, Ni, Pb and Zn than those from the E1 site.

Table 3. Geochemical analyses of S1 and E1 sediment cores (ANOC-16).

		Major Elements (dw %)		
	S1		**E1**	
	Min–max	Mean	Min–max	Mean
Al_2O_3	13.36–13.94	13.64 ± 0.17	11.90–14.81	12.93 ± 0.89
MgO	3.98–4.75	4.45 ± 0.20	3.49–3.85	3.72 ± 0.13
CaO	8.77–9.79	9.49 ± 0.30	9.67–12.73	11.34 ± 0.90
		Trace elements (mg kg^{-1})		
	S1		**E1**	
	Min–max	Mean	Min–max	Mean
Cr	176–190	183 ± 4	132–146	139 ± 4
Cu	28–40	34 ± 3	15–23	18 ± 3
Ni	120–128	128 ± 3	71–95	80 ± 7
Pb	32–39	36 ± 3	18–29	22 ± 4
Zn	114–138	129 ± 8	71–92	79 ± 7

Major elements are reported in percent dry weight. Trace elements are reported in mg kg^{-1}. Min, max, mean and standard deviation values of all sediment layers.

Geochemical analyses show that Pb and Zn average concentrations of the S1 sediment core are, respectively, 1.5 and 0.5 times higher than the background levels of the site (Table 2), while the E1 average concentrations and the background values are similar.

Depth profiles of cores from the S1 site show an increasing trend of the MgO contents upwards, relatively stationary concentrations of Cr and Ni and gradual decreasing upward trend of Pb (from 39 to 33 mg kg^{-1}), whereas Zn shows a slight increase between −16 and −8 cm (from 132 to 138 mg kg^{-1}) and then a decrease to 114 mg kg^{-1} Zn at the top (Figure 5). In cores from E1 site, MgO and Ni contents are relatively uniform, Cr reaches a peak at ~5 cm of 146 mg kg^{-1} and Zn and Pb contents show a moderate decreasing trend between 16 and 7 cm, a sharp increase between 7 and 2 cm and peaks at ~3 cm reaching 91 and 29 mg kg^{-1}, respectively.

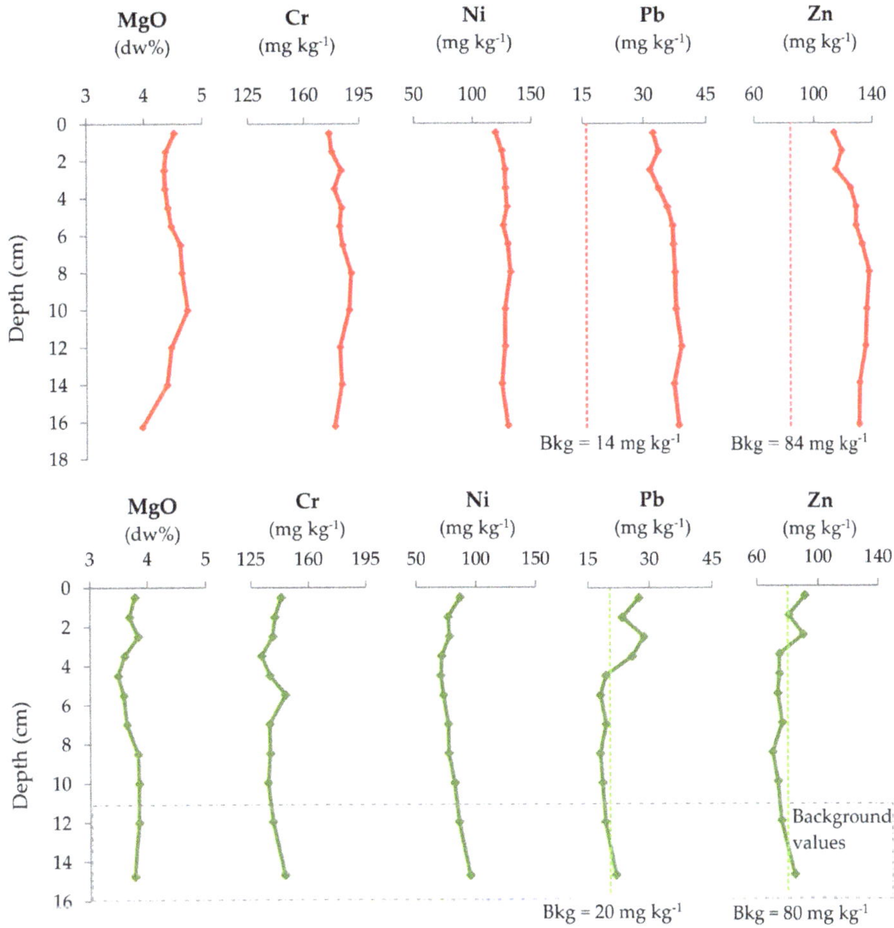

Figure 5. Depth profiles of selected major (MgO) and trace elements (Cr, Ni, Pb and Zn) in S1 (red line) and E1 (green line) sediment cores (ANOC-16).

4. Discussion

4.1. Dating

Activity profiles of ^{137}Cs in the sediment cores were used to estimate the Sediment Accumulation Rates (SAR) of the two investigated areas. Unfortunately, the short time interval recorded by our samples (max 18 cm, Table 1) due to the local high accumulation rate at the S1 site [15] did not allow us to identify the ^{137}Cs reference peak corresponding to the 1986 Chernobyl fallout (Figure 3). Consequently, for this area, we used the most recent available value of 19.4 mm/year [15,46]. Instead, we related the highest value in the ^{137}Cs distribution, detected at 3.5 cm in the E1 cores (EL14 core, Figure 3) to the 1986 event estimating a SAR of 1.2 mm/year for this site, a value comparable with those from the literature [15,38,43,47]. These results indicate that the stratigraphic succession sampled in S1 represents approximately nine years of history, while the sedimentary record recovered in E1 extends to the early 1900s (Figure 7).

4.2. Geochemical Signature of Marine Sediments

As reported in several studies [17,48–50], Cr, Ni and Mg concentrations can be used to discriminate the source areas of sediments. We compared Cr, Ni and Mg contents in the sediments from the S1 and E1 sites with those from the literature. In detail, we considered the geochemistry of deposits from the Po and Apennines rivers [30] and from boreholes in the inland areas [27].

The comparison shows that the concentrations of Cr and Ni in sediments from the Po River delta (S1 site) are very similar to those detected in sediments recovered along the Po River [30], suggesting for this area a prevailing to exclusive sediments supply from the Po River (Figure 6a). On the contrary, Cr and Ni concentrations measured in sediments from the E1 site show a wider dispersion with values comparable to those of the Northern Apennines rivers (Reno, Lamone, Fiumi Uniti, Bevano, Savio Rubicone, Uso and Marecchia), indicating an Apenninic sedimentary source for this area. These different sources of sedimentary material are in agreement with previous investigations [17,27]. It is noteworthy that the Cr and Ni concentrations of S1 site exceed the maximum permissible concentrations, according to the Italian environmental legislation, for unpolluted sites [51] (Table 4). These have a natural origin, derived from the outcrops of ultramafic rocks in the Western Alps and in the Northern Apennines, as documented by previous studies on both river and marine sediments [8,30,49].

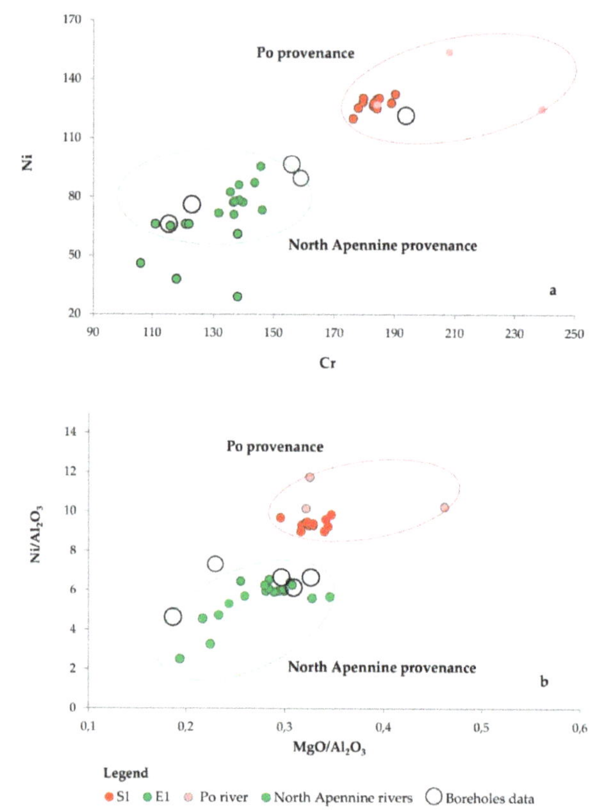

Figure 6. (**a**) Binary plots of Ni vs. Cr and (**b**) Ni/Al2O3 vs. MgO/Al2O3 ratios. Our samples are indicated as green and red dots. Reference data for rivers (light green, Apennines; light red, Po River) derive from tables presented in [30]. Reference data for boreholes (empty big circles) represent the average values presented in [27]. Provenances fields (green and red circles) are drawn in accordance with literature data [17,26–28,50,51].

Table 4. Mean metal concentrations (this work).

S1	Cr	Ni	Pb	Zn
mean	183	128	36	129
min	179	120	32	114
max	190	133	39	138
E1				
mean	139	80	22	79
min	132	71	18	71
max	149	95	29	92
Csc D.lgs 152/2006 *	150	120	100	150
LCB **	100	70	40	100
LCL **	360	75	70	170

* Italian Sediment Quality benchmarks, ** APAT and ICRAM [52]. The LCB represent the mean values of chemical content in the Italian marine sediments, whereas LCL refers to the highest values in Italian marine sediments. Values are in mg kg^{-1}.

In addition, Ni and Mg concentrations also highlight different input sources for our samples (Figure 6b). The Po River is the main source of Cr and Ni for the Adriatic Sea, accounting for ~50% of total Cr and Ni fluvial inputs [8]. According to a recent study [17], Ni/Al_2O_3 and MgO/Al_2O_3 ratios close to 6 and 0.3 mg kg^{-1}, respectively, reflect an Apennine signature of the E1 samples, whereas values of 10 and 0.32 mg kg^{-1} document the Po River provenance for the S1 samples (Figure 6b). Samples collected in the 1980s [53] were characterized by a southward decrease in magnesium rates in the northern-central Adriatic Sea, which is probably due to a higher content of dolomite (rich in Mg) brought by the Po and Alpine rivers. Based on the observations reported by Amorosi et al. [28], a value of Cr/Al_2O_3 ratio > 11.5 can be considered a clear signal of the sediment provenance dominated by the Po River supply. These data are consistent with the relatively high Cr concentrations detected in S1 sediments (Table 3 and Figures 5 and 7). The upward increasing trend of Cr/Al_2O_3 in the E1 depth profile (Figure 7) may be interpreted as a gradual increase of the Po River contribution in the E1 site over the years, in particular after the 1940s. This trend seems in contrast with the decrease of sediment supply by the Po River recorded after 1945 (with a generalized phase of degradation and partial retreat of the entire Po River delta) caused by the construction of a dam [21,54], riverbed mining activities and the channelization of watercourses [55]. The most significant retreat phase, with rates in the order of tens of meters per year, reached its peak between 1954 and 1978 [56] when the E1 core recorded the highest Cr/Al_2O_3 ratio (Figure 7). The reduction in sediment supply, combined with the recent increase in storm events produced an increase of coastal erosion in the area [18–20,57]. Sediment transport is enhanced in the northern Adriatic by the northeasterly Bora wind that intensifies waves and currents, especially in winter, with an average southward transport along the shelf and little across-shelf transport [58]. The Cr increase in the surficial sediments was recorded not only in the E1 site, but also further south, in the AN2 core offshore Ancona (Figure 1) [10]. We speculate that the high Cr values of sediments from the stations south of the Po River prodelta result from the southerly deposition of the eroded sediments of the delta wedge.

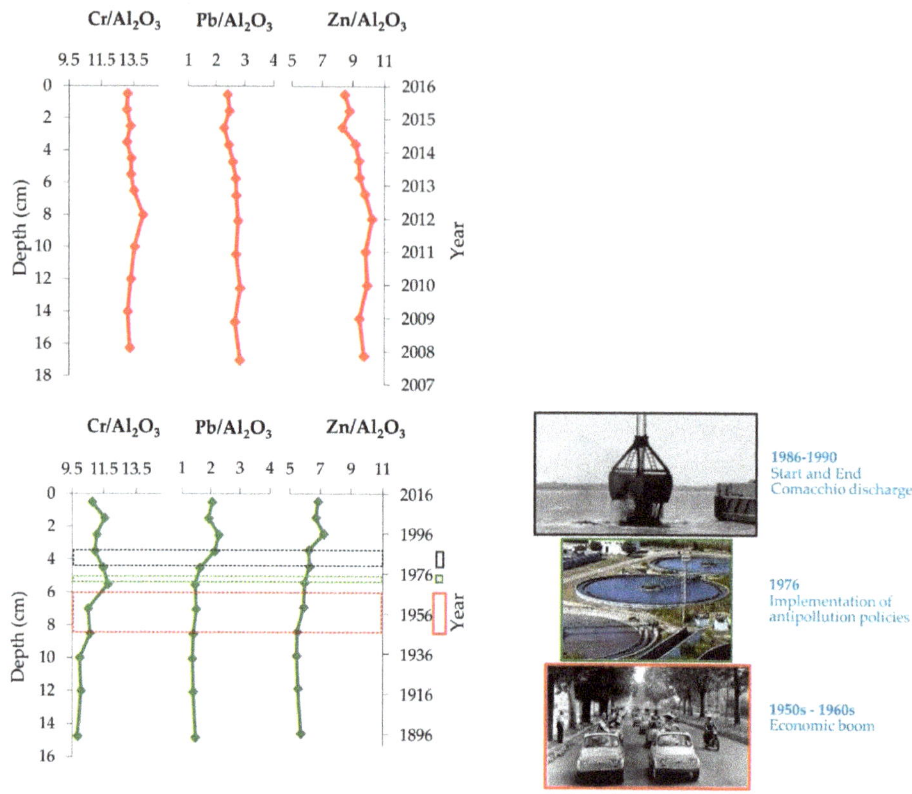

Figure 7. Down-core variability of geochemical proxies (Pb/Al_2O_3 and Zn/Al_2O_3) in S1 (red line) and E1 (green line) measured in ANOC16 cores.

4.3. Anthropic Signal

The Zn and Pb concentrations found in well dated sediment cores collected at several locations along the western Adriatic Sea were successfully used to reconstruct the anthropogenic influence in the past [7,8,10].

In samples from S1 site, Zn and Pb values fall far beyond the Holocene reference field, corresponding to the pre-industrial concentrations [10,41,42] (Table 2). The investigate time interval is limited (only the last nine years), however it is interesting to evidence that the observed decline over recent time of Pb and Zn concentrations (Figure 7) is in line with the historical trend reconstructed using data reported in the literature [59,60] (Table 5 and Appendix B) in the same area. Perhaps this is the result of the application of the Merli Italian Law 319, adopted in 1976, that regulates the concentrations of contaminants in wastewaters for environmental protection. The same assumption was suggested to explain the low content of trace metals observed in the sediments from the River Po delta [10]. Additionally, the Pb content reduction may be in response to the implementation of anti-pollution policies on automotive Pb in the second half of the 1980s in Western Europe [61].

At the E1 site, the longer time interval sampled (see Section 4.1) provides the history of contamination during the last 100 years. The E1 normalized depth profiles of Pb and Zn (Figure 7) display a gradual increasing accumulation trend from the 1940s with a peak from 1980 to 2000 in particular for Pb concentration.

Table 5. Trace metals concentration detected in the S1 area over the time (data from this work and from literature).

Year	Zn	Pb	References
2016	114	33	This work
2007	132	39	This work
1996	162	44.7	Ianni et al., 2000 [59] (site 103)
1977	189	68.3	Frascari et al., 1984 [60] (site 30)

Values are reported in mg kg^{-1} and not normalized to Al_2O_3 in order to perform the comparison between different datasets.

The detected major presence of these elements in the area around the mid-1940s is probably related to the gradual industrialization of the region after World War II. It is worth noting that, in Italy, the time interval of 1945–1963, known as "Economic Miracle", experienced a strong industrial development with an increase of the mechanical, chemical and ceramics industries [7].

The observed increase in trace metals is consistent with what was found in previous investigations carried out along the coast between Cattolica and the mouth of the Rubicone River [62]. The strong increase from the 1980s (Figure 7, E1 site) correlates in time with marine dumping off Comacchio from 1986 to 1990, used as the disposal sites for dredged materials from the Ravenna port [7,63]. In our opinion, the dumping influenced directly or indirectly the surrounding environment since the activity can destroy and/or degrade habitats for species as well as cause coastal pollution. The documented high accumulation of Pb and Zn may be explained with the southward dispersion of contaminated sediments under the influence of the WAC (Figure 1) from the dumping site to the E1 site located approximately 60 nautical miles away. The recent decrease of Zn and Pb observed since the early 2000s may be related to the application of the Italian Law 319/76 as mentioned above for S1 site.

5. Conclusions

Geochemical and distribution analyses of trace elements were performed on the marine sediments from two sites (E1 and S1) located along the western Adriatic coast.

The concentration of some main and trace elements (MgO, Cr and Ni) compared with the literature data allowed us to highlight that the S1 sediments composition largely reflects the discharges of the Po River, whereas sedimentation at the E1 site results mostly controlled by the supplies from the northern Apennines rivers, with only a marginal contribution from the Po River system. The results confirm the presence of high concentrations of Cr and Ni related to the discharge of the Po River and due to the natural presence of ultramafic rocks in its drainage basin. The high Cr values observed in the sediments of E1 station, at the south of the Po prodelta, indicate a southern deposition of the recently eroded sediments of the delta wedge. The accumulation values of trace metals in the two sites show a significant decrease over time, especially near the prodelta sediments of the Po River. Finally, using the elements Pb and Zn as diagnostic of anthropogenic impact, we reconstructed the historical trends of pollution sources. In particular, the results of the E1 site highlight the impact due to the increase in the industry after World War II, the implementation of environmental policy in 1976 and the presence of the Comacchio dumping at the end of 1980.

Author Contributions: E.B. wrote the preliminary draft of the manuscript with the support of the other authors; samples collection, analysis and data validation were carried out by S.A., E.D., F.R., P.G., L.C. and M.R.; S.A. and F.R. edited the figures; all authors discussed the results and contributed to the manuscript; L.C. and F.R. revised the final version; and M.R. was the P.I. of the scientific project and provided financial support for the publication. All authors have read and agreed to the published version of the manuscript.

Funding: This work was supported by the Jerico-NEXT project (H2020-INFRAIA-2014-2015 Research and Innovation action) and RITMARE Italy Flagship Project (SP5-WP3). The research is part of the LTER Network (https://deims.org/6869436a-80f4-4c6d-954b-a730b348d7ce) (Sites S1-GB and E1). This is contribution No. 2032 of the of the Marine Science Institute Consiglio Nazionale delle Ricerche, Bologna.

Acknowledgments: This article is in memory of Giovanni Bortoluzzi, a friend and colleague who dedicated his work to the development and support of marine observatories. We are very grateful to the technical staff of DALLAPORTA research vessel and to Mauro Bastianini cruise leader of the LTER-ANOC16, INTERNOS17 oceanographic cruises. We thank the CNR DCSR—Ufficio Programmazione e Grant Office for the technical and logistical support. We also thank Andrea Gallerani and Valerio Funari for their help in preparing the samples for analysis and Michael Marani and Marco Ligi for the helpful suggestions.

Conflicts of Interest: The authors declare no conflict of interest. The funders had no role in the design of the study, in collection, analyses, or interpretation of the data; in the writing of the manuscript, or in the decision to publish the results.

Appendix A

Table A1. Geochemical data from LTER-ANOC16 cores.

	Major elements (dw %)			
	S1		E1	
	Min–max	**Mean**	**Min–max**	**Mean**
SiO2	40.21–42.82	41.64 ± 0.88	41.81–45.24	43.63 ± 0.92
TiO2	0.61–0.64	0.62 ± 0.01	0.58–0.67	0.62 ± 0.03
Al2O3	13.36–13.94	13.64 ± 0.17	11.90–14.81	12.93 ± 0.89
Fe2O3	5.83–6.21	6.01 ± 0.13	4.91–6.43	5.48 ± 0.45
MnO	0.10–0.13	0.11 ± 0.01	0.10–0.12	0.11 ± 0.01
MgO	3.98–4.75	4.45 ± 0.20	3.49–3.85	3.72 ± 0.13
CaO	8.77–9.79	9.49 ± 0.30	9.67–12.73	11.34 ± 0.90
Na2O	1.17–4.93	1.54 ± 1.07	1.16–1.81	1.41 ± 0.21
K2O	2.58–2.79	2.72 ± 0.06	2.28–3.09	2.54 ± 0.24
P2O5	0.12–0.15	0.14 ± 0.01	0.12–0.14	0.13 ± 0.01
LOI	18.91–21.11	19.63 ± 0.73	16.15–20.67	18.08 ± 1.63
	Trace elements (mg kg^{-1})			
	S1		**E1**	
	Min-max	**Mean**	**Min-max**	**Mean**
As	8–12	10 ± 1	7–12	9 ± 2
Ba	272–320	291 ± 13	255–338	278 ± 2 4
Ce	47–73	63 ± 7	35–63	52 ± 9
Co	12–14	13 ± 1	10–13	11 ± 1
Cr	176–190	183 ± 4	132–146	139 ± 4
Cu	28–40	34 ± 3	15–23	18 ± 3
La	26–34	30 ± 2	19–29	23 ± 3
Nb	13–15	14 ± 1	13–14	14 ± 1
Ni	120–128	128 ± 3	71–95	80 ± 7
Pb	32–39	36 ± 3	18–29	22 ± 4
Rb	32–39	36 ± 3	88–120	100 ± 9
Sr	257–278	267±8	295–415	339 ± 40
V	148–161	157 ± 4	138–159	151 ± 7
Y	24–26	25 ± 1	24–27	25 ± 1
Zn	114–138	129 ± 8	71–92	79 ± 7
Zr	100–118	104 ± 5	127–162	145 ± 12

Appendix B

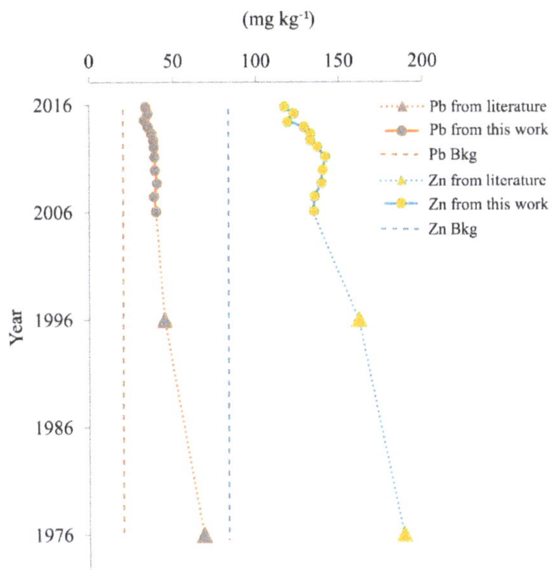

Figure A1. Pb and Zn historical concentration trends. Data from the literature [59,60] and this work (see Table 2).

References

1. Heim, S.; Schwarzbauer, J. Pollution history revealed by sedimentary records: A review. *Environ. Chem. Lett.* **2013**, *11*, 255–270. [CrossRef]
2. Lima, A. Evaluation of geochemical background at regional and local scales by fractal filtering technique: Case studies in selected Italian areas. In *Environmental Geochemistry. Site Characterization, Data Analysis and Case Stories*; de Vivo, B., Belkin, H.E., Lima, A., Eds.; Elsevier: Amsterdam, The Netherlands, 2008; pp. 135–153.
3. Matteucci, G.; Rossini, P.; Guerzoni, S.; Arcangeli, A.; Fonti, P.; Langone, L. Recent evolution of sedimentary heavy metals in a coastal lagoon contaminated by industrial wastewaters (Pialassa Baiona, Ravenna, Italy). *Hydrobiology* **2005**, *550*, 167–173. [CrossRef]
4. Pirrone, N.; Trombino, G.; Cinnirella, S.; Algieri, A.; Bendoricchio, G.; Palmeri, L. The Driver-Pressure-State-Impact-Response (DPSIR) approach for integrated catchment-coastal zone management: Preliminary application to the Po catchment Adriatic Sea coastal zone system. Reg. *Environ. Chang.* **2005**, *5*, 111–137. [CrossRef]
5. Abbiati, M.; Mistri, M.; Bartoli, M.; Ceccherelli, V.U.; Colangelo, M.A.; Ferrari, C.R.; Giordani, G.; Munari, C.; Nizzoli, D.; Ponti, M.; et al. Trade-off between conservation and exploitation of the transitional water ecosystems of the northern Adriatic Sea. *Chem. Ecol.* **2010**, *26*, 105–119. [CrossRef]
6. Zonta, R.; Cassin, D.; Pini, R.; Dominik, J. Assessment of heavy metal and as contamination in the surface sediments of Po delta lagoons (Italy). *Estuar. Coast. Shelf Sci.* **2019**, *225*, 106235. [CrossRef]
7. Romano, S.; Langone, L.; Frignani, M.; Albertazzi, S.; Focaccia, P.; Bellucci, L.G.; Ravaioli, M. Historical pattern and mass balance of trace metals in sediments of the northwestern Adriatic Sea Shelf. *Mar. Poll. Bull* **2013**, *76*, 32–41. [CrossRef]
8. Lopes-Rocha, M.; Langone, L.; Miserocchi, S.; Giordano, P.; Guerra, R. Spatial patterns and temporal trends of trace metal mass budgets in the western Adriatic sediments (Mediterranean Sea). *Sci. Tot. Environ.* **2017**, *599–600*, 1022–1033. [CrossRef]

9. Poulain, P. Adriatic Sea surface circulation as derived from drifter data between 1990 and 1999. *J. Mar. Syst.* **2001**, *29*, 3–32. [CrossRef]
10. Lopes-Rocha, M.; Langone, L.; Miserocchi, S.; Giordano, P.; Guerra, R. Detecting long-term temporal trends in sediment-bound metals in the western Adriatic (Mediterranean Sea). *Mar. Poll. Bull.* **2017**, *124*, 270–285. [CrossRef]
11. Wang, X.H.; Pinardi, N. Modeling the dynamics of sediment transport and resuspension in the northern Adriatic Sea. *J. Geophys. Res. Space Phys.* **2002**, *107*, 3225. [CrossRef]
12. Davolio, S.; Stocchi, P.; Benetazzo, A.; Bohm, E.; Riminucci, F.; Ravaioli, M.; Li, X.; Carniel, S. Exceptional Bora outbreak in winter 2012: Validation and analysis of high-resolution atmospheric model simulations in the northern Adriatic area. *Dyn. Atmos.Oceans* **2015**, *71*, 1–20. [CrossRef]
13. Boldrin, A.; Langone, L.; Miserocchi, S.; Turchetto, M.; Acri, F. Po River plume on the Adriatic continental shelf: Dispersion and sedimentation of dissolved and suspended matter during different river discharge rates. *Mar. Geol.* **2005**, *222*, 135–158. [CrossRef]
14. Milliman, J.D.; Syvitski, J.P.M. Geomorphic/tectonic control of sediment discharge to the ocean: The importance of small mountainous rivers. *J. Geol.* **1992**, *100*, 525–544. [CrossRef]
15. Frignani, M.; Langone, L.; Ravaioli, M.; Sorgente, D.; Alvisi, F.; Albertazzi, S. Fine-sediment mass balance in the western Adriatic continental shelf over a century time scale. *Mar. Geol.* **2005**, *222*, 113–133. [CrossRef]
16. Cattaneo, A.; Correggiari, A.; Langone, L.; Trincardi, F. The late-Holocene Gargano subaqueous delta, Adriatic shelf: Sediment pathways and supply fluctuations. *Mar. Geol.* **2003**, *193*, 61–91. [CrossRef]
17. Greggio, N.; Giambastiani, B.M.; Campo, B.; Dinelli, E.; Amorosi, A. Sediment composition, provenance, and Holocene paleoenvironmental evolution of the Southern Po River coastal plain (Italy). *Geol. J.* **2018**, *53*, 914–928. [CrossRef]
18. Montanari, A. Hydrology of the Po River: Looking for changing patterns in river discharge. *Hydrol. Earth Syst. Sci.* **2012**, *16*, 3739–3747. [CrossRef]
19. Ciavola, P.; Salemi, E.; Billi, P. *Sediment Supply and Morphological Evolution of a Small River Mouth (Fiumi Uniti, Ravenna, Italy): Should River Management be Storm-Driven*; Barazzutti, M., Marabini, F., Eds.; China-Italy Bilateral Symposium on the Coastal Zone and Continental Shelf Evolution Trend: Ravenna, Italy, 2010.
20. Pavanelli, D.; Cavazza, C.; Lavrnić, S.; Toscano, A. The Long-Term Effects of Land Use and Climate Changes on the Hydro-Morphology of the Reno River Catchment (Northern Italy). *Water* **2019**, *11*, 1831. [CrossRef]
21. Maselli, V.; Pellegrini, C.; Del Bianco, F.; Mercorella, A.; Nones, M.; Crose, L.; Guerrero, M.; Nittrouer, J.A. River morphodynamic evolution under dam-induced backwater: An example from the Po River (Italy). *J. Sediment. Res.* **2018**, *88*, 1190–1204. [CrossRef]
22. Correggiari, A.; Cattaneo, A.; Carrà, D.; Penitenti, D.; Preti, M.; Trincardi, F. Offshore sand for beach restoration: North Adriatic shelf examples. *CIESM Workshop* **2002**, *18*, 79–82.
23. Matteucci, G.; Riccio, S.; Rossini, P.; Sisti, E.; Bernucci, M.E.; Pari, P.; Benedettini, M.; Stanley, C.C. Shoreline evolution trend connected to progressive construction of segmented defense structures (Rimini, North Adriatic Sea, Italy). *GeoActa* **2010**, *SI 3*, 135–141.
24. Di Matteo, L.; Dragoni, W.; Maccari, D.; Piacentini, S.M. Climate change, water supply and environmental problems of headwaters: The paradigmatic case of the Tiber, Savio and Marecchia rivers (Central Italy). *Sci. Total Environ.* **2017**, *598*, 733–748. [CrossRef] [PubMed]
25. Laghi, M.; Mollema, P.; Antonellini, M. The Influence of River Bottom Topography on Salt Water Encroachment along the Lamone River (Ravenna, Italy), and Implications for the Salinization of the Adjacent Coastal Aquifer. In Proceedings of the World Environmental and Water Resources Congress 2010: Challenges of Change, Providence, RI, USA, 16–20 May 2010; pp. 1124–1135.
26. Amorosi, A.; Sammartino, I. Influence of sediment provenance on background values of potentially toxic metals from near-surface sediments of Po coastal plain (Italy). *Intern. J. Earth Sci.* **2007**, *96*, 389–396. [CrossRef]
27. Amorosi, A.; Centineo, M.C.; Dinelli, E.; Lucchini, F.; Tateo, F. Geochemical and mineralogical variations as indicators of provenance changes in Late Quaternary deposits of SE Po Plain. *Sediment. Geol.* **2002**, *151*, 273–292. [CrossRef]
28. Amorosi, A. Chromium and nickel as indicators of source-to-sink sediment transfer in a Holocene alluvial and coastal system (Po Plain, Italy). *Sediment. Geol.* **2012**, *280*, 260–269. [CrossRef]

29. Ravaioli, M.; Alvisi, F.; Vitturi, L.M. Dolomite as a tracer for sediment transport and deposition on the northwestern Adriatic continental shelf (Adriatic Sea, Italy). *Cont. Shelf Res.* **2003**, *23*, 1359–1377. [CrossRef]
30. Dinelli, E.; Lucchini, F. Sediment supply to the Adriatic Sea basin from the Italian rivers: Geochemical features and environmental constraints. *Giorn. Geol.* **1999**, *61*, 121–132.
31. Marchesini, L.; Amorosi, A.; Cibin, U.; Zuffa, G.G.; Spadafora, E.; Preti, D. Sand composition and sedimentary evolution of a Late Quaternary depositional sequence, northwestern Adriatic Coast, Italy. *J. Sediment. Res.* **2000**, *70*, 829–838. [CrossRef]
32. Böhm, E.; Riminucci, F.; Bortoluzzi, G.; Colella, S.; Acri, F.; Santoleri, R.; Ravaioli, M. Operational use of continuous surface fluorescence measurements offshore Rimini to validate satellite-derived chlorophyll observations. *J. Oper. Oceanogr.* **2016**, *9*, 167–175. [CrossRef]
33. Ravaioli, M.; Bergami, C.; Riminucci, F.; Langone, L.; Cardin, V.; Di Sarra, A.; Aracri, S.; Bastianini, M.; Bensi, M.; Bergamasco, A.; et al. The RITMARE Italian Fixed-Point Observatory Network (IFON) for marine environmental monitoring: A case study. *J. Oper. Oceanogr.* **2016**, *9*, 202–214. [CrossRef]
34. Capotondi, L.; Mancin, N.; Cesari, V.; Dinelli, E.; Ravaioli, M.; Riminucci, F. Recent agglutinated foraminifera from the North Adriatic Sea: What the agglutinated tests can tell. *Mar. Micropal.* **2019**, *147*, 25–42. [CrossRef]
35. Bastianini, M.; Riminucci, F.; Capondi, L.; Barra, E.; Pasqual, S.; Casotti, R.; Trano, A.C.; Van Dijk, M.; Mauro, C.; Fabbro, C. Rapporto sulle attività oceanografiche, biologiche, geologiche e di manutenzione della stazione meda S1-GB svolte durante la campagna oceanografica LTER-ANOC16 (26–30 aprile 2016) con N/O Dallaporta nel Mare Adriatico settentrionale. *Rapp. Tec. CNR-ISMAR* **2017**, *145*, 1–27.
36. Bastianini, M.; Riminucci, F.; Pansera, M.; Coluccelli, A.; Casotti, R.; Dal Passo, E.; Dametto, L.; Van Dijk, M.; Russo, E.; Titocci, J.; et al. Rapporto sulle attività biologiche, oceanografiche, geologiche e di manutenzione della stazione Boa E1 svolte durante la campagna INTERNOS17 (14–21 marzo 2017) con N/O Minerva Uno nel Mare Adriatico centro-settentrionale. *Rapp. Tec. CNR-ISMAR* **2017**, *146*, 1–37.
37. Berner, R.A. *Principles of Chemical Sedimentology*; McGraw-Hill: New York, NY, USA, 1971; p. 240.
38. Albertazzi, S.; Bopp, R.F.; Frignani, M.; Merlin, O.H.; Vitturi, L.M.; Ravaioli, M.; Triulzi, C. Cs-137 as a tracer for processes of marine sedimentation in the vicinity of the Po River Delta (Northern Adriatic Sea). *Mem. Soc. Geol. Italy* **1984**, *27*, 447–459.
39. Frignani, M.; Langone, L.; Ravaioli, M.; Sticchi, A. Cronologia di sedimenti marini artificiali mediante spettrometria gamma. *IGM-CNR Tech. Rep.* **1991**, 24–32.
40. Covelli, S.; Fontolan, G. Application of a Normalization Procedure in Determining Regional Geochemical Baselines. *Environ. Geol.* **1997**, *30*, 34–45. [CrossRef]
41. Correggiari, A.; Bastianini, M.; Miserocchi, S. Caratterizzazione dei sedimenti dell'Alto Adriatico nell'ambito del Monitoraggio delle microalghe potenzialmente tossiche nelle aree marino-costiere del Veneto con una particolare attenzione a Ostreopsis Ovata. *Rapp. Tec. Cnr. Ismar. Bologna* **2016**, 1–30.
42. Morelli, G.; Gasparon, M.; Fierro, D.; Hu, W.P.; Zawadzki, A. Historical trends in trace metal and sediment accumulation in intertidal sediments of Moreton Bay, southeast Queensland, Australia. *Chem. Geol.* **2012**, *300–301*, 152–164. [CrossRef]
43. Frignani, M.; Sorgente, D.; Langone, L.; Albertazzi, S.; Ravaioli, M. Behavior of Chernobyl radiocesium in sediments of the Adriatic Sea off the Po River delta and the Emilia-Romagna coast. *J. Environ. Radioact.* **2004**, *71*, 299–312. [CrossRef]
44. Giani, M.; Boldrin, A.; Matteucci, G.; Frascari, F.; Gismondi, M.; Rabitti, S. Downward fluxes of particulate carbon, nitrogen, and phosphorus in the north-western Adriatic Sea. *Sci. Total Environ.* **2001**, *266*, 125–134. [CrossRef]
45. Alvisi, F. A simplified approach to evaluate sedimentary organic matter fluxes and accumulation on the NW Adriatic Shelf (Italy). *Chem. Ecol.* **2009**, *25*, 119–134. [CrossRef]
46. Rossi, A.; Roveri, N.; Albertazzi, S.; Focaccia, P.; Ravaioli, M. Elaborazione ed Interpretazione di dati Geochimici di Sedimenti Marini dell'area Adriatica Centro Settentrionale. Master's Thesis, University of Bologna, Bologna, Italy, 2011.
47. Giordani, P.; Hammond, D.E.; Berelson, W.M.; Montanari, G.; Poletti, R.; Milandri, A.; Rabbi, E. Benthic fluxes and nutrient budgets for sediments in the Northern Adriatic Sea: Burial and recycling efficiencies. *Mar. Coas. Eutroph.* **1992**, 251–275. [CrossRef]

48. Picone, S.; Alvisi, F.; Dinelli, E.; Morigi, C.; Negri, A.; Ravaioli, M.; Vaccaro, C. New insights on late Quaternary palaeogeographic setting in the Northern Adriatic Sea (Italy). *J. Quat. Sci.* **2008**, *23*, 489–501. [CrossRef]
49. Spagnoli, F.; Dinelli, E.; Giordano, P.; Marcaccio, M.; Zaffagnini, F.; Frascari, F. Sedimentological, biogeochemical and mineralogical facies of Northern and Central Western Adriatic Sea. *J. Mar. Syst.* **2014**, *139*, 183–203. [CrossRef]
50. Colantoni, P.; Gallignani, P.; Lenaz, R. Late Pleistocene and Holocene evolution of the North Adriatic continental shelf (Italy). *Mar. Geol.* **1979**, *33*, 41–50. [CrossRef]
51. Amorosi, A.; Guermandi, M.; Marchi, N.; Sammartino, I. Fingerprinting sedimentary and soil units by their natural metal contents: A new approach to assess metal contamination. *Sci. Total. Environ.* **2014**, *500–501*, 361–372. [CrossRef]
52. APAT; ICRAM. *Manuale per la Movimentazione di Sedimenti Marini*; Corsini, S., Onorati, F., Pellegrini, D., Eds.; Ministero per l'ambiente e la tutela del territorio e del mare: Rome, Italy, 2006; p. 72.
53. Price, B.N.; Mowbray, S.; Giordani, P. *Sedimentation and Heavy Metal Input Changes of the Northwest Adriatic Shelf: A Consequence of Anthropogenic Activity*; Atti 108 Congresso AIOL.: Alassio, Italy, 1994; pp. 23–33.
54. Maselli, V.; Trincardi, F. Man made deltas. *Sci. Rep.* **2013**, *3*, 1926. [CrossRef]
55. Stefani, M.; Vincenzi, S. The interplay of eustasy, climate and human activity in the late Quaternary depositional evolution and sedimentary architecture of the Po Delta system. *Mar. Geol.* **2005**, *222*, 19–48. [CrossRef]
56. Trincardi, F.; Amorosi, A.; Bosman, A.; Correggiari, A.; Madricardo, F.; Pellegrini, C. Ephemeral rollover points and clinothem evolution in the modern Po Delta based on repeated bathymetric surveys. *Basin Res.* **2020**, *32*, 402–418. [CrossRef]
57. Billi, P.; Salemi, E. Misura del Trasporto Solido del Fiume Reno. In 2° *Giornata di Studio—Il Monitoraggio Idrotorbidimetrico dei Corsi D'acqua per la Stima dei Processi Erosivi e il Bilancio dei solidi Sospesi*; University of Ferrara: Ferrara, Italy, 2004.
58. Lee, C.M.; Askari, F.; Book, J.; Carniel, S.; Cushman-Roisin, B.; Dorman, C.; Kuzmic, M. Northern Adriatic response to a wintertime bora wind event. *Eos Trans. Amer. Geoph. Union* **2005**, *86*, 157–165. [CrossRef]
59. Ianni, C.; Magi, E.; Rivaro, P.; Ruggieri, N. Trace metals in Adriatic coastal sediments: Distribution and speciation pattern. *Toxic. Environ. Chem.* **2000**, *78*, 73–92. [CrossRef]
60. Frascari, F.; Frignani, M.; Giordani, P.; Guerzoni, S.; Ravaioli, M. Sedimentological and geochemical behavior of heavy metals in the area near the Po river delta. *Mem. Soc. Geol. Italy* **1984**, *27*, 469–481.
61. Annibaldi, A.; Truzzi, C.; Illuminati, S.; Scarponi, G. Recent sudden decrease of lead in Adriatic coastal seawater during the years 2000–2004 in parallel with the phasing out of leaded gasoline in Italy. *Mar. Chem.* **2009**, *113*, 238–249. [CrossRef]
62. Guerzoni, S.; Frignani, M.; Giordani, P.; Frascari, F. Heavy metals in sediments from different environments of a Northern Adriatic Sea area, Italy. *Environ. Geol. Water Sci.* **1984**, *6*, 111–119. [CrossRef]
63. Giani, M.; Gabellini, M.; Pellegrini, D.; Costantin, S.; Beccaloni, E.; Giordano, R. Concentration and partitioning of Hg, Cr and Pb in sediments of dredge and disposal sites of the northern Adriatic Sea. *Sci. Total Environ.* **1994**, *158*, 97–112. [CrossRef]

applied sciences

MDPI

Article

Sediment Dynamics of the Neretva Channel (Croatia Coast) Inferred by Chemical and Physical Proxies

Federico Giglio [1], Stefania Romano [2,*], Sonia Albertazzi [2], Francesca Chiarini [2], Mariangela Ravaioli [2], Marco Ligi [2] and Lucilla Capotondi [2]

1 Istituto di Scienze Polari - Consiglio Nazionale delle Ricerche ISP-CNR, Via P. Gobetti 101, 40129 Bologna, Italy; federico.giglio@cnr.it

2 Istituto di Scienze Marine - Consiglio Nazionale delle Ricerche ISMAR-CNR, Via P. Gobetti 101, 40129 Bologna, Italy; sonia.albertazzi@bo.ismar.cnr.it (S.A.); francesca.chiarini@bo.ismar.cnr.it (F.C.); mariangela.ravaioli@bo.ismar.cnr.it (M.R.); marco.ligi@bo.ismar.cnr.it (M.L.); lucilla.capotondi@bo.ismar.cnr.it (L.C.)

* Correspondence: stefania.romano@bo.ismar.cnr.it; Tel.: +39-051-639-8897; Fax: +39-051-639-8940

Received: 12 December 2019; Accepted: 21 January 2020; Published: 23 January 2020

Abstract: We examined the transport of sediments and their surficial pathways from the mouth of Neretva River, through the Neretva Channel, toward the Adriatic Sea. This research was based on twelve box-cores and five grab samples collected within the Neretva Channel. Sediment dynamics were evaluated using several proxies, such as organic matter, radiochemical isotopes and select metal concentrations and physical parameters. The data analysis showed that the influence of the river on particle distribution along the Neretva Channel decreases northward, with an estimated sediment accumulation rate ranging from 1.9 to 8.5 mm/yr. The lowest accumulation rate was found in the sector not influenced by river inflow, whereas the preferential sediment accumulation area is in the center of the basin. We speculate that dispersion and accumulation of sediments are both driven by an eddy in the waters of the Neretva Channel triggered/or intensified seasonally by the interaction of karstic springs, river input and Adriatic Sea waters. Our results indicate that the anthropogenic factor does not affect the concentration of metals within the channel and that the river particles dynamics determine the Pb areal distribution, while Cr and Ni have a possible source located to the northwest of the river-mouth.

Keywords: Adriatic Sea; Neretva Channel; sediment dynamics; age model; metal concentrations

1. Introduction

The legal instruments of EU's Environmental policies and their regional and local application levels provide an interrelated regulatory framework for protection, preservation and prevention of the European oceans [1–4]. According to these European directives, they are intended to apply an ecosystem-based approach to manage human activities whilst ensuring sustainable use of marine goods and services (respectively) [4,5]. In particular, the Marine Strategy Framework Directive (MSFD) [3] uses eleven descriptors with several indicators covering ecological, physical, chemical and anthropogenic components of the ecosystems that need to be integrated to achieve a good environmental status [6]. In this context, the study of sediment dynamics plays an important role, as it provides the basic knowledge for a correct evaluation of the environmental quality.

The Adriatic Sea is an important sub-region of the Mediterranean marine area and has been proposed as Ecologically or Biologically Significant Areas in the Mediterranean (EBSAs) from the Barcelona Convention, UNEP/MAP [7]. Several studies were carried out on sediment dynamics and geochemistry in the Adriatic Sea, but they were mainly focused on its western side (Italian Economic Water Zone; see [8–14] and reference therein). Hence, few geochemical data are available from sedimentary deposits of the eastern coast [15–17] and potential input of contaminants is still little known.

Regional geology largely controls the composition of marine sediments; however, in densely populated regions, such as coastal areas, the anthropogenic influence may strongly affect the dispersion and concentration of organic matter and contaminants, modifying the river discharge through a combination of factors such as urban settlements and roads, runoff of agricultural soils, and dry and wet atmospheric deposition [18]. In this work, the influence of the Neretva River (NR) on the southern Croatian Adriatic coastal system was investigated in order to gain information on sediment dynamics of the area (Figure 1).

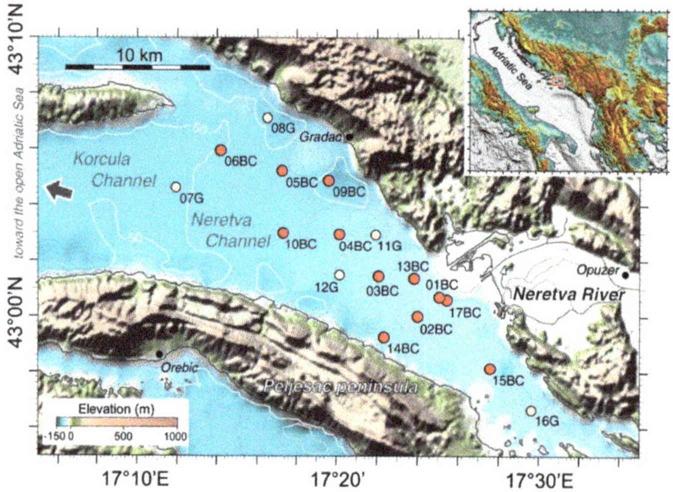

Figure 1. Study area and sampling locations. Shaded relief image derived from bathymetric data with sun illumination from NW, 45° over the horizon and no vertical exaggeration (grid resolution 10 m). Elevation and bathymetry data from EMODnet gridded data (www.emodnet.eu/bathymetry). Filled red and pink circles indicate box corer and grab samples locations, respectively. Bathymetry contour interval is 10 m. Red box in the inset indicates the study area.

The major aim of this work was to determine sediment pathways and dynamics of particles that bond metals in the Neretva Channel (NC) by combining sediment geochemistry, geochronological data and physical parameters of the water column to obtain basic information required for an assessment of the health and quality of the environment to preserve the marine life [19]. These results may provide a useful support for evaluation and decision-making processes aiming to achieve the environmental goals proposed by the Water Framework Directive [2] and the MSFD [3].

2. Study Area

The NC (Figure 1) is a narrow, semi-enclosed basin located along the southernmost part of the Croatian coast. Oriented SE–NW, it is bounded to the NE by the Croatian coastline, to the NW by the open Adriatic Sea through the Korcula Channel, and to the SW by the Peljesac Peninsula. The NC is a microtidal, low-wave energy environment with river-dominated sedimentation processes [15,17]. The river has an average annual water inflow of 332 m^3/s with peaks in December and April, and with a long-term seasonal variability in terms of minimum and maximum monthly discharge from low to moderate, respectively [20]. The NR mouth, located ≈20 km to the east of the town of Metković, is characterized by a reclaimed alluvial plain and forms the largest depositional system in the southern Croatian coast, covering an area of ≈246 km^2.

The NR, despite high annual precipitation in the catchment area, has a low stream density because part of its water is collected into karstic aquifers [20]. Furthermore, five hydropower plants in the Neretva catchment impound a total area of 36 km^2, storing a 1.1 km^3 water reserve [21]. As a consequence, the present-day delta is not associated with the large volumes of deposits that other river systems along the western Adriatic coast are. Finally, water regime in the lower course of the NR is complicated by the interaction with seawater ingression. For instance, deepening of the NR riverbed may cause higher salinity of both water and soil due to a drop in the ground water level [17,22]. Salt water constantly penetrates into underground waterways, leading to the salinity of the soil, particularly during the dry season when the river flow is reduced [23]. Hence, the natural environment of the lower river course may be strongly threatened in terms of chemical, physical and biological changes [17], especially where human activities cause pressure, such as road construction sites, urbanization, hunting and mining [22,24,25]. Moreover, the NR is moderately affected by untreated municipal and industrial (metallurgy and other lighter activities) wastewaters.

3. Materials and Methods

During the oceanographic cruise NERES06 onboard the R/V Bios DVA in May 2006, twelve box cores (BC) and five grab samples (G) were collected in the study area (Figure 1). Sampling sites were selected along three transects parallel to the NW–SE-oriented axis of the channel in order to obtain an exhaustive coverage of the investigated area.

The BC short cores (≈20 cm long) were radiographed via a directional X-ray tube (M60 Gilardoni) using an aluminum and PVC filter, allowing the identification of the sedimentary structures before sub-sampling. Once the cores were opened, sediments were described for visual characteristics, and then, one-half of each was stored for the historical archive and the other half was sub-sampled with a frequency of 2 cm for sediment grain size and tracer determinations. Before analyses, sediments were dried at 60 °C in order to calculate the sediment porosity assuming a particle density of 2.5 g cm^{-3} [26].

Total carbon (TC) and total nitrogen (TN) contents, and the stable isotopic composition, were determined on the surficial samples by using a FINNIGAN Delta Plus mass spectrometer, directly coupled to the FISONS NA2000 EA (for further details see [27]). The total organic carbon (TOC) was measured by a pre-treatment with 1.5 M HCl to remove the carbonates. TOC contents are reported as weight percent (wt%) on dry weight. Moreover, C/N was calculated as molar ratio between TOC% and TN%.

Grain size analysis was carried out on separates following wet sieving on a 63 µm mesh-size sieve in order to separate sand from finer fractions, after a pre-treatment with H_2O_2 to remove the organic matter and to favor disaggregation between sediment particles. Silt and clay fractions were determined with an X-ray Sedigraph Malvern Mastersizer 2000s.

The time framework was based on ^{137}Cs and ^{210}Pb radionuclides. ^{137}Cs activity was measured by non-destructive gamma spectrometry (see [28] and references therein) using coaxial intrinsic germanium detectors (Ortec HPGe GMX-20195P and GEM 20200), while ^{210}Pb was determined by chemical extraction of its daughter ^{210}Pb, assuming secular equilibrium between the two isotopes [11,29].

In order to determine the metal fraction of surficial sediment particles and/or dissolved in the interstitial water, 0.5 g of de-frozen wet sediment was leached with HNO_3 and H_2O_2 (10:3) under reflux [30].

Cr, Ni and Pb concentrations were determined by furnace atomic absorption spectrophotometers and the results were normalized to the sediment dry weight. Accuracy and precision were tested through repeated analysis of certified reference material NIST 2709, and in comparison with the reported values for the determination of labile or extractable elements. Results fell within the range of certified values. Accuracies (as %RDS), estimated by replicate analyses of NIST 2709, were 4% (for Ni and Cr) and 3% (for Pb). The leaching recoveries, calculated on the basis of certified values for total concentrations, were 66%, 63% and 93% for Pb, Cr and Ni, respectively.

Conductivity and temperature (CTD) data collections were carried out during the oceanographic cruise by correspondence with sampling stations with a Sea-Bird SBE 25 CTD equipped with temperature and conductivity sensors. The CTD data were processed according to UNESCO standards [31], and pressure values were averaged at 0.5 dbar intervals. The Ocean Data View software was used to interpolate spatially the CTD vertical profiles [32].

The areal distributions of metal contents, sand fraction and sediment accumulation rate (SAR) were computed by universal kriging interpolation, a method for which the values are modelled by a Gaussian process governed by prior covariance. Under suitable assumptions on the spatial continuity of the variable to interpolate, kriging gives the best linear unbiased prediction of the interpolated values [33]. Spatial continuity parameters (range, nugget and sill) were evaluated using theoretical models fitting the experimental variograms calculated for each variable [34,35]. Universal kriging assumes that a continuous property called "regionalized variable" consists of two parts: a drift, or expected value, and a residual, or deviation from the drift. The drift is modelled by a polynomial function within a given neighborhood. The residual surface obtained by drift removal can be regarded as first-order stationary in a statistical sense. However, the effectiveness of this technique depends on the correct specification of several parameters that describe the semivariogram and the model of drift. We assumed a linear model of the semivariogram implying that estimation error increases without bounds with increasing distance from the control point.

Spatial analysis and mapping were performed using the PLOTMAP [36] and the GMT [37] software packages adopting the WGS84 datum and a Mercator projection with standard parallel at the latitude of 42.5° N.

Data relationships were statistically analyzed through the Pearson correlation coefficient, using the STATISTICA software package.

4. Results

4.1. Sediment Features

Sediment grain size analyses indicate that the NC deposits are mostly clayey silts with a silty dominant fraction and a low-fraction of sand. In particular, on surficial samples, the coarser materials were found in the southernmost zone (sand content of 31% in site 15BC) and in the north-western sector of the channel (sand fractions of 51%, 39% and 31% at sites 07BC, 06BC and 10BC, respectively; Figure 2a).

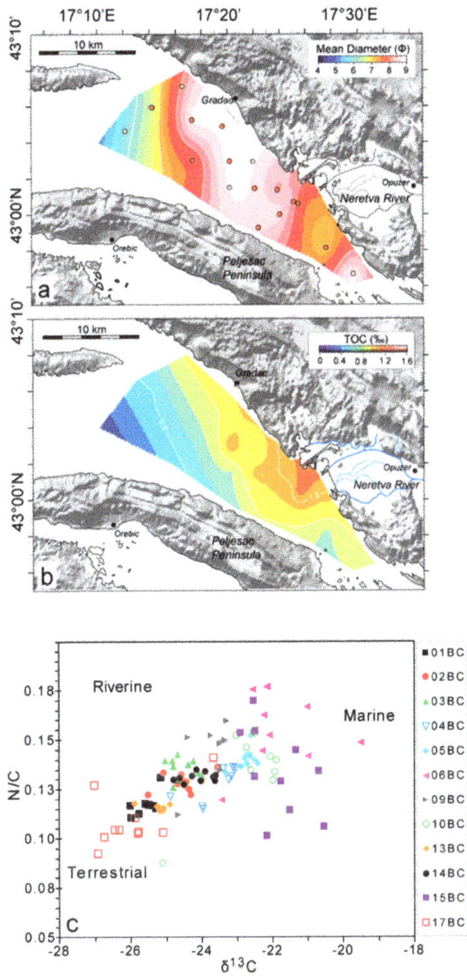

Figure 2. (**a**) Log-normal sediment mean diameter [38] areal distribution and (**b**) organic carbon surficial areal distribution in the Neretva Channel with (**c**) covariation of N/C versus δ^{13}C.

Surficial TOC (Figure 2b) varies from 0.6% (06BC) to 1.18% (17BC) and δ^{13}C from -22.3‰ (06BC) to -27.0‰ (17BC). Total nitrogen (TN) content ranges from 0.15% (17BC) to 0.09% (06BC), resulting in C/N ratios from 5.9 to 8.5.

X-radiograph images, together with depth distributions of sediment porosity and sand fractions of selected BC samples, are shown in Figure 3. X-ray images of BCs do not display evident inner sedimentary structures; instead, bioturbations are observed in several sediment intervals; in particular, in cores from the area close to the river mouth (i.e. site 01BC). Sample porosities of the surficial level along the entire channel range between 0.62 and 0.71. In addition, at each sampling site porosity slightly decreases with depth (toward the bottom of the BC) without abrupt changes through the sedimentary sequence (Figure 3).

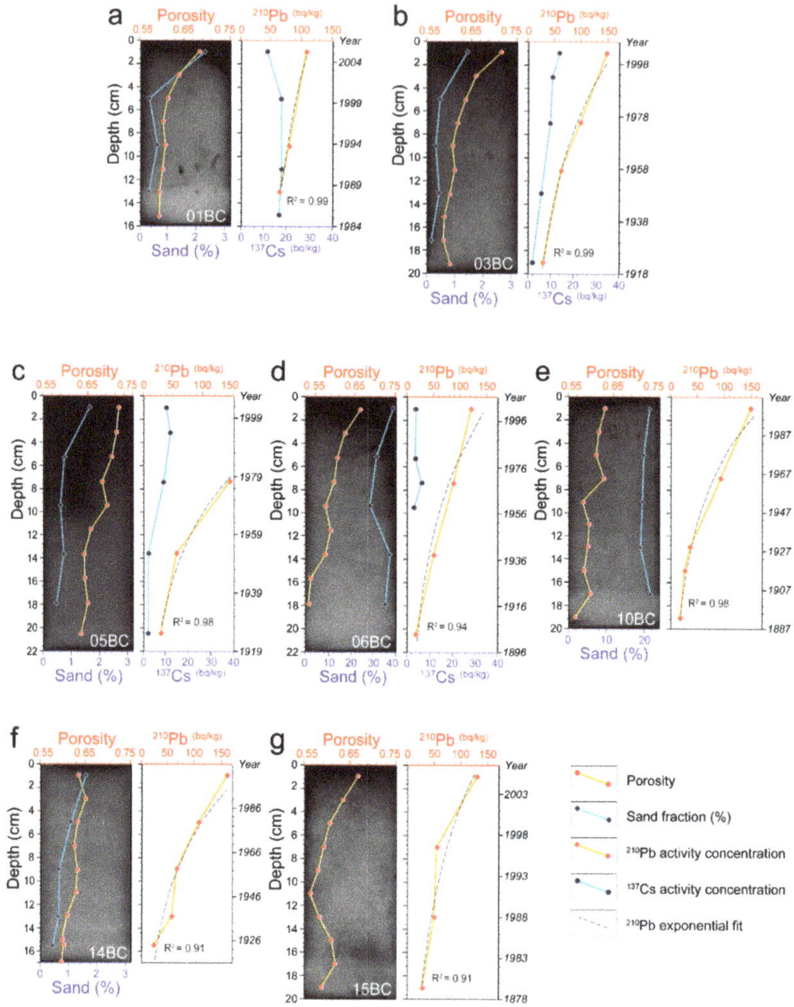

Figure 3. Examples of X-ray images, porosity and sand content depth profiles (left panel), and ^{210}Pb and ^{137}Cs concentration activity variations with depth (right panel) from selected box cores (BCs): (**a**) 01BC; (**b**) 03BC; (**c**) 05BC; (**d**) 06BC; (**e**) 10BC; (**f**) 14BC and (**g**) 15BC.

Vegetal remains are found mostly in sediments near the river mouth (sites 01BC, 02BC, 09BC and 15BC), whereas fragments of molluscs, bryozoans and ostracods are widespread in the surficial sediments of the NC, with the exception of the nearest site to the river mouth (01BC), where only fragments of molluscs are found, and of the deepest site (07G) where the benthic macrofauna is almost absent.

4.2. Activity-Depth Profile of ^{210}Pb and ^{137}Cs Radionuclides

^{137}Cs and ^{210}Pb content variations with depth of selected BC are shown in Figure 3. ^{137}Cs measured only in three BC (Figure 3a–d), displays maximum activity concentration at the top of BCs (i.e. 01BC and 03BC or at sub-surficial levels as in 02BC and 05BC; Figure 3a,c) with values ranging from the detection limit to 18 Bq/kg (Figure 3d,e). Unfortunately, the 1986 peak of ^{137}Cs relative to the Chernobyl accident cannot be resolved in our samples.

^{210}Pb contents in cores generally decrease with depth following a regular exponential trend (Figure 3). Since cores 04BC, 09BC, 13BC, 14BC and 15BC are strongly affected by physical or biological mixing (see X-radiograph; i.e. Figure 3f,g), a different ^{210}Pb pattern is observed.

4.3. Regional Water Circulation and Water-Column Structure

During the sampling survey, temperature (T) and salinity (S) values in the water column ranged from 12.7 to 17.1 °C and from 19.2 to 38.0 psu, respectively (Figure 4a,b). At sites 01BC and 17BC, in front of the river mouth, recorded values of T and S at sea surface were ≈ 16 and 15.2 °C, and ≈ 34 and 19.3 psu respectively, whereas values at the sea bottom displayed lower temperature and higher salinity (≈ 13 and 13.5 °C, and ≈ 37.5 and 37.6 psu respectively).

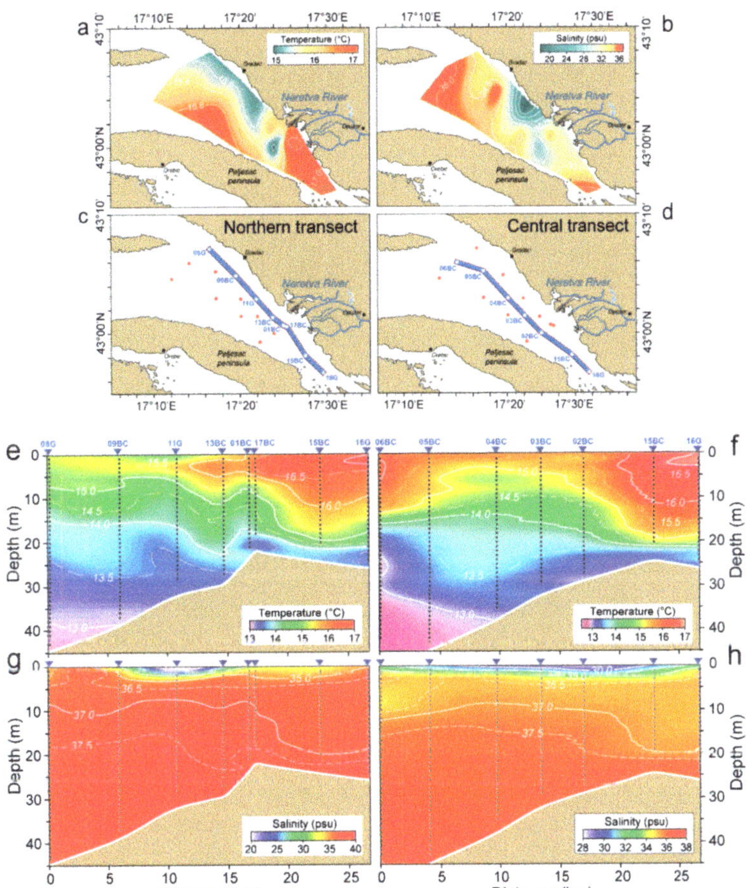

Figure 4. (**a**) Temperature and (**b**) salinity spatial distribution. Temperature (**e**,**f**) and salinity (**g**,**h**) vertical cross sections along the northern (**c**) and the central (**d**) transects running parallel to the NE coast of the Neretva Channel.

T and S profiles across the water column are shown in Figure 4e–h. Observed values reveal water stratification with the highest temperatures at the surface in the southeast sector of the NC. The northern transect (Figure 4e,f) and surficial areal data (Figure 4a,b) suggest water inflow with lower temperature and salinity localized along the coast to the north of the main Neretva river-mouth in the vicinity of station 11G, where a local decrease of salinity has been also recorded at 0 and 5 m water-depth intervals (Figure 4e,h). The water mass with lower T and S moves deeper toward the center of the channel.

4.4. Metal Distributions

Among the several metals, we focused on Cr, Ni and Pb in this study, because their origin may be due to both natural and anthropogenic causes. The spatial distributions of their concentrations (μg/g) allowed identifying zones of provenance (source areas) and areas at critical levels (Figure 5). Results shown in Figure 5 indicate that site 08G, in the northern part of the investigated area, displays the maximum contents of Ni and Cr (84.0 and 71.5 μg/g respectively), whereas the minimum contents were recorded at site 06BC (28.6 and 28.7 μg/g, respectively) located in center of the northern sector of the NC (Figure 5b,c). In particular, surficial Cr and Ni concentrations decrease toward the river mouth. Differently, the Pb concentrations show two different areas of local maxima: the first (36 μg/g) is located in front of the river mouth at site 01BC; the second (67.9 μg/g) in the center of the NC at site 10BC.

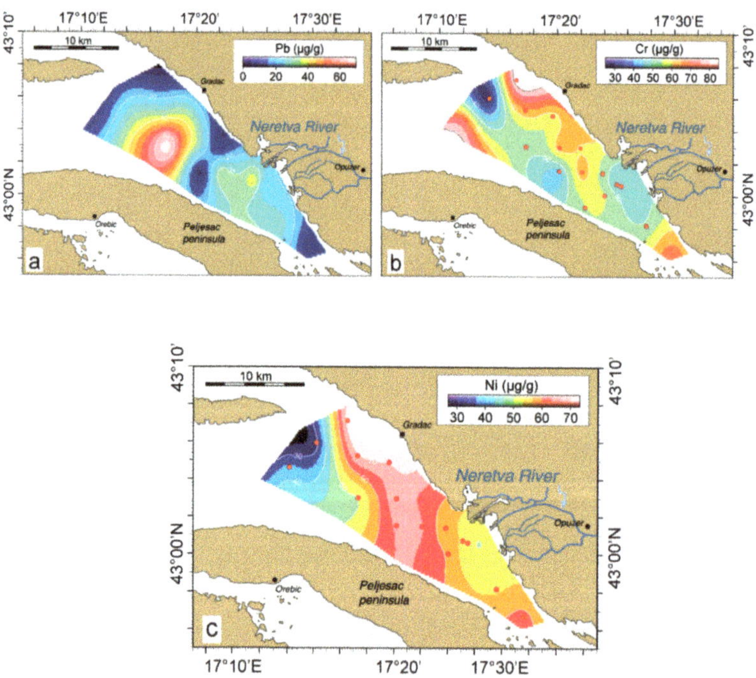

Figure 5. Surficial distribution of (**a**) Pb, (**b**) Ni and (**c**) Cr concentrations.

5. Discussions

5.1. Surface Sediment Distribution

The Neretva's surficial sediment is generally composed of clayey silt. The distribution suggests an influence of the river plume decreasing along the SE–NW direction (Figure 2a). In particular, accumulation of fine-grained particles (mean grain size > 7.5 φ [38]) occurs mainly in two areas: in front of the NR mouth and in the central/northeastern part of the NC (Figure 2). This accumulation pattern is in agreement with the observed water circulation, where a hypopycnal river plume formed at the mouth distributes fine-grained particles over the entire channel area [16,17]. These observations suggest a different nature of the two observed coarser-grains depositional zones. The southernmost zone may only receive river sediments sporadically during events of strong hydro-dynamism in the channel and may be considered a non-depositional area characterized by relict coarser sediment. This silt-sandy accumulation zone was previously reported by [16]. Instead, the northernmost sandy zone may be related to the influence of high energy currents from the open Adriatic Sea that remove the finest part of the sediments.

The observed porosities and particle grain size in the BCs (Figure 3) suggest that over the time recorded by our samples, the NC sedimentation was continuous in a low-energy environment. Moreover, the chaotic textures of sediments in the cores nearest to the river pro-delta area (BC01-17 and 13) suggest also that these deposits are affected by several post-depositional processes, including physical mixing and biological reworking due to microbial activity. Both of those processes may be related to the seasonal river input variability or episodic flooding events from the NR.

5.2. Organic Carbon

Relationships between C/N ratios and the $\delta^{13}C$ contents are commonly used as proxies to determine the source of organic matter in coastal areas [26,39,40]. In addition, the $\delta^{13}C$ and $\delta^{15}N$ isotopic compositions and the N/C ratio help to discriminate the marine versus terrestrial origins of the particles [41]. Since the photosynthetic processes and carbon sources are different between marine organisms and terrestrial plants, there is an inverse correlation between the $\delta^{13}C$ and C/N ratio: A high $\delta^{13}C$ values suggests a predominant influence of marine phytoplankton [42]. On the contrary, low $\delta^{13}C$ values (lower than -27‰) are indicative of vegetable sources (C3 plants) from inland; i.e., proximity to the shoreline and/or proximity to a source of organic matter of continental origin [43]. Here, we considered the N/C ratio instead of the C/N, because it behaves linearly in a mixing model [43,44] and an estimate of the sedimentary organic carbon fraction is more reliable [45]. The observed N/C and $\delta^{13}C$ distributions (Figure 2c) indicate that the sites close to the river mouth (01BC and 17BC) have a clear terrestrial signature due to the strong fluvial input. The marine input increases northward, and the transition between terrestrial and marine regimes may be located along the theoretical line joining sites 12G, 4BC and 9BC (Figure 2b). In addition, a strong negative correlation (−0.81) is observed between TOC and $\delta^{13}C$, a result in agreement with that of [17]. These data together with a lack of correlation with the clay fraction (Table 1), suggest riverine ^{13}C depleted terrestrial organic matter as the main source of sedimentary organic carbon [27].

Table 1. Linear correlation coefficients between Pb, Cr and Ni concentrations (μg/g), porosity, organic carbon (TOC, %), total carbon (TC, %), inorganic carbon (TIC, %), C/N ratio, carbon stable isotope ($\delta^{13}C$, ‰) and sediment compositions (fines, sand, clay, silt contents, %). Statistically relevant coefficients are reported in bold and highly significant one in bold red.

	Pb	Cr	Ni	Porosity	TOC	TC	TIC	C/N	$\delta^{13}C$	Fines	Sand	Silt	Clay
Pb	1.00												
Cr	−0.04	1.00											
Ni	0.12	**0.67**	1.00										
Porosity	**−0.69**	**0.43**	0.06	1.00									
TOC	0.10	0.09	0.30	−0.10	1.00								
TC	−0.04	−0.16	**−0.72**	0.09	**−0.52**	1.00							
TIC	−0.06	−0.16	**−0.70**	0.10	**−0.64**	**0.99**	1.00						
C/N	−0.03	0.04	0.07	0.12	**0.59**	−0.29	**−0.36**	1.00					
$\delta^{13}C$	−0.13	0.07	−0.17	0.20	**−0.82**	**0.40**	**0.50**	**−0.56**	1.00				
fines	0.09	0.19	**0.68**	−0.11	**0.70**	**−0.89**	**−0.92**	**0.45**	**−0.59**	1.00			
Sand	−0.10	−0.21	**−0.67**	0.08	**−0.73**	**0.87**	**0.91**	**−0.48**	**0.62**	**−1.00**	1.00		
Silt	0.12	−0.17	0.31	**−0.39**	0.24	**−0.60**	**−0.58**	0.18	**−0.45**	**0.67**	**−0.63**	1.00	
Clay	−0.05	**0.42**	0.24	**0.42**	**0.40**	−0.07	−0.13	0.23	−0.02	**0.10**	−0.15	**−0.67**	1.00

These results allow us to hypothesize that the input material from the NR is quickly moved northward by marine currents and released not far from the river mouth, particularly in the area near and along the coast (north of the river mouth). The observed organic carbon distribution highlights a reduced fluvial influence moving seaward. However, surficial sediments may be affected by seasonal conditions and their distribution may be different in other periods of the year due to different amounts of input material from the river.

5.3. Age Model and Sediment Accumulation Rates

Activity-depth profiles of ^{137}Cs and ^{210}Pb were evaluated along the BCs to define an age model of the NC recent sedimentary sequence (Figure 3).

Correlations between ^{210}Pb age determinations and depth are strongly model-dependent in cores with non-exponential ^{210}Pb profiles [46]. Assuming a constant ^{210}Pb flux below the surficial mixed layer, a "constant flux–constant sedimentation" model has been adopted to obtain an estimate of sedimentation accumulation rates (SARs) for the last hundred years. However, since the main assumptions of most common conceptual models in sediment chronologies refer to the inputs of particles and/or a radiotracer onto the sediment, in those samples where ^{210}Pb profiles were not suitable for model calculations (such as in samples 06BC and 10BC), rough estimates of sediment accumulation rates were obtained by assuming an age of 100 years at the depth where the ^{210}Pb background value of ≈18 Bq/kg is found (Figure 3d,e). Since physical mixing and bioturbation were neglected in the calculations, these apparent average rates have to be considered upper limits.

Since the ^{210}Pb dating model requires validation by a second, independent stratigraphic tracer [47,48], ^{137}Cs activity-depth profiles were considered in the calculations. Unfortunately, as mentioned above, the 1986 ^{137}Cs peak (Chernobyl accident) is not cleatracer [47,48]rly identifiable in the investigated sedimentary intervals. This is probably due to the strong bioturbation and sediment mixing (see X-ray images in Figure 3f,g) as suggested before by [16]. Moreover, since the half-life of ^{137}Cs is 30.17 years, it could not be excluded that the concentration is below the experimental sensitivity. This effect may be enhanced in case of high contents of carbonate (such as in karstic areas) that are less reactive to radiotracers [49]. In order to overcome validation troubleshooting, the core depth where ^{137}Cs reached the instrumental detection limit was used as reference age, assuming it corresponded to the early 1960s, when a large number of nuclear bombs were exploded in the atmosphere.

Estimated sediment accumulation rates (SARs; Figure 6) range from 1.9 to 8.5 mm/yr. These values differ partially from those previously estimated by other authors [16], ranging from 4 to 6 mm/yr and based only on the distribution of ^{137}Cs in core sediments. In particular, the lowest SAR value of 1.9 mm/yr was found at site 06BC (Figure 6), located to the NW of the channel at the boundary with the Korcula Channel, a location comparable to that of sample K3 of [16] where a SAR of 4 mm/yr was estimated. In the sectors of the channel less influenced by the river outflow, SARs range from 1.9 to 2.3 mm/yr, while in the areas where the river influence is high, SARs range from 4.0 to 8.5 mm/yr. The highest accumulation rates are observed in front of the river mouth and along the northeastern coast with a peak of localized preferential accumulation at site 04BC (8.5 mm/yr; Figure 6). Estimated accumulation rates are in agreement with the observed water circulation, where a hysopycnal river plume formed at the river mouth distributes particles toward the open sea.

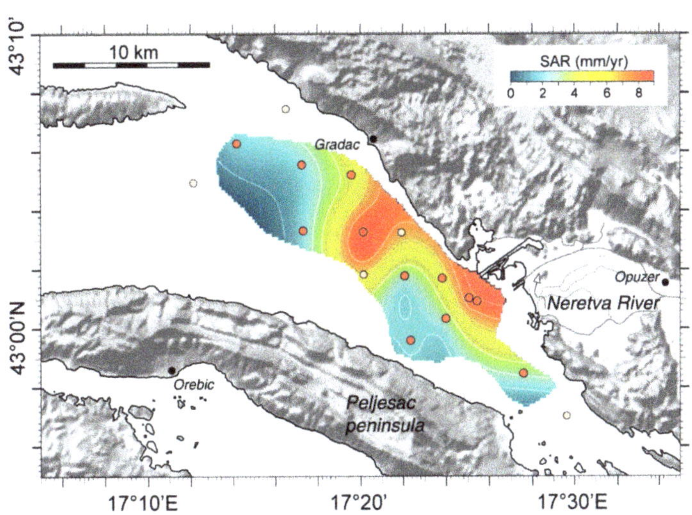

Figure 6. Apparent sediment accumulation rates (SAR) over the study area.

Apparent accumulation rates (Figures 3 and 6) suggest that most surficial sediments in the NC result from the present day seasonally river-dominated coastal sedimentary regime, with few relict deposits in its southernmost sector, south of the river mouth. According to the calculated SARs, our BCs record a time interval ranging from 18 (04BC) to 100 years (06BC) ago.

5.4. Regional Water Circulation and Water-Column Physical Properties

Surficial currents along the eastern Adriatic in the proximity of the NC are always oriented from SE to NW following the overall cyclonic circulation of the basin [50]. Local data on annual mean values of water circulation within the NC are not available in the literature, except for a little punctual information. Given that the NC represents a semi-closed marine environment, it is reasonable to assume that channel hydrodynamics are the result of a balance between tidal currents and fluvial influence, and seasonal variations in precipitation may act as a driving factor for the regional water exchange with the Adriatic open sea.

Generally, river currents do not exceed the velocity of 0.1 m/s and are higher during the winter season [51,52]. In particular, in 2004, daily outflows (m^3/s) during winter and spring periods ranged from 300 to 400 m^3/s, with short time intervals of peak flow (>500 m^3/s) or reduced flow (< 200 m^3/s). However, during the summer season, the river outflow is drastically reduced (up to 50 m^3/s), as reported by UNEP/MAP [53]. In addition, horizontal distribution of freshwater into the channel may depend on the direction of local winds and may vary on a daily basis [54].

Based on CTD data, sediment grain size and SAR surficial distributions, we suggest the presence of an eddy in the NC waters located at the center of the channel that may strongly influence sediment dispersion and determine areas of localized preferential sediment accumulation not necessarily in correspondence with the river mouth (e.g., 10BC, Figure 4). Moreover, a considerable amount of freshwater is discharged into the channel by several submarine karstic springs [55], suggesting high inflow of cold water that lacks sediment load during periods of intense rainfall.

The observed water stratification along the northern transect may represent a seasonal snapshot of the hydrologic conditions of the study area. Mixed-water conditions may dominate through the entire NC area during flooding events in autumn. However, for the objectives of this study, spring conditions may be considered the optimal hydrological situation. Close to the river mouth, less salty and colder water was detected. This could be due to a reduction of the salt-wedge during low tide conditions and/or intense river outflow, or the karstic characteristics of the region. Karstic fresh waters have generally poor or null suspended material contents, explaining why these inflows do not affect organic matter or metal distributions.

5.5. Metal Distributions

Metal concentrations suggest that the river outflow plays only a marginal role in Pb, Cr and Ni surficial distributions (Figure 5). In detail, Pb shows the highest concentration at site 10BC located in the center of the channel and far from possible direct sources. This may be due to an event of local discharge or to a pulse in the transport of metal enriched sediment, as suggested by the higher sand content at the top of the core and by the N/C values (Figure 2b,c and Figure 3). A second local maximum of Pb content is located in front of the river mouth following a distribution inversely correlated with the distance from the mouth (Figure 5a).

Cr and Ni contents (Figure 5b,c) are affected by a source located along the coast to the NW of the river mouth (close to station 08G). Moreover, Cr content shows another local maximum at site 07G (70.4 μg/g), where an increase in the sand fraction is observed, suggesting an additional source. On the other hand, the minimum values of Ni and Cr contents at site 06BC may indicate a scouring resuspension or not sedimentation, as suggested by the observed N/C values. In the southern part of the NC, where coarser sediments are located, an increase of Cr and Ni contents associated to a low organic carbon and to a low sediment accumulation rate of 2.3 mm/yr (15BC; Figure 3) suggests that this zone acts as a trap for metal bearing particles. Alternatively, the Cr and Ni abundances in this area may be influenced by sediment inputs from Peljesac Peninsula.

Some authors indicate the ratio Cr/Ni as a possible tracer of geo-genic versus anthropogenic influences [56–58] and suggest it as suitable to determining geochemical baselines in the case of high natural concentrations [34,35,59]. In particular, Cr concentrations in south Dalmatia show a mean value of 126 μg/g; that is the highest value found in karstic regions, probably due to the presence of chromite-bearing ultramafic rocks and/or of clastic deposits derived from older mafic magmatic rocks. Similarly, Ni levels in the same area show a mean value as high as 84 μg/g as a consequence of the presence of mafic and ultramafic rock outcroppings in the region. These rocks are probably a source of Ni in alluvial soils of the floodplains of rivers draining this region [57]. The expected Cr/Ni value reported in literature [56–58] is about 1.5, whereas in the study area Cr/Ni ranges from 0.7 (site 12G) to 1.9 (site 07G) with a mean value of 0.9. This discrepancy may be due to a natural variability but also due to the extraction efficiency of the leaching method. However, ratio values are generally uniform in the area supporting the hypothesis of Cr and Ni background concentrations.

According to previous data [17], heavy metals (such as Pb, Cr and Ni) in surficial sediments are tightly linked to fine grain size fractions, and in particular, to clay minerals coated by iron oxides/oxy-hydroxides that may trap metal cations within their crystal-chemical structures. Univariate statistical analysis shown in Table 1 confirms the previous observations for Ni and Cr: that they correlate positively with each other, and with clay and porosity, while they correlate negatively with total and inorganic carbon. In addition, organic carbon shows a positive correlation with fine grain size, in particular, with the clay fraction, and negative ones with total carbon and inorganic carbon. These results may support the hypothesis of clay minerals and organic matter as driving factors of Cr and Ni areal distribution. Pb shows a different behavior, characterized by a negative significant correlation with porosity. Since porosity is correlated negatively to silt and positively to clay, this suggests that Pb is not immobilized in the clay fraction of the sediments having higher affinity with the silt fraction.

A preliminary evaluation of environmental risks due to contaminant loads within the channel has been carried out, comparing surficial concentration values with quality criteria for marine sediments (Table 2)—threshold effect level (TEL), effect range low (ERL), probable effect level (PEL) and effect range median (ERM) [60–62]—and with benchmarks (LCB and LCL) for Italian coastal areas [63]. These guidelines are screening tools to predict potential sediment toxicity, linking sediment metal concentrations to the adverse biological effects that may result from exposure to chemicals.

Table 2. Minimum, maximum and average concentrations (in µg/g d.w.) of heavy metals in box-core sediments. Comparison with Italian and international sediment quality guidelines benchmarks.

	Cr	Ni	Pb
Mean surficial concentration	56.3	58.2	25
Minimum	26.9	28.7	4.5
Maximum	84	86.8	67.9
BG	45	51.8	24
WA	100	40	10
LCB *	100	70	40
LCL *	360	75	70
ERL **	81	20.9	46.7
ERM **	370	51.6	218
ISQGs **	52.3	15.9	30.2
PEL **	160	42.8	112.2

Mean metal concentrations pre-seventies (BG); world average (WA) [64]; * Italian Sediment Quality benchmark: APAT and ICRAM [63]; ** International Sediment Quality Guidelines [60–62].

The Pb and Cr levels in surficial samples are below the benchmarks of PEL, ERM, and the Italian LCL guidelines, except at sites 10BC and 14BC for Pb (ERL and LCB) and 08G for Cr (ERL). However, the incidence of the effect for these samples may be considered low or scarcely probable (8%–30% for Pb and 2.9%–21.1% for Cr; [62–64]).

The Ni surficial concentration exceeds, at several sites, both PEL and ERM guidelines, whereas stations 06BC and 07G and 10BC comply under PEL; however, the Italian benchmark LCB is exceeded only at site 08G. In general, the effect incidence for exceeding ERM is common, although for Ni it is only 16.9% [62–64]. The Pb, Cr and Ni concentrations in surficial sediments of the Neretva Channel fall into an unpolluted class with exception of Pb content of sample 10BC falling into a moderately polluted class. All this highlights a general unpolluted situation of the channel.

The sediment flux (carried out where BCs were available; Figure 7) provides a quantitative estimation of the amount of material that deposits in the channel. Two areas of higher flux are identified: in front the river mouth and in the center of the NC, at site 04BC. The latter location is not related to any particular morphologic setting of the channel and coincides with the area where we suggest the presence of an eddy in the water circulation, probably favoring a critical area of accumulation. TOC, following a similar distribution, tells the same story (Figure 2b), given that TOC input depends mostly from the river inflow because springs water may be assumed containing very few particles. These observations document that local hydrology plays a key role in controlling the sediment distribution in the area.

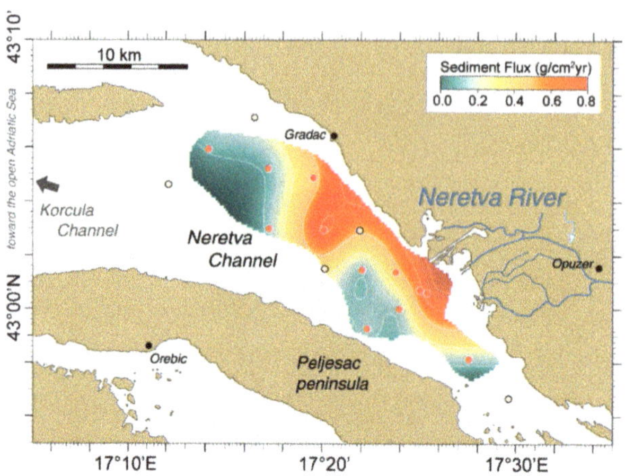

Figure 7. Areal distribution of sediment flux (g/cm^2 yr) in surficial sediment of the Neretva Channel.

6. Conclusions

In this study, a multi-proxy approach on surficial sediment samples allowed us to determine sediment dynamics of the Neretva delta system, the largest modern siliciclastic depositional system on the eastern Adriatic coast.

Samples from 12 box-corer and five grab stations document that surficial sediments are predominantly clayey silt and coarse silt with a distribution of finer grained particles decreasing southward moving away from the river mouth.

Estimated accumulation rates, based on ^{210}Pb and ^{137}Cs activity versus depth profiles, range from 1.9 to 8.4 mm/yr, suggesting the center of the channel and the river mouth as the sectors with the highest sediment deposition rates. In particular, two areas of higher sediment accumulation were identified: (i) close to the river mouth; and (ii) in the central-northern sector of the Neretva Channel along the axis of the basin.

Temperature and salinity vertical distributions indicate stratified waters along the axis of the basin close to the Croatian coast, and the presence of less salty and colder water in the proximity of the river mouth. The latter consideration, together with the observation of strong salt intrusions in the NR delta, suggests a salt wedge reduction during low tide conditions and submerged fresh water inputs. Moreover, relationships between TOC and TN and their spatial distribution, in particular, covariations between N/C and δ^{13}C, indicate a terrestrial origin with prevalent accumulation in the vicinity of the river mouth that decreases northward towards the open sea. The scenario depicted above agrees with temperature and salinity data of the water column and their surficial distribution in the channel.

Finally, based on CTD data, grain size, TOC and SAR distributions, we suggest the presence of an eddy in the NC waters near the center of the channel that may be triggered and/or intensified seasonally. This may strongly affect sediment dispersion favoring areas of localized preferential sediment accumulation not necessarily close to the river mouth (e.g., 10BC, Figure 4).

The distribution of Pb contents in the surficial sediments follows a general trend inversely proportional to the distance of the river mouth with two local zones of accumulation: in front of the river mouth and in the center of the channel. Univariate statistics indicate that Pb is not immobilized in the clay fraction. On the contrary, Ni and Cr distributions indicate that they follow the clay fraction and that they account for other sources besides the river, such as sediment input from the Peljesac peninsula. The comparison with marine sediment guidelines shows that surficial concentrations of the examined metals have a low toxicity.

Author Contributions: F.G. had the role of project administrator and investigator in this study. S.R. wrote the preliminary version of the manuscript with the support of F.G. and the other authors. The methodology and validation of data were carried out by S.A., L.C. and M.L. F.C. reviewed and edited text and figures, and curated data. All authors discussed the results and contributed to the final manuscript. M.R. supervised the scientific project and provided financial support for the scientific work and publication. All authors have read and agreed to the published version of the manuscript.

Funding: Funds were provided, within the framework of ADRICOSM-NERES project (2006–2007) (Environmental regeneration and sustainable development of the delta of the NEretva river), by the Italian Ministry of Foreign Affairs and the Ecoss project Italy-Croatia (2019).

Acknowledgments: Thanks to Frano Matic, Grozdan Kuspilic and Slavica Matijevc of the Institute of Oceanography and Fisheries, Split Croatia; Gabriele Marozzi and Costante Luttazzi, for sample collection, sampling and treatment on R/V Bios; Mauro Frignani for helpful suggestions; and Nadia Pinardi for project coordination. Special thanks to Prof. Enrico Bonatti of Earth Institute Lamont-Doherty Earth Observatory, Columbia University for the careful english language revision of the manuscript. This is contribution number 2022 of the Istituto Di Scienze Marine of the Consiglio Nazionale delle Ricerche, Bologna.

Conflicts of Interest: The authors declare no conflict of interest. The funders had no role in the design of the study, in collection, analyses, or interpretation of the data; in the writing of the manuscript, or in the decision to publish the results.

References

1. UNEP/MAP 1995. Convention for the Protection of the Mediterranean Sea Against Pollution (Barcelona Convention) modified by Conference of Plenipotentiaries on the Convention for the Protection of the Mediterranean Sea against Pollution and its Protocols, held in Barcelona on 9 and 10 June 1995 (UNEP(OCA)/MED IG.6/7). Available online: https://ec.europa.eu/environment/marine/international-cooperation/regional-sea-conventions/barcelona-convention/index_en.htm (accessed on 22 January 2020).
2. Directive 2000/60/EC of the European Parliament and of the Council of 23 October 2000 establishing a framework for Community action in the field of water policy (Water Framework Directive). OJ L 327, 22.12.2000, p. 1–73. Available online: http://data.europa.eu/eli/dir/2000/60/oj (accessed on 22 January 2020).
3. Directive 2008/56/EC of the European Parliament and of the Council of of 17 June 2008 establishing a framework for community action in the field of marine environmental policy (Marine Strategy Framework Directive). OJ L 164, 25.6.2008, p. 19–40. Available online: http://data.europa.eu/eli/dir/2008/56/oj (accessed on 22 January 2020).
4. Directive 92/43/EEC of 21 May 1992 on the conservation of natural habitats and of wild fauna and flora (Habitat Directive). Official Journal L 206, 22/07/1992 P. 0007 – 0050. Available online: http://data.europa.eu/eli/dir/1992/43/oj (accessed on 22 January 2020).
5. Directive 2009/147/EC of the European Parliament and of the Council of 30 November 2009 on the conservation of wild birds (Bird Directive). OJ L 20, 26.1.2010, p. 7–25. Available online: http://data.europa.eu/eli/dir/2009/147/oj (accessed on 22 January 2020).
6. Van Hoey, G.; Borja, A.; Birchenough, S.; Buhl-Mortensen, L.; Degraer, S.; Fleischer, D.; Kerckhof, F.; Magni, P.; Muxika, I.; Reiss, H.; et al. The use of benthic indicators in Europe: From the Water Framework Directive to the Marine Strategy Framework Directive. *Mar. Pollut. Bull.* **2010**, *60*, 2187–2196. [CrossRef] [PubMed]
7. Micheli, F.; Levin, N.; Giakoumi, S.; Katsanevakis, S.; Abdulla, A.; Coll, M.; Fraschetti, S.; Kark, S.; Koutsoubas, D.; Mackelworth, P.; et al. Setting Priorities for Regional Conservation Planning in the Mediterranean Sea. *PLoS ONE* **2013**, *8*, e59038. [CrossRef] [PubMed]
8. Cattaneo, A.; Correggiari, A.; Langone, L.; Trincardi, F. The Late-Holocene Gargano subaqueous delta, Adriatic shelf: Sediment pathways and supply fluctuations. *Mar. Geol.* **2003**, *193*, 61–91. [CrossRef]
9. Sherwood, C.S.; Carniel, S.; Cavaleri, L.; Chiggiato, J.; Das, H.; Doyle, J.D.; Harris, C.K.; Niedoroda, A.W.; Pullen, J.; Reed, C.W.; et al. Sediment dynamics in the Adriatic Sea investigated with coupled models. *Oceanography* **2004**, *17*, 58–69. [CrossRef]
10. De Lazzari, A.; Rampazzo, G.; Pavoni, B. Geochemistry of sediments in the Northern and Central Adriatic Sea. *Estuar. Coast. Shelf Sci.* **2004**, *59*, 429–440. [CrossRef]
11. Frignani, M.; Langone, L.; Ravaioli, M.; Sorgente, D.; Alvisi, F.; Albertazzi, S. Fine-sediment mass balance in the western Adriatic continental shelf over a century time scale. *Mar. Geol.* **2005**, *222*, 113–133. [CrossRef]

12. Romano, S.; Langone, L.; Frignani, M.; Albertazzi, S.; Focaccia, P.; Bellucci, L.G.; Ravaioli, M. Historical pattern and mass balance of trace metals in sediments of the northwestern Adriatic Sea Shelf. *Mar. Pollut. Bull.* **2013**, *76*, 32–41. [CrossRef]
13. Goudeau, M.-L.S.; Grauel, A.-L.; Bernasconi, S.; De Lange, G.J. Provenance of surface sediments along the southeastern Adriatic coastoff Italy: An overview. *Estuar. Coast. Shelf Sci.* **2013**, *134*, 45–56. [CrossRef]
14. Lopes-Rocha, M.; Langone, L.; Miserocchi, S.; Giordano, P.; Guerra, R. Spatial patterns and temporal trends of trace metal mass budgets in the western Adriatic sediments (Mediterranean Sea). *Sci. Tot. Environ.* **2018**, *599*, 1022–1033. [CrossRef]
15. Jurina, I.; Ivanić, M.; Vdović, N.; Mikac, N.; Sondi, I. Mechanism of the land-sea interactions in the Neretva River delta (Croatia): The distribution pattern of sediments and trace element. *Rapp. Comm. Int. Mer. Medit.* **2010**, *39*, 35.
16. Jurina, I.; Ivanic, M.; Troskot-Corbic, T.; Barisic, D.; Vdovic, N.; Sondi, I. Activity concentrations and distribution of radionuclides in surface and core sediments of the Neretva Channel (Adriatic Sea, Croatia). *Geol. Croat.* **2013**, *66*, 143–150. [CrossRef]
17. Jurina, I.; Ivanic, M.; Vdovic, N.; Troskot-Corbic, T.; Lojen, S.; Mikac, N.; Sondi, I. Deposition of trace metals in sediments of the deltaic plain and adjacent coastal area (the Neretva River, Adriatic Sea). *J. Geochem. Explor.* **2015**, *157*, 120–131. [CrossRef]
18. Kljakovic-Gaspic, Z.; Bogner, D.; Ujevic, I. Trace metals (Cd, Pb, Cu, Zn and Ni) in sediment of the submarine pit Dragon ear (Soline Bay, Rogoznica, Croatia). *Environ. Geol.* **2009**, *58*, 751–760. [CrossRef]
19. Law, G.T.W.; Geissler, A.; Boothman, C.; Burke, I.T.; Livens, F.R.; Lloyd, J.R.; Morris, K. Role of nitrate in conditioning aquifer sediments for technetium bioreduction. *Environ. Sci. Technol.* **2010**, *44*, 150–155. [CrossRef] [PubMed]
20. Orlic, M.; Dadic, V.; Grbec, B.; Leder, N.; Marki, A.; Matic, F.; Mihanovic, H.; Paklar, G.B.; Pasaric, M.; Pasaric, Z.; et al. Wintertime buoyancy forcing, changing seawater properties, and two different circulation systems produced in the Adriatic. *J. Geophys. Res. Ocean.* **2006**, *111*, C03S07. [CrossRef]
21. Skoulikidis, N.T. The environmental state of rivers in the Balkans-A review within the DPSIR framework. *Sci. Total Environ.* **2009**, *407*, 2501–2516. [CrossRef]
22. EIA R E P O R T. Environmental Impact Assessment in the Neretva and Trebišnjica River Basin (NTRB); No. TF052845/GE-P084608: Sarajevo / Banja Luka, August 2006; p. 285. Available online: http://documents.banquemondiale.org/curated/fr/793041468104056146/pdf/E14680REVISED0ECA1EA1P084608.pdf (accessed on 22 January 2020).
23. Ivanković, A.; Petrović, D.; Ivanković, P.; Majstorović, J. Monitoring of water salinity of the lower river Neretva. *Int. J. Ecosyst. Ecol. Sci.* **2017**, *2*, 136–143.
24. Kralj, D.; Romic, D.; Romic, M.; Cukrov, N.; Mlakar, M.; Kontrec, J.; Barisic, D.; Sirac, S. Geochemistry of stream sediments within the reclaimed coastal floodplain as indicator of anthropogenic impact (River Neretva, Croatia). *J. Soils Sediment.* **2015**, *16*, 1150–1167. [CrossRef]
25. Calo, F.; Parise, M. Waste management and problems of groundwater pollution in karst environments in the context of a post-conflict scenario: The case of Mostar (Bosnia Herzegovina). *Habitat Int.* **2009**, *33*, 63–72. [CrossRef]
26. Berner, R.A. Authigenic mineral formation resulting from organic matter decomposition in modern sediments. *Fortschr. Mineral.* **1981**, *59*, 117–135.
27. Tesi, T.; Langone, L.; Goni, M.A.; Miserocchi, S.; Bertasi, F. Changes in the composition of organic matter from prodeltaic sediments after a large flood event (Po River, Italy). *Geochim. Cosmochim. Acta* **2008**, *72*, 2100–2114. [CrossRef]
28. Bellucci, L.G.; Frignani, M.; Cochran, J.K.; Albertazzi, S.; Zaggia, L.; Cecconi, G.; Hopkins, H. ^{210}Pb and ^{137}Cs as chronometers for salt marsh accretion in the Venice Lagoon and links to flooding frequency and climate change. *J. Environ. Radioactiv.* **2007**, *97*, 85–102. [CrossRef] [PubMed]
29. Frignani, M.; Langone, L. Accumulation rates and ^{137}Cs distribution in sediments off the Po River Delta and the Emilia Romagna coast (north-western Adriatic Sea, Italy). *Cont. Shelf Res.* **1991**, *11*, 525–542. [CrossRef]
30. Bellucci, L.G.; Frignani, M.; Paolucci, D.; Ravanelli, M. Distribution of heavy metals in sediments of the Venice Lagoon: The role of the industrial area. *Sci. Total Environ.* **2002**, *295*, 35–49. [CrossRef]

31. UNESCO. *The Acquisition, Calibration and Analysis of CTD Data: Unesco Technical Papers in Marine Science 54; A Report of SCOR Working Group 51*; United Nation Educational, Scientific and Cultural Organization: Paris, France, 1988; p. 102, ISSN 0503-4299.
32. Schlitzer, R. Ocean Data View. 2018. Available online: https://odv.awi.de (accessed on 22 October 2019).
33. Olea, R.A. Optimal contour mapping using universal kriging. *J. Geophys. Res.* **1974**, *79*, 695–702. [CrossRef]
34. Cabral Pinto, M.M.S.; Ferreira da Silva, E.A.; Silva, M.M.V.G.; Melo-Gonçalves, P. Heavy metals of Santiago Island (Cape Verde) top soils: Estimated background value maps and environmental risk assessment. *J. Afr. Earth Sci.* **2015**, *101*, 162–176. [CrossRef]
35. Cabral Pinto, M.M.S.; Ferreira da Silva, E.A.; Silva, M.M.V.G.; Melo-Gonçalves, P.; Candeias, C. Environmental risk assessment based on high-resolution spatial maps of potentially toxic elements sampled on stream sediments of Santiago, Cape Verde. *Geosciences* **2014**, *4*, 297–315. [CrossRef]
36. Ligi, M.; Bortoluzzi, G. PLOTMAP: Geophysical and geological applications of good standard quality cartographic software. *Comput. Geosci.* **1989**, *15*, 519–585. [CrossRef]
37. Wessel, P.; Smith, W.H.F. New, improved version of Generic Mapping Tools released. *AGU Eos Trans.* **2006**, *79*, 579. [CrossRef]
38. Folk, R.L.; Ward, W.C. Brazos River bar: A study in the significance of grain size parameters. *J. Sediment. Petrol.* **1957**, *27*, 3–26. [CrossRef]
39. Graham, M.C.; Eaves, M.A.; Farmer, J.G.; Dobson, J.; Fallick, A.E. A study of carbon and nitrogen stable isotope and elemental ratios as potential indicators of source and fate of organic matter in sediments of the Forth Estuary, Scotland. *Estuar. Coast. Shelf Sci.* **2001**, *52*, 375–380. [CrossRef]
40. Middelburg, J.J.; Nieuwenhuize, J. Carbon and nitrogen stable isotopes in suspended matter and sediments from the Schelde. *Estuary Mar. Chem.* **1998**, *60*, 217–225.
41. Ogrinc, N.; Fontolan, G.; Faganeli, J.; Covelli, S. Carbon and nitrogen isotope compositions of organic matter in coastal marine sediments (the Gulf of Trieste, N Adriatic Sea): Indicators of sources and preservation. *Mar. Chem.* **2005**, *95*, 163–181. [CrossRef]
42. Lamb, A.L.; Wilson, G.P.; Leng, M.J. A review of coastal palaeoclimate and relative sea-level reconstructions using delta C-13 and C/N ratios in organic material. *Earth Sci. Rev.* **2006**, *75*, 29–57. [CrossRef]
43. Gõni, M.A.; Gardner, L.R. Seasonal dynamics in dissolved organic carbon concentrations in a coastal water-table aquifer at the forest-marsh interface. *Aquat. Geochem.* **2003**, *9*, 209–232. [CrossRef]
44. Garcia-Garcia, A.; Orange, D.; Lorenson, T.; Radakovitch, O.; Tesi, T.; Miserocchi, S.; Berne, S.; Friend, P.L.; Nittrouer, C.; Normand, A. Shallow gas off the Rhone prodelta, Gulf of Lions. *Mar. Geol.* **2006**, *234*, 215–231. [CrossRef]
45. Perdue, E.M.; Koprivnjak, J.F. Using the C/N ratio to estimate terrigenous inputs of organic matter to aquatic environments. *Estuar. Coast. Shelf Sci.* **2007**, *73*, 65–72. [CrossRef]
46. Robbins, J.A. Geochemical and geophysical application of radioactive lead. In *The Biogeochemistry of Lead in the Environment*; Nriagu, J.O., Ed.; Elsevier: Amsterdam, The Netherlands, 1978; pp. 285–393.
47. Smith, J.N. Why should we believe ^{210}Pb sediment geochronologies? *J. Environ. Radioactiv.* **2001**, *55*, 121–123. [CrossRef]
48. Kirchner, G. ^{210}Pb as a tool for establishing sediment chronologies: Examples of potentials and limitations of conventional dating models. *J. Environ. Radioact.* **2011**, *102*, 490–494. [CrossRef]
49. Bellucci, L.G.; Frignani, M.; Lin, S.; Muntau, H. Accumulation and metal fluxes in the central Venice Lagoon during the last century. *Chem. Ecol.* **2005**, *21*, 425–439. [CrossRef]
50. Vilibic, I.; Book, J.W.; Paklar, G.B.; Orlic, M.; Dadic, V.; Tudor, M.; Martin, P.J.; Pasaric, M.; Grbec, B.; Matic, F.; et al. West Adriatic coastal water excursions into the East Adriatic. *J. Mar. Syst.* **2009**, *78*, S132–S156. [CrossRef]
51. Poulain, P.M. Adriatic Sea surface circulation as derived from drifter data between 1990 and 1999. *J. Mar. Syst.* **2001**, *29*, 3–32. [CrossRef]
52. Poulain, P.M. Tidal currents in the Adriatic as measured by surface drifters. *J. Geophys. Res. Oceans* **2013**, *118*, 1434–1444. [CrossRef]
53. UNEP/MAP. Riverine transport of sediments and pollutants to the Mediterranean Sea. *MAP Tech. Ser.* **2003**, *141*, 1–111.
54. Matic, F. Air-Sea Interaction in Split, Brac and Neretva Channel. Master's Thesis, University of Zagreb, Zagreb, Croatia, 2005.

55. Vidjak, O.; Bojanic, N.; Kuspilic, G.; Gladan, Z.N.; Ticina, V. Zooplankton community and hydrographical properties of the Neretva channel (eastern Adriatic Sea). *Helgol. Mar. Res.* **2007**, *61*, 267–282. [CrossRef]
56. Amorosi, A.; Sammartino, I. Influence of sediment provenance on background values of potentially toxic metals from near-surface sediments of Po coastal plain (Italy). *Int. J. Earth Sci.* **2007**, *96*, 389–396. [CrossRef]
57. Miko, S.; Peh, Z.; Parika, M.A. *Geokemijski atlas zapadne Hrvatske*; Institut za Geologska Istrazivanja Fond strucne dok: Zagreb, Croatia, 2001; p. 22.
58. Mugnai, C.; Bertolotto, R.M.; Gaino, F.; Tiberiade, C.; Bellucci, L.G.; Giuliani, S.; Romano, S.; Frignani, M.; Albertazzi, S.; Galazzo, D. History and Trends of Sediment Contamination by Heavy Metals Within and Close to a Marine Area of National Interest: The Ligurian Sea off Cogoleto-Stoppani (Genoa, Italy). *Water Air Soil Pollut.* **2010**, *211*, 69–77. [CrossRef]
59. Cabral Pinto, M.M.S.; Silva, M.M.V.G.; Ferreira da Silva, E.A.; Dinis, P.A.; Rocha, F. Transfer processes of potentially toxic elements (PTE) from rocks to soils and the origin of PTE in soils: A case study on the island of Santiago (Cape Verde). *J. Geochem. Explor.* **2017**, *183*, 140–151. [CrossRef]
60. Long, E.R.; Macdonald, D.D.; Smith, S.L.; Calder, F.D. Incidence of Adverse Biological Effects within Ranges of Chemical Concentrations in Marine and Estuarine Sediments. *Environ. Manag.* **1995**, *19*, 81–97. [CrossRef]
61. Long, E.R.; Field, L.J.; MacDonald, D.D. Predicting toxicity in marine sediments with numerical sediment quality guidelines. *Environ. Toxicol. Chem.* **1998**, *17*, 714–727. [CrossRef]
62. Grimwood, M.J.; Dixon, E. Assessment of Risks Posed by List II Metals to Sensitive Marine Areas (SMAs) and Adequacy of Existing Environmental Quality Standards (EQSs) for SMA Protection. *WRc Rep. Engl. Nat.* **1997**.
63. APAT; ICRAM. *Manuale per la Movimentazione di Sedimenti Marini*; Corsini, S., Onorati, F., Pellegrini, D., Eds.; Ministero per l'ambiente e la tutela del territorio e del mare: Rome, Italy, 2006; p. 72.
64. Förstner, U.; Wittmann, G.T.W. *Metal Pollution in the Aquatic Environment*, 2nd ed.; Springer: Berlin/Heidelberg, Germany, 1981; p. 486.

applied sciences

MDPI

Article

Impact of Dredged Material Disposal on Heavy Metal Concentrations and Benthic Communities in Huangmao Island Marine Dumping Area near Pearl River Estuary

Wei Tao [1,†], Zhongchen Jiang [1,†], Xiaojuan Peng [1], Zhenxiong Yang [1], Weixu Cai [1], Huili Yu [2] and Jianjun Ye [1,*]

1 South China Sea Environment Monitoring Center, State Oceanic Administration (SOA), Guangzhou 510300, China; twei527@163.com (W.T.); jcc1967@163.com (Z.J.); xiaojuanpeng@163.com (X.P.); snrdfdcw@163.com (Z.Y.); scsemc@163.com (W.C.)

2 National Center of Ocean Standards and Metrology, Tianjin 300112, China; yuhuili2003@sina.com

* Correspondence: jjye@live.com; Tel.: +86-134-8025-9901

† These authors contributed equally to this work.

Abstract: The Huangmao Island dumping area is adjacent to the Pearl River Estuary in the South China Sea. From its first dumping activity in 1986 to 2017, 6750×10^4 m^3 dredged materials were dumped in this dumping area. Sediment pollution levels, ecological risk, and benthic communities in 2011–2017 were evaluated; the results showed that the concentrations of the heavy metals (HMs; except Hg) in surface sediments of the dumping area met the class I standard of marine sediment quality (GB 18668-2002). HMs in the surface sediments were relatively high in the northern and central areas but relatively low in the south of the dumping area. Speculation was that the spatial variation in HM concentrations might be caused by dumping activities. The Nemerow index implied that the contaminated area was mainly in the north of the dumping area (S1, S2, and S3), where the dumping amount was the largest. The potential ecological risk (E^i_r) indices of Zn, As, Cu, and Pb indicate that these metals posed a low risk to the ecosystem of the dumping area, whereas Cd and Hg posed a high risk at some stations. The geoaccumulation indices (I_{geo}) of Zn, As, Cu, and Pb specified no pollution or light pollution in the study area, whereas those of Cd and Hg in most years indicated mild contamination levels. Benthic organisms in the study area were arthropods, chordates, annelids, mollusks, echinoderms, nemertinean, coelenterate, and echiuran, among which arthropods were the most abundant. The abundance of taxa and density of benthic organisms had a little difference among the stations within the dumping area, but were significantly lower than those of the stations outside the dumping area. In addition, non-metric multidimensional scaling analysis confirmed that the observed patterns separated the stations within the dumping area from stations outside the dumping area. The evaluation results of the HMs revealed that the dumping area with a large dumping amount was more severely polluted. Dumping dredged materials seemed to have a negative impact on the benthic community in the dumping area.

Keywords: marine dumping area; heavy metal; dredged material; benthic community; Huangmao Island

Citation: Tao, W.; Jiang, Z.; Peng, X.; Yang, Z.; Cai, W.; Yu, H.; Ye, J. Impact of Dredged Material Disposal on Heavy Metal Concentrations and Benthic Communities in Huangmao Island Marine Dumping Area near Pearl River Estuary. *Appl. Sci.* **2021**, *11*, 9412. https://doi.org/10.3390/app11209412

Academic Editor: Mauro Marini

Received: 3 September 2021
Accepted: 29 September 2021
Published: 11 October 2021

1. Introduction

In urbanization and the rapid development of the marine transportation industry and coastal engineering construction projects, many dredged materials are produced [1,2]. Many disposal methods are available for dredged materials, but land disposal is a priority, such as reclaiming land from the sea and making solidified materials [3]. For cost and complexity reasons, another common disposal method for dredged materials is marine dumping [4]. From 2011 to 2017, the volume of dumping materials in China was estimated to be 11.2×10^8 m^3 [5]. Dredging and disposal are serious environmental concerns in coastal management [6,7].

Dredged materials mainly originate from harbors and channels. These areas are characterized by low hydrodynamic force, poor automatic purification capacity, low dissolved oxygen levels, intensive human activities, and high HM contents [8]. HMs cannot be biodegraded, accumulate rapidly, and reach a toxic level within a short time. During dumping, these toxic, persistent, and bioaccumulative HMs enter sediments through the decomposition, sedimentation, absorption, and formation of complexes and eventually harm benthic ecosystems [9,10]. Meanwhile, their removal is difficult and sometimes impossible [11,12]. Therefore, sediments are usually considered indicators of HM pollution [13].

Dumping dredged materials can cause the resuspension of marine sediments [14], resulting in the release of HMs from sediments into overlying seawater. In addition, this practice leads to an increase in seawater turbidity and a decrease in seawater depth, which interferes with the respiration and feeding of marine organisms and affects the regional hydrodynamic force [3]. Studies have indicated that dumping can have various impacts, ranging from obvious and long-term [4,15] to unobvious and short-term damage in the benthic community [16,17]. Harvey et al. [4] and Roberts et al. [18] suggested that the diversity of communities in the dumping area might be reduced, dominant patterns within the community might be altered, the abundance of some species might decrease, and the abundance of opportunistic species might increase.

The Huangmao Island marine dumping area is near the Pearl River Estuary in the South China Sea. The dumping area began to receive dumping materials in 1986. As of 2017, this dumping area had received 6750×10^4 m^3 of dredged materials. Dumping materials in the Huangmao Island marine dumping area were mainly dredged materials from wharves, harbor basins, and channels: Most were cleaning dredged materials (Class I standard for dredged materials for dumping in dumping area) [19], a small amount was contaminated dredged materials (Class II), and none were polluted dredged materials (Class III).

To alleviate the marine environment during the use of the dumping area, monitoring and evaluating the dumping area is necessary, especially in terms of sediments and benthos. A few studies have been carried out to evaluate the effect of dumping activities to marine sediments and benthic community [3,17,20,21], but few studies have been conducted in the South China Sea, where there are 25 dumping areas as of 2017. Among these dumping areas, the Huangmao Island dumping area is the earliest to received dredged materials. Therefore, it is highly necessary to perform assessment of the dredged material disposal in this dumping area. Based on years of monitoring of the Huangmao Island marine dumping area, an evaluation of the pollution status of sediments in the Huangmao Island marine dumping area was carried out by the geoaccumulation index and ecological risk index. The effects of dumping on the benthic community were investigated using non-metric multidimensional scaling analysis (n-MDS). According to the research results, suggestions and measures are presented to provide technical support for the sustainable use of the dumping area.

2. Materials and Methods

2.1. Description of the Study Area

The Huangmao Island marine dumping area is adjacent to the Pearl River Estuary, southeast of Hengqin Island, south of Huangmao Island, and northwest of the Wanshan Islands (Figure 1). The geographical coordinates are 113°38′30″–113°40′30″ E and 21°58′00″–22°01′00″ N, and the dumping area is approximately 20.6 km^2. The Huangmao Island marine dumping area was approved in 1986. The dumping volume in the early stage was small but increased significantly since 2011. From 2011 to 2017, a total volume of 4762×10^4 m^3 of dredged materials was dumped in this area. The dumping volume was the largest (1607×10^4 m^3) in 2016 and the smallest (134×10^4 m^3) in 2017. The sediment type in the dumping area was mainly silt, followed by sand and clay. The water depth of the dumping area gradually increased from the northwest to the southeast. In 2011, the water depth of the dumping area ranged from 8.2 to 14.4 m, with an average of 12.2 m. In 2017, the water depth ranged from 9.4 to 14.5 m, and the average value decreased by 0.3 m, of which the maximum decrease (1.2 m) was in the northwest of the dumping area.

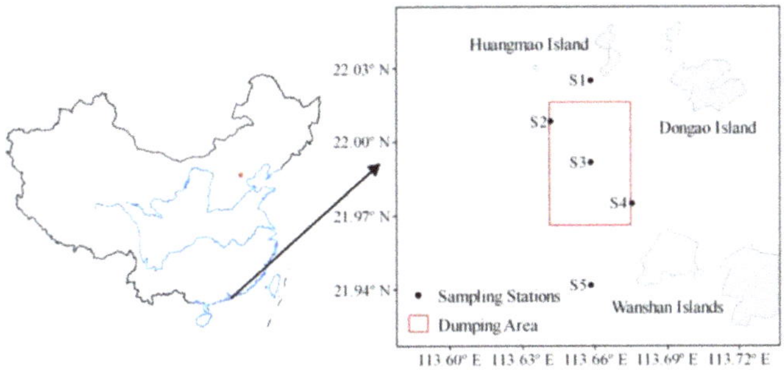

Figure 1. Sampling stations at the dumping area and adjacent sea areas.

2.2. Sampling

Surface sediment samples (0–5 cm) were collected from five sampling stations in October each year from 2011 to 2017. Because only a small amount of data was collected in 2015, the situation in 2015 was not included in this study. In addition, a benthic community investigation was conducted in the study area in 2017. Sediments were collected 2–3 times at each site by a grab sampler and then mixed and packed into pre-cleaned glass jars and frozen at −20 °C until further treatment. Benthos samples were collected using qualitative and quantitative methods. Quantitative samples of benthos were collected using a grab sampler. The samples were washed with seawater through a sieve with a diameter of 1 mm, and all biological samples were collected and transferred to a plastic container. Qualitative samples were collected with a 1.0 m wide Agassiz trawl. After dragging slowly for 10 min (approximately 200 m) at each station, all benthic organisms from the trawl were collected. Samples were bottled, numbered, and fixed with a 5% neutral formalin solution. Species were identified, counted, and weighed in the laboratory.

2.3. Laboratory Analysis

Before chemical analysis, sediments were freeze-dried to a constant weight. After removing the gravel and shells, the dried sediments were ground and sieved through a 96 μm stainless-steel sieve. Sediment samples (0.2000 g) for the measurement of Hg and As were digested with 10.0 mL of a mixture of acid (HNO_3+HCl). Sediment samples (0.1000 g) for the measurement of Cu, Zn, Cd, and Pb were digested with a mixture of concentrated HNO_3 (1.0 mL) and $HClO_4$ (2.0 mL). HMs (Cu, Pb, Zn, and Cd) in the sediment samples were determined using a flame atomic absorption spectrometer (Analytik Jenna ContrAA 700). As and Hg in the sediment samples were tested using an atomic fluorescence spectrometer (Beijing Haiguang AFS 9560). Sulfide and TOC in sediment samples were determined by methylene blue spectrophotometry and potassium dichromate volumetric method, respectively. Oils were analyzed using a UV spectrophotometer (Shimadzu UV-2450). Eh and pH values of sediments were measured with a potentiometer and a pH meter, respectively. Although As is a metalloid that exhibits intermediate properties between those of metals and non-metals, this text refers to it as a metal. Organisms were sorted, counted, and identified to species level.

Quality assurance and quality control were evaluated using duplicates, blanks, and standard reference material (GB W07333) from the National Research Center for Standard of China. The detection limits of Hg, As, Cu, Pb, Cd, and Zn were 0.002, 0.06, 0.1, 0.1, 0.02, and 0.2 mg/Kg, respectively. All chemicals used for the analysis were of analytical grade or above. Blanks and duplicates were run for each batch of 10 samples. The blank values are below the detection limit. From the values of the duplicates, the relative errors of Hg, As, Cu, Pb, Cd, and Zn were below 6.5%, 1.8%, 2.5%, 3.7%, 5.9%, and 2.1%, respectively. The measured values of the standard reference material were within the error allowed. The results of quality assurance and quality control indicate that the accuracy and precision are acceptable.

2.4. Pollution and Ecological Risk Assessment Methods

2.4.1. Nemerow Pollution Index (P_i)

The Nemerow pollution index (P_i) was used to evaluate the sediments, as follows [22]:

$$P_{ij} = C_{ij}/S_i$$

$$P_{ijave} = \frac{1}{\mathrm{m}} \sum_{i=1}^{\mathrm{m}} P_{ij}$$

$$P_i = \left\{ \left[\left(P_{ijmax}\right)^2 + \left(P_{ijave}\right)^2 \right] /2 \right\}^{1/2}$$

where P_{ij} is the standard index of the ith evaluation factor, C_{ij} is the measured concentration of the ith evaluation factor, S_i is the evaluation standard of the ith evaluation factor, m represents the number of evaluation factors, and P_{ijave} and P_{ijmax} refer to the average and maximum single factor pollution indices, respectively. The Nemerow pollution index was divided into five zones to describe pollution levels: unpolluted ($\leq$0.7), lightly polluted ($0.7 < P_i \leq 1$), mildly polluted ($1 < P_i \leq 2$), middle-level polluted ($2 < P_i \leq 3$), and seriously polluted ($P_i > 3$).

2.4.2. Geochemical Accumulation Index (I_{geo})

The pollution status of HMs in sediments was evaluated using the geochemical accumulation index (I_{geo}) as follows [23]:

$$I_{geo} = \log_2 \frac{c_i}{1.5 \times c_{Bi}}$$

where C_i is the measured concentration of element i in the sediment, and C_{Bi} refers to the geochemical background value of an element. The geochemical accumulation index is divided into six zones to describe pollution levels: clean ($I_{geo} < 0$), light pollution ($0 \leq I_{geo} < 1$), mild contamination ($1 \leq I_{geo} < 2$), middle-level pollution ($2 \leq I_{geo} < 3$), and serious contamination ($I_{geo} \geq 3$).

2.4.3. Integrated Potential Ecological Risk Index (*RI*)

The ecological risk was evaluated using the potential ecological hazard index (*RI*) as follows [24]:

$$E_r^i = T_r^i \times C_f^i = T_r^i \times \frac{C_i}{C_{Bi}}$$
$$RI = \sum_{i=1}^{n} E_r^i$$

where C_f^i is the accumulation factor of metal i, expressed as $C_f^i = \frac{C_i}{C_{Bi}}$; C_i is the concentration of metal i in the sample; C_{Bi} is the geochemical background value of metal i in the sediments; and $T_r^{\,i}$ is the toxicity coefficient of metal i. $E^i{}_r$ is the individual ecological risk of metal i, and *RI* represents the potential ecological risk caused by overall contamination, which is the sum of all risk coefficients for metals.

Based on $E^i{}_r$ value, the potential risk is classified into five categories: low risk ($E^i{}_r \leq 40$), middle risk ($40 < E^i{}_r \leq 80$), relatively high risk ($80 < E^i{}_r \leq 160$), high risk ($160 < E^i{}_r \leq 320$), and extra-high risk ($E^i{}_r > 320$). *RI* is classified into four categories to describe integrated potential ecological risk: low risk ($RI < 150$), middle risk ($150 \leq RI < 300$), relatively high risk ($300 \leq RI < 600$), and high risk ($RI > 600$).

2.5. Statistical Analysis

SPSS 25 was used to evaluate the correlation between HMs and environmental factors. Multivariate analysis was performed after the fourth root transformation of the abundance data from each sampling station. Outputs from n-MDS ordination models of the Bray–Curtis similarity matrix were obtained. For benthos data, n-MDS was conducted based on the abundance of taxa, density, and diversity of each sampling station. Multivariate analysis was conducted using PRIMER V5 software [25].

3. Results and Discussion

3.1. HMs in the Sediments

The concentrations of Hg, As, Cu, Pb, Cd, and Zn in surface sediments of the dumping area were 0.030–0.26, 7.90–19.8, 13.1–36.3, 17.4–54.7, 0.11–0.34, and 50.0–122 mg/kg, respectively, with an average of 0.080, 15.1, 24.5, 30.8, 0.21, and 97.6 mg/kg, respectively (Table 1).

Table 1. The concentrations of HMs in the dumping area (unit: mg/kg).

	Hg	As	Cu	Pb	Cd	Zn	References
2011	0.050–0.090	14.2–17.4	20.2–31.5	29.7–32.3	0.15–0.21	108–117	This study
2012	0.056–0.14	16.4–19.6	21.1–29.6	23.0–54.7	0.16–0.27	97.4–122	
2013	0.054–0.086	16.1–17.8	19.8–36.3	19.9–36.0	0.16–0.34	66.1–116	
2014	0.052–0.12	14.2–17.9	20.3–26.7	21.6–33.4	0.16–0.28	62.2–122	
2016	0.053–0.26	14.2–19.8	21.8–25.4	25.4–32.8	0.22–0.32	88.6–119	
2017	0.030–0.050	7.90–10.9	13.1–21.7	17.4–35.6	0.11–0.24	50.0–83.0	
Average [a] ± standard deviation	0.080 ± 0.041	15.1 ± 3.20	24.5 ± 4.40	30.8 ± 7.70	0.21 ± 0.061	97.6 ± 20.5	
Coefficient of variation (CV%) [b]	0.51	0.21	0.18	0.25	0.29	0.21	
Class I [c]	0.2	20	35	60	0.5	150	
Background	0.03	7.7	15	20	0.07	65	[26]
TEL [d]	0.13	7.24	18.7	30.2	0.68	124	[27]
PEL [e]	0.7	41.6	108	112	4.21	271	[27]
ERL [f]	0.15	8.2	34	46.7	1.2	150	[28]
ERM [j]	0.71	70	270	218	9.6	410	[28]

[a] Average value of all the samples over six years. [b] CV%=average/standard deviation. [c] Standard of the Marine Sediment Quality Standard (MSQS) of China (GB 18668-2002). [d] Threshold effect level. [e] Probable effect level. [f] Effect range low from the National Oceanic and Atmospheric Administration (NOAA). [j] Effect range medium from the National Oceanic and Atmospheric Administration (NOAA).

If the concentrations of pollutants in the sediments are lower than the threshold effect level (TEL), adverse biological effects are expected to rarely occur. If the concentrations of pollutants in the sediments are higher than the probable effect level (PEL), adverse effects are expected to frequently occur. If the concentrations of pollutants in the sediment are between the TEL and PEL, the probability of adverse biological effects is close to that of no adverse biological effects. In this study, the concentration of As in the sediments was higher than the TEL in each year, that of Cu was higher than the TEL except in 2013 and 2017, and that of Pb was lower than the TEL except in 2011 and 2012. The concentrations of Hg, Zn, and Cd were lower than the TEL. The concentrations of all HMs were lower than the PEL. The HMs in sediments from the dumping area may have adverse biological effects. The effect range low (ERL) and effect range medium (ERM) were proposed by the National Oceanic and Atmospheric Administration [27]. The results showed that the concentration of As was higher than ERL, and that of other HMs was lower than ERL and ERM.

The concentrations of HMs in the surface sediments of all stations exceeded the background values of coastal sediments [26]. Hg, As, Cu, Pb, Cd, and Zn were approximately 2.7-, 1.9-, 1.6-, 1.5-, 3-, and 1.5-times higher than the coastal background values, respectively, indicating that the surface sediments of the dumping area have been slightly polluted over time. Although the concentrations of HMs in the sediments of the dumping area were higher than the background value, all the HMs except Hg fulfilled the class I standard of the Marine Sediment Quality Standard (MSQS) of China (GB 18668-2002). The quality of sediments in the marine dumping area was required to fulfill the class III standard of the MSQS. Therefore, the HM pollution was within an acceptable range. However, compliance with the sediment quality standards does not mean that HMs have no harm or risk, because the concentrations of HMs cannot fully reflect the pollution state [29], especially, the bioaccumulation and amplification of HMs in marine organisms pose potential health risks to human beings [30,31].

The coefficient of variation (CV%) was used to explain the changes in HM concentrations in different years. It reflects the average degree of variation of each sampling station in the samples. If the variation is greater than 0.5, the spatial distribution of HM concentrations in the sediments is uneven, and local point source pollution may exist [32]. The CV% of Hg in the sediments of the Huangmao Island dumping area was greater than 0.5, indicating that Hg entering the dumping area was mainly sourced from external input, that is, dumping. The coefficients of variation of Pb, Cu, As, and Zn were all smaller than 0.5, indicating that these HMs were relatively stable. There is a dynamic process between the release and absorption of HMs in marine sediments. In this process, different physicochemical characteristics (e.g., pH, redox potential, temperature, particle size, and salinity), ocean current, and dumping intensity will cause changes in the CV% values of HMs in sediments [33].

Compared with other dumping areas, HM concentrations in the Huangmao Island dumping area were generally at an intermediate level (Table 2). For individual HMs, the Hg concentration was close to the levels recorded in the Jinzhou Port Dumping Area [34] and Marine Dumping Area outside Jiaozhou Bay [35]. The As concentration was close to the levels recorded in the Hulu Island Port Dumping Area [34] and Tianjin Dumping Area [34]. The Zn concentration was close to the levels recorded in the Hulu Island Port Dumping Area [34], Fangchenggang Dumping Area [36], and Dumping Site at the Taiwan Shelf [37]. The Cd concentration was close to the levels recorded in the Jinzhou Port Dumping Area [34], Huangye Port Dumping [34], Lanshan Port Temporary Marine Dumping Area [38], and Marine Dumping Area outside Jiaozhou Bay [35]. Last, the Pb concentration was similar to those recorded in the Huangye Port Dumping Area [34], Laizhou Port Dumping Area [34], and Fangchenggang Dumping Area [36].

Table 2. The concentrations of HMs in sediments from other dumping areas.

Locate	Hg	As	Cu	Pb	Cd	Zn	Reference
	mg/kg						
Huangmao Island Dumping Area	0.03–0.26	7.9–19.8	13.1–36.3	17.4–54.7	0.10–0.34	50–122	This study
Hulu Island Port Dumping Area	0.31–0.36	6.07–10.5	15.6–20.7	12–18.5	0.17–0.63	63.1–126	[34]
Jinzhou Port Dumping Area	0.10–0.26	4.97–8.62	0.15–0.45	10.7–17.5	0.15–0.45	22.8–234	
Tianjin Dumping Area	0.016–0.023	16–25	0.112–0.149	20.5–23.4	0.112–0.149	53.9–66.5	
Huangye Port Dumping Area	0.016–0.019	50.9–73.1	0.167–0.23	17.2–23.7	0.156–0.201	6.37–7.71	
Laizhou Port Dumping Area	0.015–0.061	6.91–10.71	0.22–0.24	19.1–23.5	0.22–0.24	58.6–64.2	
Marine Dumping Area outside Jiaozhou Bay	0.038–0.082	NA	2.79–3.84	0.99–1.32	0.09–0.13	3.73–16.16	[35]
Fangchenggang Dumping Area	0.063–0.071	9.02–9.89	13.8–16.7	26.2–32.1	0.02–0.11	31.7–79.5	[36]
Dumping Site at Taiwan Shelf	NA	NA	12.5–15.6	15.6–18.5	NA	101.3–124.7	[37]
Lanshan Port Temporary Marine Dumping Area	0.038–0.082	NA	10.32–13.67	7–9.15	0.15–0.17	22.59–28.15	[38]

3.2. *Temporal and Spatial Distribution Characteristics of HMs*

Changes in the concentrations of HMs in the surface sediments in different years are shown in Figure 2. The highest concentrations of Hg, As, Cu, Pb, Cd, and Zn were observed in 2016, 2012, 2013, 2012, 2016, and 2011, respectively. In 2017, the concentrations of all HMs were the lowest, which might be related to the significant reduction in the dumping amount in 2017. From 2011 to 2016, the concentrations of As, Cu, and Pb did not change significantly, that of Zn first decreased and then increased, and those of Hg and Cd increased. Changes in HM concentrations in sediments were affected by multiple factors, such as the pollution level of dumped dredged materials, dumping volume, and hydrodynamic characteristics of the dumping area.

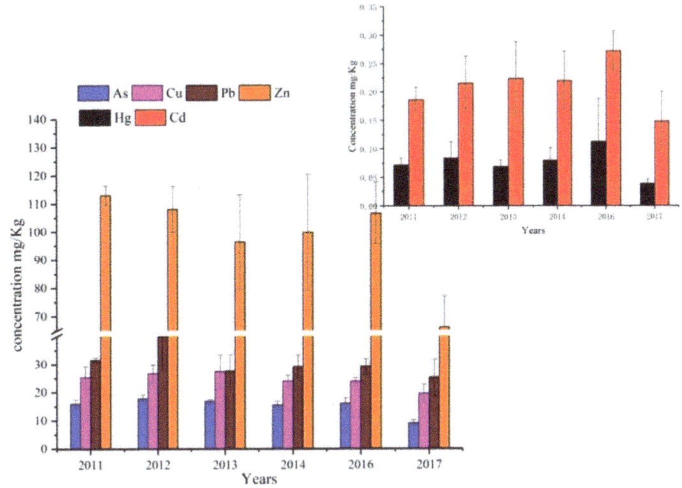

Figure 2. Average concentrations of HMs in surface sediments in different years.

Spatial distribution profiles of the HMs in the sediments from the study area are shown in Figure 3. Concentrations of HMs in surface sediments generally showed a high trend in the north and central areas of the dumping area and a low trend in the south of the dumping area. The concentration of Cd was higher in the dumping area but lower outside the dumping area; it reached its highest value at station S3 (in the center of the dumping area) and the lowest at station S5 (in the south outside the dumping area). The concentration of Pb was higher in the central and south of the dumping area but lower in the northwest; it reached its highest value at station S3 and the lowest at station S2 (at the edge of the dumping area). The concentrations of Hg and As gradually decreased from northwest to southeast, reaching the highest at station S2 and the lowest at station S5. The concentration of Cu gradually decreased from north to south, reaching its highest at station S1 and lowest at station S4. The concentration of Zn gradually decreased from north to south in the dumping area, reaching its highest at station S5 and lowest at station S2.

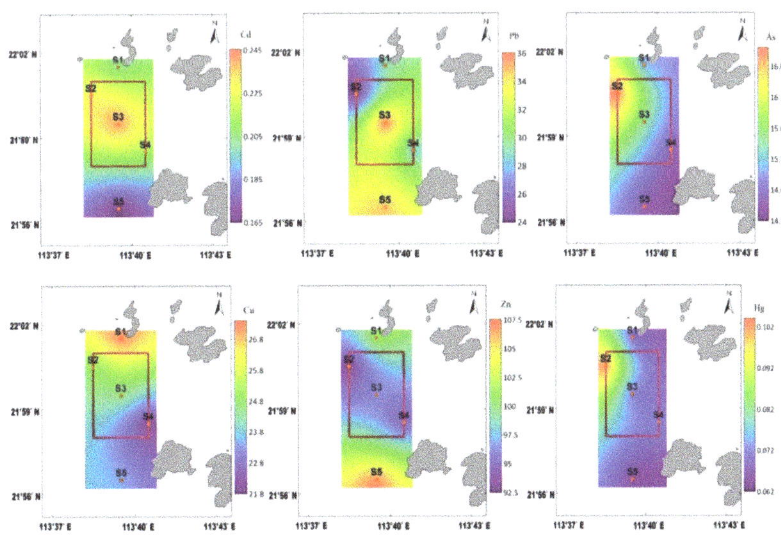

Figure 3. Spatial distribution of HMs in sediments.

3.3. Correlation Analysis between HMs and Other Environmental Factors

The concentrations of HMs in surface sediments are affected by various conditions, including environmental background, marine physical and chemical properties, biological effects, and human activities [39]. Analyzing the correlation between HMs and environmental factors (e.g., oils, TOC, sulfides) is helpful in determining the possible sources of HMs [40].

In this study, the data followed a normal distribution, and the Pearson coefficient analysis was performed using SPSS 25 software. In Table 3, organic carbon had a significant positive correlation with Cu and Zn ($p < 0.01$), indicating that Cu and Zn were closely related to organic matter, which was a main carrier. Additionally, organic carbon and HMs easily form organic matter complexes. Therefore, organic carbon plays an important role in the distribution of HMs. Among the six HMs, Zn was significantly correlated with Cu ($p< 0.01$) and moderately correlated with As and Pb ($p < 0.05$). Hg and Cd were significantly correlated with As but not with the other three HMs, implying that Hg and Cd might have sources similar to As, which differed from those of other metals. Oils and sulfides were not correlated with the concentrations of the six HMs, indicating that they had no effect on HMs in the dumping area.

Table 3. Correlation analysis for HMs, TOC, Oils, and sulfides in sediments.

	TOC	Hg	As	Cu	Pb	Cd	Zn	Oils	Sulfides
TOC	1								
Hg	0.139	1							
As	0.493	0.657 **	1						
Cu	0.663 **	0.245	0.484	1					
Pb	0.438	0.06	0.103	0.44	1				
Cd	0.017	0.419	0.723 **	0.179	0.187	1			
Zn	0.661 **	0.269	0.549 *	0.766 **	0.515 *	0.254	1		
Oils	0.363	−0.016	0.16	0.164	0.45	0.243	0.451	1	
Sulfides	0.213	0.017	−0.249	0.052	−0.002	−0.556 *	−0.164	0.012	1

* Correlation is significant at the 0.05 level (two-tailed). ** Correlation is significant at the 0.01 level (two-tailed).

3.4. Pollution Assessment of HMs in Sediments

Pollution assessments for HMs in sediments have been conducted widely based on several indices, such as the Nemerow pollution index, $E^i{}_r$, RI, and I_{geo} [22,41,42].

The Nemerow index was used to evaluate the pollution of HMs in the sediments in the dumping area over different years (Figure 4). In 2012, 2013, and 2016, the Nemerow index exceeded 0.7 but was less than 1. This finding indicated that the sediments in the dumping area had reached light pollution levels in 2012, 2013, and 2016. Sediments in other years were at a clean level. The pollution level in 2016 was slightly higher, which might be related to larger quantity of dumping materials in this year. According to the pollution distribution, the polluted areas were mainly in the northwest (S1 and S2) and central areas (S3) of the dumping area. Additionally, the change in water depth showed that these stations were areas with large dumping quantities, indicating that dumping might have a negative impact on sediments.

The I_{geo} values of As, Cu, Zn, and Pb ranged from 0 to 1 from 2011 to 2016 (Figure 5), indicating that the pollution levels of these four HMs were light pollution. Notably, in 2017, the I_{geo} values of As, Cu, Zn, and Pb were lower than 0, implying that the pollution levels of these four metals were not polluted in 2017, which may be related to the least dumping amount in this year. In most years, the I_{geo} values of Hg and Cd were between 1.0 and 2.0, indicating that Hg and Cd had reached a mild contamination level. From 2011 to 2016, the pollution levels of Hg and Cd showed an increasing trend and were higher than those of the other four HMs. Even in 2017, when the dumping amount was the lowest, the pollution levels of Hg and Cd were rated as light pollution, and the pollution from the other four HMs was not observed. Therefore, special attention should be paid to Hg and Cd before dumping dredged materials.

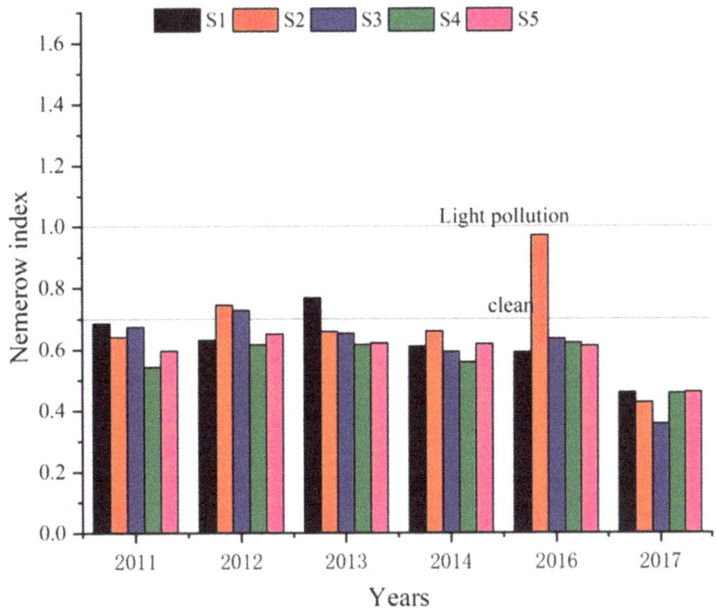

Figure 4. The Nemerow pollution index of HMs in different years.

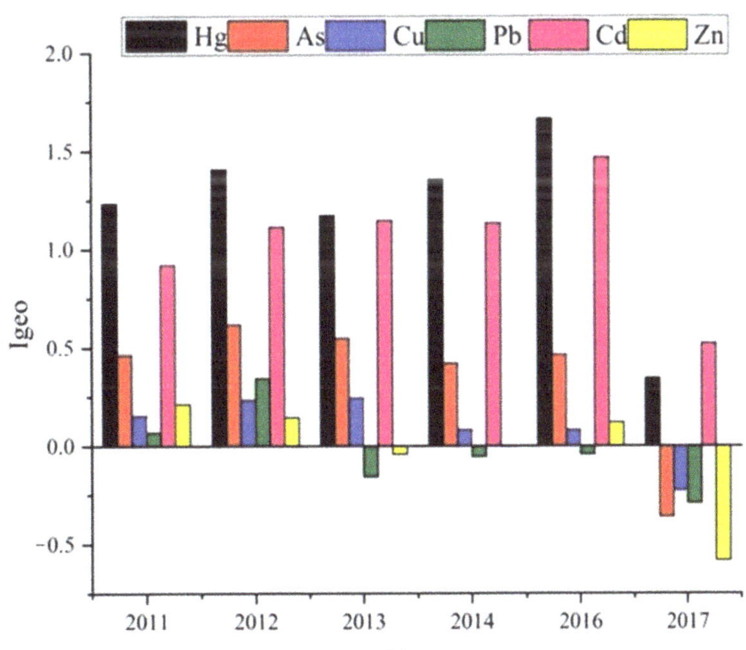

Figure 5. The I_{geo} values of HMs in different years.

Figure 6 shows the integrated potential ecological risk index (*RI*) and potential ecological risk index (E^i_r) of an individual HM. The *RI* values of HMs in the sediments of the dumping area were higher than 150 in 2011, 2013, 2014, and 2017, suggesting that HMs had a middle ecological risk. The *RI* values in 2012 and 2016 were 310 and 388, respectively, indicating that the ecological risk of sediments was relatively high. The *RI* values of the dumping area were greater than 150 every year, implying that the potential ecological risk of the dumping area must be paid more attention.

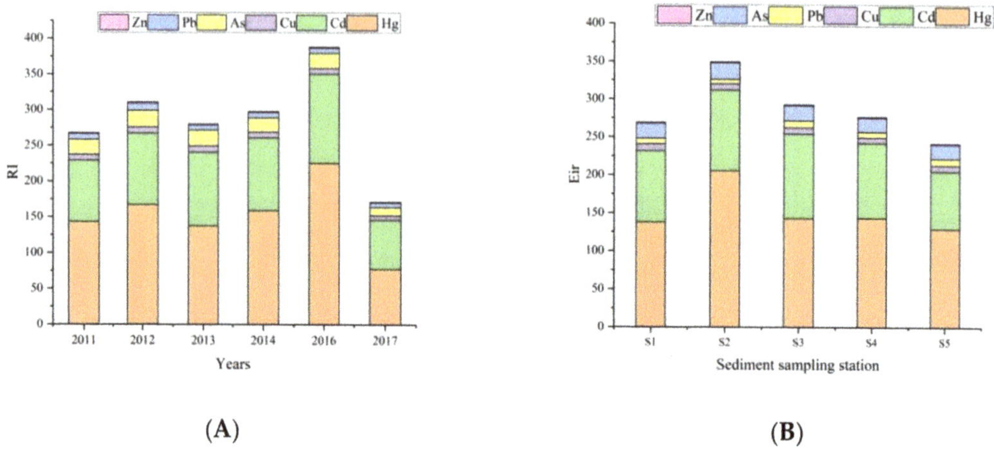

Figure 6. *RI* of HMs in different years (**A**) and E^i_r of HMs in different stations (**B**).

For individual metals, Hg and Cd posed more severe ecological risk than the other HMs, with E^i_r values higher than 40 in each year from 2011 to 2017, especially in 2016, a high risk level ($E^i_r \geq 160$). The E^i_r values of other HMs were less than 40, indicating that the potential ecological risk of other HMs was low. In general, Hg and Cd were the main factors causing the potential ecological risks of sediments in the study area.

At present, extensive ecological risk assessments have been conducted on HMs in sediments based on *RI* [31,43]. In these studies, contamination levels or risk contributions of HMs varied across study areas. For example, Cd was the largest contributor to the ecological risk of sediments [43]. Similar to the results of this study, Lao et al. [31] observed that the ecological risk of Hg in the Beibu Gulf was relatively high, and Tang et al. [42] found that the ecological risk of Hg and Cd in HMs in Daya Bay was also high. Liu et al. [33] found that the ecological risk of Hg in the sediments of the Maowei River aquaculture area was high. Hg and Cd had stronger toxic effects on organisms [44]. The toxicity coefficients ($T_r{}^i$) of Hg and Cd were 40 and 30, respectively, much higher than those of the other four HMs, which led to a higher potential risk in sediments of the dumping area. Hg and Cd have strong biological toxicity and high bioaccumulation potential, which seriously threaten marine ecosystems and human health. Therefore, to reduce the ecological risk caused by dumping, the monitoring of Hg and Cd in dredged materials should be strengthened before ocean dumping.

The E^i_r values of HMs at the different stations are shown in Figure 6. In terms of spatial distribution, the order of the potential ecological risk of each monitoring station was S2 > S3 > S4 > S1 > S5. Areas with high ecological risk were mostly located in the northwest of the dumping area, and the ecological risk in the south of the dumping area are low. In the north of the dumping area, the results of ecological risk are consistent with the concentration distribution and the Nemerow evaluation results. In areas (S1, S2, and S3) with frequent dumping, higher concentrations of HMs in sediments were detected, indicating that dumping activities had a significant impact on the marine ecological environment.

3.5. Characteristics of Benthic Community in the Dumping Area

Fifty-two species of benthos were identified and characterized by the presence of the following groups: arthropods, chordates, annelids, mollusks, echinoderms, nemertinean, coelenterate, and echiuran. Among these groups, arthropods were the most abundant, accounting for 38.5% of all benthos, and chordate, annelid, and mollusks accounted for 25.0%, 15.4%, and 13.5%, respectively. Benthic communities differ in their substrate requirements, and different species dwell indifferent ecosystems [45,46]. After investigating the benthos in Riga Bay in the eastern Baltic Sea, Pallo et al. [47] found that arthropods were more suitable for coarse sand sediment environments. In this study, the similar result was found. Sediments in the Huangmao Island dumping area were mainly silt and the most abundant benthos in this area was arthropods.

A general belief is that filter-feeding benthos are more likely to enrich HMs [48]. However, Quan et al. [49] found that arthropods can easily affect the biological community structure and biomass amount by adsorbing HMs from sediments.

The abundance of taxa, individual density of benthos, and diversity of the benthic fauna are shown in Figure 7. There was little difference in the abundance of taxa, density, and diversity between the central area (S3) and the northwest area (S1 and S2). The abundance of taxa and density of S4 in the southeast corner of the dumping area was low, and those of S5 located to the south of the dumping area were significantly higher than those at the other stations. The response of benthic communities to cumulative dumping was analyzed based on the abundance of taxa, density, and diversity of each station in the dumping area. The abundance of taxa, density, and diversity of benthos at each station were transformed by logarithmic conversion to create a Bray–Curtis similarity matrix and to conduct clustering and MDS sequencing. The results are presented in Figure 8. The benthic fauna community in the study area was divided into four groups at a similar level of 92.9%. S1 and S3 belong to group 1, in the central area and to the northern part of the dumping area. S2 belongs to group 2, in the northwest corner of the dumping area. Groups 1 and 2 had a high similarity. S4 belongs to group 3, in the southeast corner of the dumping area. S5 belongs to group 4, outside the dumping area. Stations in groups 1 and 2 were the areas with the largest dumping amount and higher ecological risk. Stations in groups 3 and 4 were the areas with less dumping amount than group 1 and 2, and pollution and ecological risk were also lower. The temporal and spatial distribution characteristics of benthic communities were highly related to the pollution levels in different dumping areas.

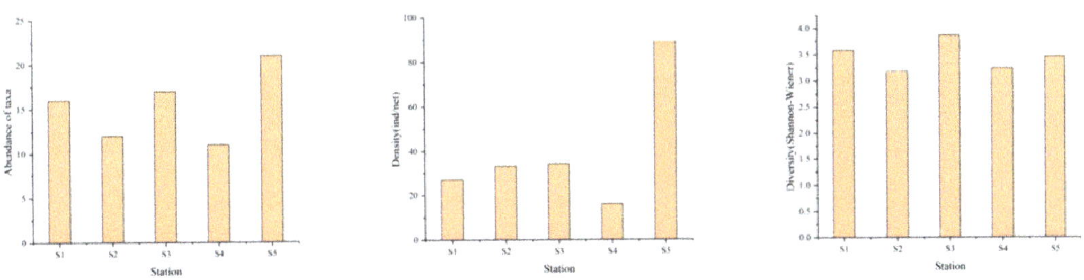

Figure 7. The abundance of taxa, individual density of benthos, and diversity for the benthic fauna (Shannon–Wiener index).

3.6. Influence of Dumping Dredged Materials and the Pollution Status

To minimize the impact of dumping on the marine ecological environment, the dumping distance, hydrodynamic conditions, water depth, and other factors were comprehensively considered during the selection of the location of the Huangmao Island dumping area. In the case of large amounts of dumping and complex types of dredged materials, benthos was affected even during short-term dumping [3]. HMs in dredged materials enter sediments in different ways, causing changes in the sediment environment and affecting the structure and composition of organisms. Jia et al. [50] found that the concentrations of

Cu, Pb, and Cd in sediments had a significant impact on the richness and equity of benthos. Li et al. [51] found that the correlation coefficient between the HM content in sediments and the community structure was the highest, and the content of HMs in sediments was the main environmental factor affecting the community structure of benthos in this area. The HMs in seawater can enter the sediment in different ways, which can change the sediment environment and affect the structure and composition of organisms.

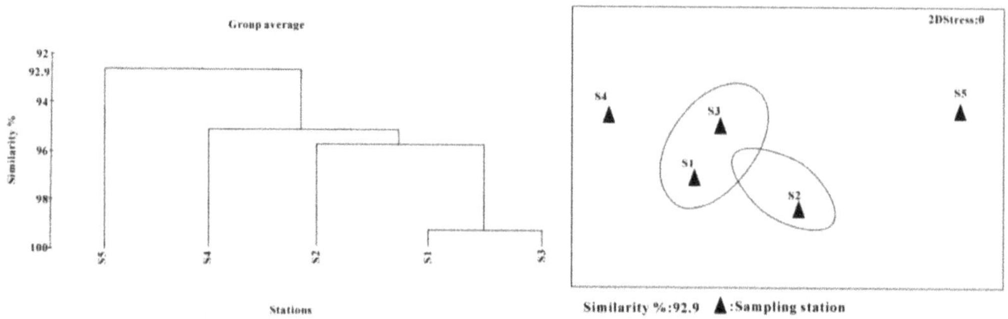

Figure 8. Cluster and n-MDS plots based on abundance of taxa, density, and diversity in each station.

The abundance of taxa, density, and diversity have been widely used in the impact assessment of benthos [52]. In this study, there were significant differences in the abundance of taxa and density between the areas inside and outside the dumping area. The abundance of taxa and density outside the dumping area were significantly higher than those inside the dumping area, indicating that dumping dredged materials might affect the benthos. The n-MDS analysis confirmed the observed patterns, separating the S5 stations from the others (S1–S4). Similar to this study, Fonseca et al. [48] also found that the area with more concentrated dumping had more severe pollution and lower richness of benthos. The Nemerow index showed that the central and northern parts of the Huangmao island dumping area (S1, S2, and S3) were polluted, and the *RI* values of HMs in this area were relatively high. Zhang et al. [38] investigated the sediments in the Lanshan Port temporary marine dumping area and found that the concentrations of HMs in the dumping area were significantly higher than those outside the dumping area, corroborating the results of this study.

The results of this study clearly indicate that the source of HMs in the dumping area are mainly from the dumped dredged materials, and dumping had affected the abundance of taxa and the density of benthos, especially in areas with a large dumping quantity and high dumping frequency. Since metal concentrations and evaluation of metal accumulation in benthos were not conducted in this study, the differences in the characteristics of the benthic community between the areas inside and outside the dumping area cannot be directly associated to HM content only. Community structure and composition can also be directly affected by the burial of dumping materials [53]. Other factors, such as dumping amount, dumping frequency, the type of dredged material, and toxicity of the dredged material also change the structure and composition of the benthic community.

4. Conclusions

The average concentrations of Hg, As, Cu, Pb, Cd, and Zn in surface sediments of the dumping area were 0.080, 15.1, 24.5, 30.8, 0.21, and 97.6 mg/kg, respectively. HMs in the surface sediments of the study area generally showed a trend of high in the northwest and central areas and low in the southeast of the dumping area. All HMs, except for Hg, met the class I standard for MSQS. The Nemerow index showed that sediments in the dumping area were at a light pollution level, and the polluted areas were mainly in the northwest and central areas with large dumping amounts. The values of I_{geo} showed that the dumping area was at a mild contamination level, and Hg and Cd were the major pollutants. The *RI* of the dumping area was relatively high. Hg and Cd are the main factors that cause the potential ecological risk of sediments. Areas with high potential ecological risk and high pollution levels were mainly those that undertake frequent and large dumping amounts of dredged materials.

Benthos identified in this study include arthropods, chordates, annelids, mollusks, echinoderms, nemertineans, coelenterates, and echiurans. Among these groups, arthropods were the most abundant, accounting for 38.5% of all benthos, and chordate, annelid, and mollusks accounted for 25.0%, 15.4%, and 13.5%, respectively. The abundance of taxa and density of benthic organisms showed little difference among the stations within the dumping area but were significantly lower than those of the stations outside the dumping area. It can be inferred that dumping dredged materials in the Huangmao Island marine dumping area had a negative impact on the benthos.

The assessment of HMs in the dumping area indicated that the pollutants in the dredged materials may have harmful effects on benthos. As, Hg, and Cd must be treated before dumping dredged materials. To study the impact of HMs more objectively in dumping dredged materials on marine organisms, metal concentration evaluation in benthic organisms and the toxicity test in sediments of dumping areas should be evaluated in further research. Additionally, the dumping area should be managed by zones to avoid intensive dumping in the northern part.

Author Contributions: Conceptualization, W.T. and Z.J.; methodology, W.T. and Z.J.; validation, X.P. and Z.Y.; investigation, Z.J.; data curation, H.Y. and J.Y.; writing—original draft preparation, W.T. and J.Y.; writing—review and editing, W.C. and W.T.; supervision, Z.J. and J.Y.; Funding acquisition, W.T. and J.Y. All authors have read and agreed to the published version of the manuscript.

Funding: This research was funded by State Oceanic Administration, grant number DOMEP-01-03.

Institutional Review Board Statement: Not applicable.

Informed Consent Statement: Not applicable.

Data Availability Statement: Data employed in this study will be available on request to the corresponding authors.

Acknowledgments: The authors would like to thank the technicians from South China Sea Environment Monitoring Center, China for sample collection and analysis.

Conflicts of Interest: The authors declare no conflict of interest.

References

1. Song, K.-H.; Choi, K.-Y.; Kim, C.-J.; Kim, Y.-I.; Chung, C.-S. Assessment of the Governance System for the Management of the East Sea-Jung Dumping Site, Korea through Analysis of Heavy Metal Concentrations in Bottom Sediments. *Ocean Sci. J.* **2015**, *50*, 721–740. [CrossRef]
2. Fatoki, O.S.; Mathabatha, S. An Assessment of Heavy Metal Pollution in the East London and Port Elizabeth Harbours. *Water Sa* **2001**, *27*, 233–240. [CrossRef]
3. Katsiaras, N.; Simboura, N.; Tsangaris, C.; Hatzianestis, I.; Pavlidou, A.; Kapsimalis, V. Impacts of Dredged-Material Disposal on the Coastal Soft-Bottom Macrofauna, Saronikos Gulf, Greece. *Sci. Total Environ.* **2015**, *508*, 320–330. [CrossRef] [PubMed]
4. Harvey, M.; Gauthier, D.; Munro, J. Temporal Changes in the Composition and Abundance of the Macro-Benthic Invertebrate Communities at Dredged Material Disposal Sites in the Anse a Beaufils, Baie Des Chaleurs, Eastern Canada. *Mar. Pollut. Bull.* **1998**, *36*, 41–55. [CrossRef]
5. State Oceanic Administration of China. *Bulletin of China Marine Ecological Environment Status*; Ocean Press: Beijing, China, 2011–2017.
6. Marmin, S.; Lesueur, P.; Dauvin, J.C.; Samson, S.; Tournier, P.; Lavanne, A.G.; Dubrulle-Brunaud, C.; Thouroude, C. An Experimental Study on Dredge Spoil of Estuarine Sediments in the Bay of Seine (France): A Morphosedimentary Assessment. *Cont. Shelf Res.* **2016**, *116*, 89–102. [CrossRef]
7. Moog, O.; Stubauer, I.; Haimann, M.; Habersack, H.; Leitner, P. Effects of Harbour Excavating and Dredged Sediment Disposal on the Benthic Invertebrate Fauna of River Danube (Austria). *Hydrobiologia* **2018**, *814*, 109–120. [CrossRef]
8. Guerra-Garcia, J.; Garcia-Gomez, J. Polychaete Assemblages and Sediment Pollution in a Harbour with Two Opposing Entrances. *Helgol. Mar. Res.* **2004**, *58*, 183–191. [CrossRef]

9. Islam, M.S.; Ahmed, M.K.; Raknuzzaman, M.; Habibullah-Al-Mamun, M.; Islam, M.K. Heavy Metal Pollution in Surface Water and Sediment: A Preliminary Assessment of an Urban River in a Developing Country. *Ecol. Indic.* **2015**, *48*, 282–291. [CrossRef]
10. Abdel-Ghani, N.T.; Elchaghaby, G.A. Influence of Operating Conditions on the Removal of Cu, Zn, Cd and Pb Ions from Wastewater by Adsorption. *Int. J. Environ. Sci. Technol.* **2007**, *4*, 451–456. [CrossRef]
11. Yuan, H.; Song, J.; Li, X.; Li, N.; Duan, L. Distribution and Contamination of Heavy Metals in Surface Sediments of the South Yellow Sea. *Mar. Pollut. Bull.* **2012**, *64*, 2151–2159. [CrossRef]
12. Gleyzes, C.; Tellier, S.; Astruc, M. Fractionation Studies of Trace Elements in Contaminated Soils and Sediments: A Review of Sequential Extraction Procedures. *Trac Trends Anal. Chem.* **2002**, *21*, 451–467. [CrossRef]
13. Peng, B.; Peng, J.X.; Sun, K.F. A review on heavy metals contamination in Daya Bay and adjacent waters. *Ecol. Sci.* **2015**, *34*, 170–180.
14. Bolam, S.G.; Barry, J.; Bolam, T.; Mason, C.; Rumney, H.S.; Thain, J.E.; Law, R.J. Impacts of Maintenance Dredged Material Disposal on Macrobenthic Structure and Secondary Productivity. *Mar. Pollut. Bull.* **2011**, *62*, 2230–2245. [CrossRef]
15. Newell, R.C.; Seiderer, L.J.; Hitchcock, D.R. The Impact of Dredging Works in Coastal Waters: A Review of the Sensitivity to Disturbance and Subsequent Recovery of Biological Resources on the Sea Bed. *Oceanogr. Mar. Biol. D* **1998**, *36*, 127–178.
16. Vandolah, R.; Calder, D.; Knott, D. Effects of Dredging and Open-Water Disposal on Benthic Macroinvertebrates in a South-Carolina Estuary. *Estuaries* **1984**, *7*, 28–37. [CrossRef]
17. Roberts, R.D.; Forrest, B.M. Minimal Impact from Long-Term Dredge Spoil Disposal at a Dispersive Site in Tasman Bay, New Zealand. *N. Z. J. Mar. Freshw. Res.* **1999**, *33*, 623–633. [CrossRef]
18. Roberts, R.D.; Gregory, M.R.; Foster, B.A. Developing an Efficient Macrofauna Monitoring Index from an Impact Study—A Dredge Spoil Example. *Mar. Pollut. Bull.* **1998**, *36*, 231–235. [CrossRef]
19. State Oceanic Administration of China. *Technical Guidelines for Selecting of Ocean Dumping Area (HY/T122-2009)*; Standard Press: Beijing, China, 2009.
20. Donazar-Aramendia, I.; Sanchez-Moyano, J.E.; Garcia-Asencio, I.; Miro, J.M.; Megina, C.; Garcia-Gomez, J.C. Impact of Dredged-Material Disposal on Soft-Bottom Communities in a Recurrent Marine Dumping Area near to Guadalquivir Estuary, Spain. *Mar. Environ. Res.* **2018**, *139*, 64–78. [CrossRef]
21. Wang, X.; Kong, L.; Cheng, J.; Zhao, D.; Chen, H.; Sun, R.; Yang, W.; Han, J. Distribution of Butyltins at Dredged Material Dumping Sites around the Coast of China and the Potential Ecological Risk. *Mar. Pollut. Bull.* **2019**, *138*, 491–500. [CrossRef]
22. Yan, N.; Liu, W.; Xie, H.; Gao, L.; Han, Y.; Wang, M.; Li, H. Distribution and Assessment of Heavy Metals in the Surface Sediment of Yellow River, China. *J. Environ. Sci.* **2016**, *39*, 45–51. [CrossRef]
23. Muller, G. Index of Geoaccumulation in Sediments of the Rhine River. *Geojournal* **1969**, *2*, 108–118.
24. Hakanson, L. An Ecological Risk Index for Aquatic Pollution-Control—A Sedimentological Approach. *Water Res.* **1980**, *14*, 975–1001. [CrossRef]
25. Clarke, K.R.; Warwick, R.M. *Change in Marine Communities: An. Approach to Statistical Analysis and Interpretation*; Plymouth Marine Laboratory: Plymouth, UK, 1994.
26. Zhang, Y. A Background Value Study on Heavy Metal Elements in the Sediments of Daya Bay. *Trop. Ocean.* **1991**, *3*, 76–80.
27. MacDonald, D.D.; Carr, R.S.; Calder, F.D.; Long, E.R.; Ingersoll, C.G. Development and Evaluation of Sediment Quality Guidelines for Florida Coastal Waters. *Ecotoxicology* **1996**, *5*, 253–278. [CrossRef]
28. Long, E.; Macdonald, D.; Smith, S.; Calder, F. Incidence of Adverse Biological Effects Within Ranges of Chemical Concentrations in Marine and Estuarine Sediments. *Environ. Manag.* **1995**, *19*, 81–97. [CrossRef]
29. Rauret, G. Extraction Procedures for the Determination of Heavy Metals in Contaminated Soil and Sediment. *Talanta* **1998**, *46*, 449–455. [CrossRef]
30. Bonsignore, M.; Manta, D.S.; Mirto, S.; Quinci, E.M.; Ape, F.; Montalto, V.; Gristina, M.; Traina, A.; Sprovieri, M. Bioaccumulation of Heavy Metals in Fish, Crustaceans, Molluscs and Echinoderms from the Tuscany Coast. *Ecotoxicol. Environ. Saf.* **2018**, *162*, 554–562. [CrossRef]
31. Lao, Q.; Su, Q.; Liu, G.; Shen, Y.; Chen, F.; Lei, X.; Qing, S.; Wei, C.; Zhang, C.; Gao, J. Spatial Distribution of and Historical Changes in Heavy Metals in the Surface Seawater and Sediments of the Beibu Gulf, China. *Mar. Pollut. Bull.* **2019**, *146*, 427–434. [CrossRef] [PubMed]
32. Adila, S.; Mamattursun, E.; Alimujiang, K. Assessment of Pollution and Ecological Risk of Heavy Metals in Surface Dust in Karamay City. *Asian J. Ecotoxicol.* **2021**, *16*, 310–322. [CrossRef]
33. Liu, J.; Zhang, J.; Lu, S.; Zhang, D.; Tong, Z.; Yan, Y.; Hu, B. Interannual Variation, Ecological Risk and Human Health Risk of Heavy Metals in Oyster-Cultured Sediments in the Maowei Estuary, China, from 2011 to 2018. *Mar. Pollut. Bull.* **2020**, *154*, 111039. [CrossRef] [PubMed]
34. Zheng, L.; Liu, Y.; Yuan, Y.; Zhou, C.Y.; Qu, L. Enrichment of Heavy Metals in the Surface Sediments from the Dumping Areas in Bohai Sea and Assessment of Their Potential Ecological Risk. *Mar. Sci. Bull.* **2014**, *33*, 340–346.
35. Zhang, N.X.; Cao, C.H.; Ren, R.Z.; Sun, X. Heavy Metals in the Surface Sediment of the Dumping Ground Outside Jiaozhou Bay and Their Potential Ecological Risk. *HuanjingKexue* **2011**, *32*, 1315–1320.
36. Liu, G.Q.; Liu, B.L.; Qing, S.M.; Xing, S.K. Assessment on Pollution and Potential Ecological Risk of Heavy Metals in the Sediments of the Temporary Marine Dumping Area of Fangchenggang. *Ecol. Sci.* **2013**, *32*, 177–182.

37. Chen, C.F.; Chen, C.W.; Ju, Y.R.; Kao, C.M.; Dong, C.D. Impact of Disposal of Dredged Material on Sediment Quality in the Kaohsiung Ocean Dredged Material Disposal Site, Taiwan. *Chemosphere* **2018**, *191*, 555–565. [CrossRef]
38. Zhang, L.; Ren, R.Z.; Wu, F.C.; Wang, J.W.; Zhang, N.X. Analysis on the Distribution of Surface Sediment Heavy Metals Contamination and the Influencing Factors of Temporary Ocean Dumping Site in Lanshan Port. *Trans. Oceanol. Limnol.* **2012**, *1*, 130–136.
39. Hilton, J.; Davison, W.; Ochsenbein, U. A Mathematical-Model for Analysis of Sediment Core Data—Implications for Enrichment Factor Calculations and Trace-Metal Transport Mechanisms. *Chem. Geol.* **1985**, *48*, 281–291. [CrossRef]
40. Sundaray, S.K.; Nayak, B.B.; Lin, S.; Bhatta, D. Geochemical Speciation and Risk Assessment of Heavy Metals in the River Estuarine Sediments-A Case Study: Mahanadi Basin, India. *J. Hazard. Mater.* **2011**, *186*, 1837–1846. [CrossRef] [PubMed]
41. Liu, B.; Wang, J.; Xu, M.; Zhao, L.; Wang, Z. Spatial Distribution, Source Apportionment and Ecological Risk Assessment of Heavy Metals in the Sediments of Haizhou Bay National Ocean Park, China. *Mar. Pollut. Bull.* **2019**, *149*, 110651. [CrossRef]
42. Tang, H.; Ke, Z.; Yan, M.; Wang, W.; Nie, H.; Li, B.; Zhang, J.; Xu, X.; Wang, J. Concentrations, Distribution, and Ecological Risk Assessment of Heavy Metals in Daya Bay, China. *Water* **2018**, *10*, 780. [CrossRef]
43. Cao, Y.; Lei, K.; Zhang, X.; Xu, L.; Lin, C.; Yang, Y. Contamination and Ecological Risks of Toxic Metals in the Hai River, China. *Ecotoxicol. Environ. Saf.* **2018**, *164*, 210–218. [CrossRef]
44. Lin, L.H.; Wei, H.J.; Huang, H.M. Contamination status and bioaccumulation of the heavy metals in the surface sediments and benthos in Daya Bay. *Ecol. Sci.* **2017**, *36*, 173–181.
45. Sarr, A.B.; Joao Benetti, C.; Fernandez-Diaz, M.; Garrido, J. The Microhabitat Preferences of Water Beetles in Four Rivers in Ourense Province, Northwest Spain. *Limnetica* **2013**, *32*, 1–9.
46. de Castro Vasconcelos, M.; Melo, A.S. An Experimental Test of the Effects of Inorganic Sediment Addition on Benthic Macroinvertebrates of a Subtropical Stream. *Hydrobiologia* **2008**, *610*, 321–329. [CrossRef]
47. Pallo, P.; Widbom, B.; Olafsson, E. A Quantitative Survey of the Benthic Meiofauna in the Gulf of Riga (Eastern Baltic Sea), with Special Reference to the Structure of Nematode Assemblages. *Ophelia* **1998**, *49*, 117–139. [CrossRef]
48. Fonseca, E.M.; Fernandes, J.R.; Lima, L.S.; Delgado, J.; Correa, T.R.; Costa, P.M.S.; Baptista Neto, J.A.; Aguiar, V.M.C. Effects of Dredged Sediment Dumping on Trace Metals Concentrations and Macro Benthic Assemblage at the Continental Shelf Adjacent to a Tropical Urbanized Estuary. *Ocean. Coast. Manag.* **2020**, *196*, 105299. [CrossRef]
49. Quan, F. Intertidal Pollution on Benthic Macro-Arthropod Community in Mangrove Forest. Master's Thesis, Hainan Normal University, Hainan, China, 2011.
50. Jia, H.B.; Hu, H.Y.; Tang, J.L.; Wang, Y.M.; Chai, H.P. Effect of Heavy Metals on Macro-Benthos in Surface Sediments in Changjiang Estuary and Adjacent Sea. *Mar. Environ. Sci.* **2011**, *30*, 809–813.
51. Li, X.Z.; Li, B.Q.; Wang, H.F. Community Structure of Macrobenthos in Coastal Water off Rushan, Southern Shandong Peninsula, and the Relationships with Environmental Factors. *Acta Oceanol. Sin.* **2009**, *28*, 81–93.
52. Xu, R.; Yang, Y.; Li, Z. Diffusion of Heavy Metals from Marine Environment to Shellfish. *Mar. Sci. Bull.* **2007**, *26*, 117–120.
53. Simonini, R.; Ansaloni, I.; Cavallini, F.; Graziosi, F.; Iotti, M.; N'Siala, G.M.; Mauri, M.; Montanari, G.; Preti, M.; Prevedelli, D. Effects of Long-Term Dumping of Harbor-Dredged Material on Macrozoobenthos at Four Disposal Sites along the Emilia-Romagna Coast (Northern Adriatic Sea, Italy). *Mar. Pollut. Bull.* **2005**, *50*, 1595–1605. [CrossRef]

applied sciences

MDPI

Article

The Impact of Coastal Geodynamic Processes on the Distribution of Trace Metal Content in Sandy Beach Sediments, South-Eastern Baltic Sea Coast (Lithuania)

Dovilė Karlonienė [1,*], Donatas Pupienis [1,2], Darius Jarmalavičius [2], Aira Dubikaltinienė [1] and Gintautas Žilinskas [2]

[1] Faculty of Chemistry and Geosciences, Naugarduko st. 24, LT 03225 Vilnius, Lithuania; donatas.pupienis@gamtc.lt (D.P.); aira.dubikaltiniene@chgf.vu.lt (A.D.)
[2] Nature Research Centre, Akademijos st. 2, LT 08412 Vilnius, Lithuania; darius.jarmalavicius@gamtc.lt (D.J.); gintautas.zilinskas@gamtc.lt (G.Ž.)
* Correspondence: dovile.karloniene@chgf.vu.lt

Featured Application: Geochemical analysis can provide valuable information about the local and regional patterns of sediment transport, distribution, provenance, and coasts' conditions.

Abstract: Sandy coasts are one of the most dynamic spheres; continuously changing due to natural processes (severe weather and rising water levels) and human activities (coastal protection or port construction). Coastal geodynamic processes lead to beach sediment erosion or accumulation. The coast's dynamic tendencies determine the changes in the volume of beach sediments; grain size; mineralogical; and geochemical composition of sediments. In addition to lithological and mineralogical analysis of sediments, geochemical analysis can provide valuable information about the local and regional patterns of sediment transport, distribution, provenance, and coasts' conditions. The study aims to assess trace metals' temporal and spatial distribution determined in the sandy beach sediments along the south-eastern Baltic Sea coast (Lithuania) during 2011–2018. The Lithuanian seacoast is divided into two parts: mainland and spit coast. Our results revealed that the dominant group of elements on the mainland includes Ca–Mg–Mn–Ti and on the Curonian Spit Fe–Pb–As–Co–Cr–Ni–Al, which remain unchanged during the years. The analysis included additional parameters such as beach volume, grain size and sorting, and heavy mineral concentration on the beach. The spatial analysis of trace elements indicated that the trace metal content depends on the coastal processes, but it differs in the mainland and spit sea coast. We identified a higher concentration of trace metals in the erosion-dominated areas in all analysed years on the mainland coast. On the spit coast, the trace metal concentration increased in areas associated with relict coarse sand and where the loading of sediments was active on the beach due to the northward along-shore transport.

Keywords: trace metals; beach sediments; coastal processes; lithology

Citation: Karlonienė, D.; Pupienis, D.; Jarmalavičius, D.; Dubikaltinienė, A.; Žilinskas, G. The Impact of Coastal Geodynamic Processes on the Distribution of Trace Metal Content in Sandy Beach Sediments, South-Eastern Baltic Sea Coast (Lithuania). *Appl. Sci.* **2021**, *11*, 1106. https://doi.org/10.3390/app11031106

Academic Editor: Mauro Marini
Received: 29 December 2020
Accepted: 20 January 2021
Published: 25 January 2021

Publisher's Note: MDPI stays neutral with regard to jurisdictional claims in published maps and institutional affiliations.

1. Introduction

Trace metals enter the coastal system from the entire Baltic Sea catchment area, which is four times larger than the sea area. The main sources of trace metals besides natural ones in this area are the combustion of fossil fuels (transport and energy production), municipal and industrial sewage management, agriculture, manufacturing processes (pulp and paper, metallurgy, etc.), and military activities (chemical ammunition buried after World War II) [1]. These metals are transported to the sea and coastal areas by rivers, deposited from the air along with precipitation and other pathways.

Sandy beaches are considered important recreational sites and less recognised as highly threatened and fragile natural ecosystems, which may work as natural barriers to pollutants transported by sea [2,3]. Approximately, 50% of coastlines globally are composed of sand and gravel, mostly due to river and land sediment transport and deposition [4]. However, the ultimate magnitude of the accumulation/erosion of beach sediments depends also on anthropogenic activities [5]. The textural, lithological, and mineralogical composition of beach sediments depend on many factors: the physical and geographical conditions of the location [6–8], geological framework [9–12], tectonic settings [13–17], provenance [18–22], climate and sea hydrodynamics (waves, tides, and currents) conditions [13,23–27], and the anthropogenic activities [27–31]. Sandy beaches are presently threatened by several forms of environmental degradation, although beach management has traditionally concentrated on geomorphic hazards [32] and recreational use of coast [33], while trace metals are rarely considered [34]. The geochemical analysis of beach sediments together with lithological and mineralogical characterisation can provide valuable information about the local and regional patterns of sediment transport, distribution, and provenance, as well as the conditions of the coast [35].

Global studies related to trace elements analysis on beaches focus on assessing the changes in trace element concentrations in the context of the influence of local pollution sources such as mining sites, urban areas, and industrial complexes [28,29,36–38] or tourism [39–41]. Several studies analysing the migration and distribution of trace elements in coastal environments have been conducted; e.g., the distribution of trace elements concentrations across (from the beach to dunes) [42] and along the coast [38,43–48] and vertical migration [49]. All overviewed studies intended to identify the source or how the particular pollution source affects beach sediments, but the coastal processes were not deeply analysed. The concentration of elements can be affected, as it is known in heavy minerals, by changes in the hydrometeorological conditions [23] or by the season [35,49]. The dominant coastal processes could affect the accumulation of trace elements, which could be of natural origin.

In Lithuania, the formation and dynamics of recent beach sediments have been analysed using lithological, mineralogical, morphometric, and other methods [12,27,50,51]. The nearshore mineralogical and geochemical composition of the Baltic Sea's Lithuanian territorial waters and distribution has already been well analysed [21,52,53]. However, the detailed geochemical composition of beach sediments in relation to active coastal processes has not been previously investigated, except for some local studies [54,55]. This study aims to assess the temporal change of the trace metal content in sandy beach sediments and the dependence on coastal lithomorphodynamical processes.

2. Study Area

The Lithuanian coast is divided by the 1.1 km wide Klaipėda Strait, where Klaipėda Port is located into two parts: the Curonian Spit (hereinafter—the spit) sea coast (51 km) and the mainland coast (39 km) (Figure 1). Beyond 1991, when the Curonian Spit National and Seaside Regional Parks were established, the major part of the Baltic Sea coast in Lithuania (about 70 km long) has acquired the status of a protected area. Several coastal sectors with different characters can be distinguished: a technogenic coast that predominates near to Klaipėda and Šventoji Ports, i.e., areas of waste water disposal from the Būtingė oil terminal, and the Mažeikiai oil processing plant, and protected areas such as the Curonian Spit National Park, the Seaside Regional Park, the Baltic Sea Talasological Reserve, and the Būtingė Geomorphological Reserve (Figure 1).

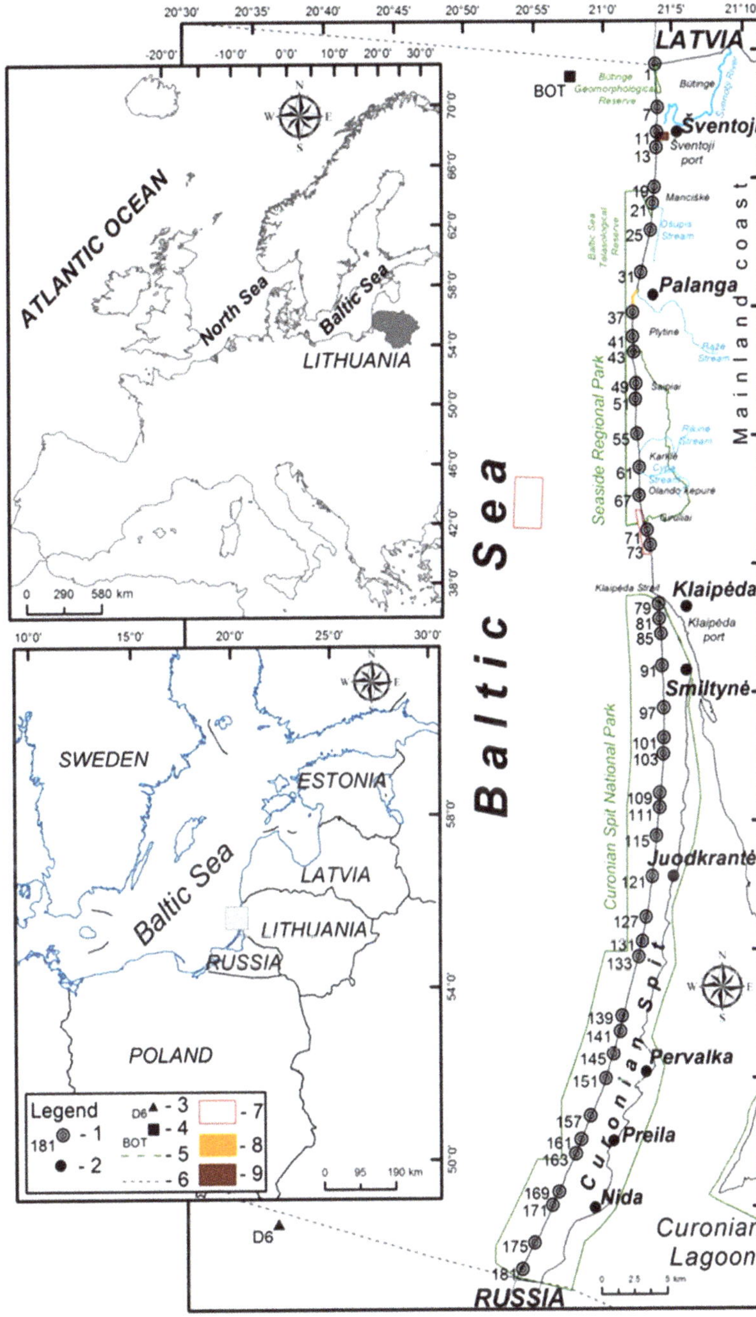

Figure 1. Study area. (**1**) surface sand sampling site, (**2**) main settlements, (**3**) D6 oil platform (Ru), (**4**) Būtingė oil terminal, (**5**) conservation areas, (**6**) borders, (**7**) offshore and nearshore dumping area, (**8**) beach nourishment site in Palanga, and (**9**) storage site of dredged bottom sediments.

The Holocene clastic deposits occur along the entire length of the Lithuanian coastal zone. The Lithuanian coastal sediments are composed of quartz, K–feldspar (orthoclase, microcline), plagioclase (albite, anorthite), carbonates (dolomite, calcite), mica (biotite, muscovite), and clay minerals (illite, chlorite, kaolinite, montmorillonite, glauconite, and vermiculite) with admixture of heavy minerals [21,50].

On the mainland coast, the average 20–100 m wide beaches are backed by foredunes 4–12 m in height or by moraine cliffs 5–24 m in height. The southern part of the mainland coast structure between the 25th and 31st km (distance from Latvia-Lithuania border) is characterised by glacial (moraine) deposits formed during the Late Pleistocene and that, in most cases, occur on the abraded cliff coast. The first moraine cliff is located between the 25th and 27th km on the mainland coast near Šaipiai (Figure 1). The cliff is 2.1 km long and 5–8 m in height. The second moraine cliff (Olando kepurė) is 950 m long and 24 m in height at its highest point [56]. The highest cliffs are between the 30th and 31st km at Karklė. On the Curonian spit sea coast, the average width of the beaches varies from 32 to 75 m, and the relative height of the foredune changes from 5 to 16 m [12].

The Baltic Sea is nontidal (amplitudes reach 3.5–4.0 cm); therefore, a wind-wave regime dominates. The cold (autumn–winter) season is marked by most days with strong winds. The annual mean wind speed is 4.7 m s^{-1} and the wave height is 0.65 m. During the strongest storms, the wave height varies from 5 to 6 m [56]. The prevailing westerly (SW, W, NW) winds and waves are dominant in the coastal zone and generate the alongshore sediment transport from the Sambian Peninsula to the end of the Curonian Spit and along the mainland coast [24,27,57]. Currently, the alongshore sediment transport is disrupted by the Klaipėda port gate [58].

3. Materials and Methods

In the analysis, we focus on trace elements which are considered of anthropogenic origin. The macroelements, lithological and morphological parameters are used for the data interpretation and definition of the causes.

Sediment samples were collected along the entire Baltic Sea coast of Lithuania in 2011, 2014, and 2018 (Figure 1). During the collection of samples, the sea and wind conditions were relatively constant. In the coastal zone, the most active processes occurred in the surf and swash zone; for that reason, it was decided to take samples from the middle of the beach [26]. Totally, 43 composite surface sand (0–5 cm) (from 5 subpoints) samples were collected at equal distances of 3 and 5 km and in plastic containers delivered for laboratory analysis [27]. In the laboratory, sand samples were air-dried and split to 100 g of sediment. The samples were mechanically sieved 15 min on a vibratory sieve shaker Fritsch Analysette 3 Spartan Pulverisette 0 using a set of 11 sieves. After determination of size fraction, statistical grain-size Falk and Ward method mean (d, mm), sorting (S_o) parameters were calculated using the GRADISTAT 8.0 software [59].

To characterize the relative concentrations of heavy minerals, a Bartington MS3 field scanning sensor was used for rapid and effective measurements of low-field volume magnetic susceptibility [27]. Sandgren and Snowball [60] point out that bulk magnetic susceptibility (MS) is a good indicator of allochthonous mineral matter in sediments. Measuring MS helps to determine the net contribution of ferromagnetic and paramagnetic minerals in sediments. Heavy mineral-rich ($\rho > 2.90$ g/cm^3) sediments have ferromagnetic and paramagnetic properties, and high magnetic susceptibility values, depending on the predominant iron content. Quartz-rich minerals ($\rho < 2.65$ g/cm^3) have diamagnetic properties. Quartz-rich sands (dominate quartz, feldspar, carbonate, and mica group minerals) have weaker positive magnetic susceptibilities values of $\kappa < 3.0$ µSI, heavy mineral-rich sand (elements like Ti, Cr, Mn, Fe, Co, Ni, and Cu can sometimes result in magnetism) κ values range from 30 to 150 µSI and higher values $\kappa > 150$ µSI are typical for heavy minerals with Fe, Ni, and Co elements [26,61,62]. MS measurement method is helpful to detect ferromagnetic minerals when their concentration in sediments is deficient [60]. Magnetic susceptibility and grain sizes of beach sediments belongs of the provenance, geologic framework, alongshore sediment transport, deposition and coastal processes (erosion/accretion), etc. The grain size composition also might have contributed to the difference in proportions of ferromagnetic, paramagnetic, and diamagnetic minerals [63]. Magnetic minerals are known as important sources of trace elements in sediments.

The beach sediment volume (Q, m^3/m) was calculated for each profile based on repeated cross-shore levelling once per year. The changes in sediment volume comprise the changes in volume of the coastal profile from the foredune lee side, where the vertical variability is negligible during observation, to the intersection with the mean sea level [12]. Total beach sediment volume was counted with the formula: $Q = (Q_i + Q_i + 1)\ L_i/2$, where Q—sediment volume (m^3) at the coastal segment; $i = 1, 2, 3, \ldots$ number of transects; Q_i—sediment volume at the separate coastal cross-section profile (m^3/m); and L_i—distance between levelling cross-section profile lines [64].

For the geochemical analysis, the dried samples were ground using an agate mortar and pestle. Prior to each sample's homogenisation, the agate mortar and pestle were washed with deionised water and dried twice. Geochemical analyses of the samples were performed at the Bureau Veritas Commodities Canada Ltd., laboratory. 0.5–2.0 g of bulk sample digested after application of modified aqua regia (1:1 HNO_3: HCl) solution for low to ultralow determination of soil and analysed with an inductively coupled plasma mass/emission spectrometer (ICP–MS/ES). In this study, we mostly analysed trace elements that might originate from the anthropogenic activities common in coastal systems as fossil fuel burning, sewage discharge, metals common in ship ports, etc. and monitored at the national monitoring program [65]. We also selected macroelements that could help to describe the origin of sediments. The results are given in ppm for elements As, Cu, Cr, Co, Mn, Ni, Pb, and Zn; in ppb for Hg; and in percentage for Ca, Mg, Fe, Al, and Ti. Analytical quality was monitored in each batch of samples by repeated analyses, recovery of spiked samples, and analysis of a certified reference material (OREAS45EA and DS11), and duplicates and blanks were used to assure the quality of the analysis. The reaction mixture was chosen to evaluate the labile trace elements dissolving sulphide/oxide type minerals to exclude the elements incorporated in the silicate lattice as metals from anthropogenic sources tend to be more mobile than those from pedogenic or lithogenic sources [43,66].

Descriptive and multivariate statistical methods were applied to analyse the results. Correlation analysis using Pearson's coefficient (linear relation, significant $p > 0.01$ or 0.05) and principal component analysis (PCA) were applied to transform the correlation matrix to identify the relationship between the metals [31]. For the PCA analysis, we used the Varimax rotation method to derive more reliable information on the distribution of the weights of the variables on a factor, and loadings higher than 0.5 were considered. Statistical programs IBM SPSS Statistics 22.0 and PAST 3.24 were used.

The concentration of trace metal in each site was compared with the median estimated of all samples at that year $K_k = (K_n)/(M_n)$, where K_n—concentration of the element n and M_n—the median concentration of the element n. Median presents the central or typical value in a set of data and is weakly dependent on minimal and maximal values and outliers, this approach has been used and developed in a few studies [48,52,67]. This analysis helps to compare the loading of analysed elements among sites; the lower limit of the anomaly is considered as 1.5 [52]. Following a multi-element index, the integrating and averaging data were estimated, $K_d = \Sigma K_k/n$, where K_k—is the concentration ratio for a specific element and n—number of elements. We assumed that K_d values higher than 2 indicate an anomaly concentration in the site [68].

Other studies use concentration ratios to determine trace element accumulation in sediments and focus on anthropogenic pollution [43,53,69–71], the determination of background concentrations in such studies is essential [71]. The main concept to identify the anthropogenic impact is to compare the concentrations of elements measured in uncontaminated sites—to establish a local baseline or, as in most studies, compare sediment element concentrations with preindustrial levels such as average shale [72] or average crustal value [73]. Using the average shale or earth's crust concentrations, the local geochemical framework is being ignored, which might lead to misinterpretation of anomaly concentrations in the analysed region. Second, the shale concentration mostly represents fine grain sediments and earth's crust—coarse sediments, meanwhile the background concentration represents sites in similar mineralogical and textural environments [71]. In this study, we analysed beach sand, dynamic environment and usage of shale or earth's crust as the background concentration might miss an important pattern in the distribution of trace elements. For that reason, we chose to use the values calculated during our study. We also considered the values as a background—calculated from 76 samples of aeolian sediments collected in the western region of Lithuania located close to our studied area [74]. However, this area is remote from the sea—where the sediments are affected by both wind and water; thus, we considered not to apply these data in the current study.

4. Results

4.1. The Lithology of Beaches

The sandy beaches on the Lithuanian sea coast are composed of fine-medium grain sediments (d = 0.27 mm, σ = 0.07 in 2011; d = 0.29 mm, σ = 0.09 in 2014; d = 0.29 mm, σ = 0.11 in 2018). Although in 2011, fine and medium sand predominated on the beaches, in 2014 and 2018, coarse sand was already detected in several places (Figure 2). In 2011, fine sand dominated in 23 and medium sand in 20 sites. In 2014, fine sand was determined in 20 sites, the medium—dominated in 21 places, and coarse sand—in two sites—73rd and 131st. In 2018, fine sand was indicated as well in 21 places, medium sand in 19 places, and coarse sand in 3 places—61st, 73rd, and 133rd. The appearance of coarse sand at Olando kepurė, north of Klaipėda port pier, is related to the intensified local erosion processes. Erosive processes are local and usually occur in specific stretches of the coast, e.g., where moraine cliffs predominate, or are affected by hydrotechnical construction.

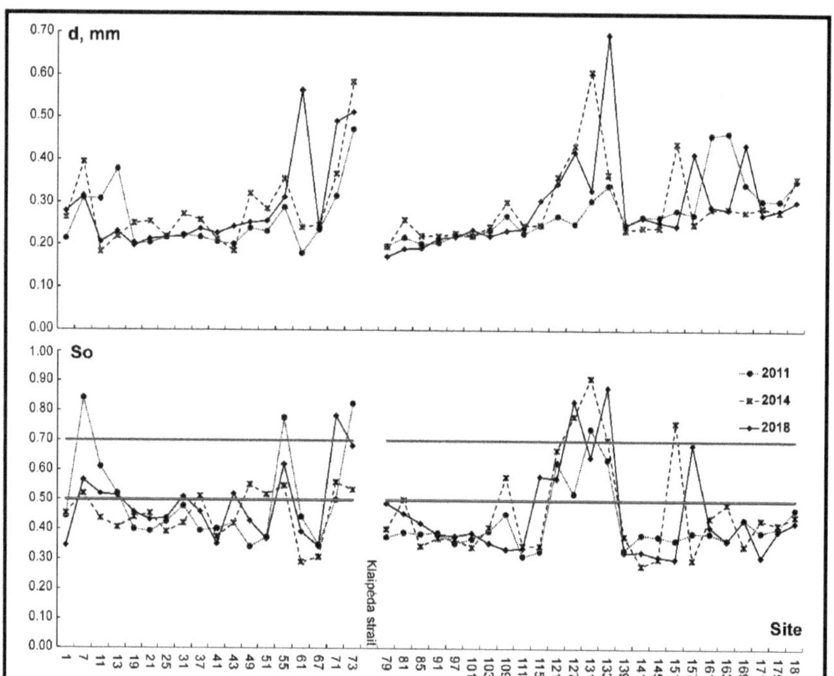

Figure 2. The distribution of grain-size (d, mm) and sorting (S_o) of the beach sediments along the coast of the south-eastern Baltic Sea coast (Lithuania) in 2011, 2014, and 2018.

In all investigated years, the surface sand on the mid-beach was very well sorted ($S_o < 0.35$) and well sorted ($0.35 < S_o > 0.50$) (Figure 2). Moderately well-sorted ($0.50 < S_o > 0.70$) or moderately sorted ($S_o > 0.70$) sand was in areas with a medium or coarse sediment fraction. In 2011, moderately sorted sand was identified in 9; in 2014—in 13, and in 2018—in 14 sites. The distribution of beach sand particles shows that the average particle diameter decreases from south to north on both the mainland and the Curonian Spit coast (Figure 2).

4.2. Magnetic Susceptibility of Beach Sediments

The bulk MS values of the sediments on the investigated coast ranged between 13.9 and 357.7 μSI (mean $\kappa = 64.4$, $\sigma = 58.7$) in 2011, between 16.5 and 804.8 μSI (mean $\kappa = 139.1$, $\sigma = 161.6$) in 2014, and between 7.9 and 271.8 μSI (mean $\kappa = 77.7$, $\sigma = 60.2$) in 2018 (Figure 3). The MS values differed between the mainland and Curonian Spit coast, i.e., in 2011 and 2014, the MS values measured higher on the mainland coast, and in 2018, the difference in the values faded. The anomalous MS values were determined at the 7th, 21st, 127th, and 181st.

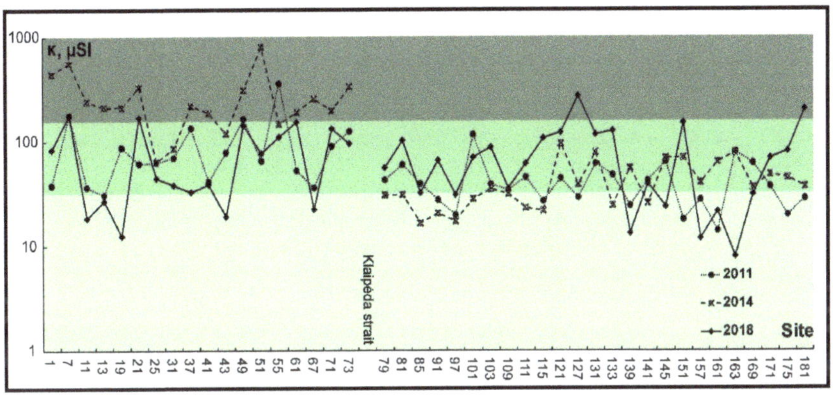

Figure 3. The distribution of magnetic susceptibility values (κ, μSI) of the beach sediments along the coast of the south-eastern Baltic Sea coast (Lithuania) in 2011, 2014, and 2018 (the dark green colour field indicate heavy minerals, green—heavy mineral-rich sand and light green—quartz-rich sands).

The MS values differed between the mainland and Curonian Spit coast. In 2011 and 2014, the MS values were higher on the mainland coast, and in 2018, the difference in the values faded. The higher values of magnetic susceptibility were detected in the beach sediments, enriched with perimagnetic and ferromagnetic minerals. The anomalous MS values were determined at the 7th, 21st, 127th, and 181st. Magnetic susceptibility reflects different coastal processes and magnetic mineral sources. Magnetic susceptibility of beach sediments increases greatly close to the provenance and decreases moving away from source.

4.3. Beach Sediment Volume

The beach sand volume (Q, m^3/m) compared between 2011 and 2014 increased by in almost all sites on the mainland coast except in the 25th, 41st, and 49th (Figure 4). However, on the Curonian Spit coast, the sediment content on the beaches in 2014 compared to 2011 decreased in the end of the spit (79th and 91st–109th) and in Juodkrantė–Pervalka section (115th, 127th, 133rd, 139th), higher in the 141st, 161st, and 175th sites, the highest increase in sediments was in the south of Nida (157th, 169th, 171st, and 181st).

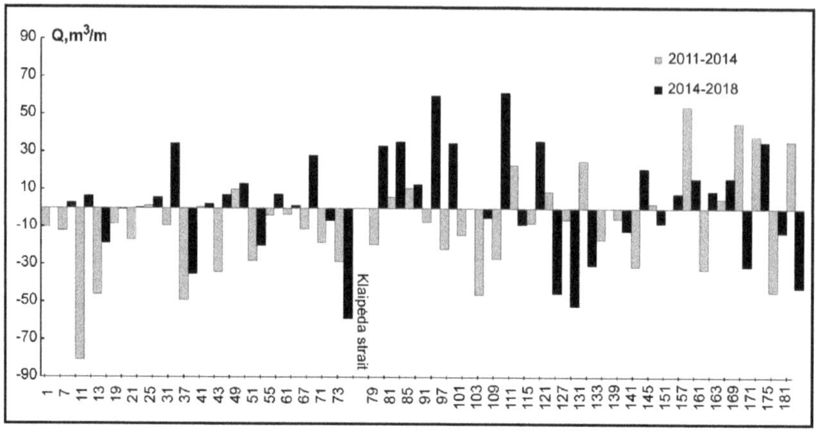

Figure 4. The changes in beach sediment volume (Q, m^3/m) distribution along the coast of the south-eastern Baltic Sea coast (Lithuania) in 2011, 2014, and 2018.

Comparing 2018 and 2014 sediment volume on the beaches, on the mainland coast the highest increase was determined in the 31st, 49th, and 61st, and the significant decrease was determined in the 13th, 37th, and 51st sites, but mostly in the 73rd. On the Curonian Spit coast, the sediment volume increased mostly at the end of the Curonian Spit (79th–97th, 109th, 115th) and 141st. The considerable decrease in sediments on the beaches of the spit was estimated on Juodkrantė–Pervalka (121st, 127th, 131st), 169th, and 181st sites.

4.4. *Descriptive Statistics of Trace Elements*

The descriptive statistics for the 3 years of the analysed trace elements (Cu, Pb, Zn, Ni, Co, Mn, As, Cr, Hg, and Cd) and macroelements (Fe, Al, Ca, Mg, and Ti) are provided in Table 1.

The comparison of the average concentrations of the trace elements in the analysed years revealed that the higher average concentrations of most elements were measured in 2014. In 2011, the higher concentrations of As and Cr compared to other years were determined, while in 2018, only for Mn. Additionally, in 2014, Hg was measured in sediments collected on 18 sites; in 2011, Hg was detected only on eight sites; and in 2018, Hg was found on six sites.

4.5. *Distribution of Trace Elements Along Coast*

We assessed the distribution of the average concentration of trace elements and the medium value ratio (K_k) along the coast. The pattern of the ratio differed among elements. The K_k of As were estimated only on the Curonian Spit coast where the ratio exceeded 1.5 in six sites on the distal end of the spit (79th, 85th, and 103rd–111st) and Pervalka–Preila section (145th) in 2011, in four sites (109th, 111th, 131st and 145th) in 2014, and in only one Juodkrantė–Pervalka section (133rd) in 2018. We estimated higher than 1.5 Kk values for Cr in 2011 in seven places, a significant majority at the distal end of the Curonian Spit (79th–111th); in 2014, already in 12 sites (79th, 91st, 111th, 131st, 145th, 157th, 171st, and 175th), and only in one site (51st) on the mainland coast (Figure 4). In 2018, as in previous years, we indicated the highest Cr concentration at the distal end of Curonian Spit (79th–109th) and in the coast stretch southern from Juodkrantė (133rd, 139th, 145th, 151st, and 181st) (Figure 5).

Table 1. Trace element concentrations of the surface sediments collected on the beach along the coast of the south-eastern Baltic Sea coast (Lithuania) in 2011–2018 (values in ppm unless otherwise indicated).

Element	Year	N	Mean	σ	Median	Min–Max
Cu	2011	42	0.32	0.13	0.30	0.17–0.70
	2014	43	0.77	1.01	0.40	0.16–4.15
	2018	42	0.44	0.54	0.28	0.14–3.10
Pb	2011	42	1.29	0.42	1.18	0.70–2.59
	2014	43	1.42	0.34	1.34	0.89–2.69
	2018	42	1.01	0.24	0.96	0.70–1.80
Zn	2011	42	3.61	1.09	3.20	2.20–6.40
	2014	43	4.48	1.73	4.30	2.70–13.00
	2018	42	3.39	1.44	2.95	1.90–10.70
Ni	2011	42	0.64	0.18	0.60	0.40–1.10
	2014	43	0.70	0.16	0.70	0.50–1.10
	2018	42	0.61	0.19	0.60	0.30–1.30
Co	2011	42	0.40	0.15	0.40	0.20–1.00
	2014	43	0.43	0.14	0.40	0.30–0.80
	2018	42	0.41	0.14	0.40	0.20–1.00
Mn	2011	42	19.88	8.06	19.00	10.00–46.00
	2014	43	23.70	12.43	20.00	10.00–66.00
	2018	42	24.12	15.14	21.00	8.00–82.00
As	2011	41	0.79	0.36	0.70	0.20–1.70
	2014	43	0.73	0.20	0.70	0.40–1.30
	2018	42	0.70	0.34	0.70	0.10–2.00
Cr	2011	43	2.62	1.62	2.05	1.00–9.00
	2014	43	2.55	1.23	2.10	1.10–6.50
	2018	42	2.15	1.04	1.70	0.08–5.30
Hg (ppb)	2011	8	8.25	1.98	7.5	6.00–12.00
	2014	18	8.89	2.42	9.00	5.00–13.00
	2018	6	8.33	2.42	8.00	5.00–11.00
Fe, %	2011	42	0.18	0.10	0.15	0.09–0.59
	2014	43	0.19	0.07	0.17	0.11–0.42
	2018	42	0.16	0.06	0.15	0.08–0.38
Al, %	2011	42	0.06	0.02	0.06	0.05–0.15
	2014	43	0.07	0.02	0.07	0.05–0.11
	2018	42	0.06	0.02	0.05	0.03–0.12
Ca, %	2011	42	0.36	0.25	0.28	0.09–1.19
	2014	43	0.44	0.39	0.33	0.11–1.78
	2018	42	0.47	0.48	0.28	0.08–2.30
Mg, %	2011	42	0.07	0.03	0.07	0.03–0.17
	2014	43	0.08	0.05	0.07	0.03–0.24
	2018	42	0.08	0.05	0.08	0.03–0.30
Ti, %	2011	42	0.004	0.002	0.003	0.002–0.01
	2014	43	0.005	0.003	0.004	0.002–0.02
	2018	42	0.004	0.002	0.003	0.001–0.01

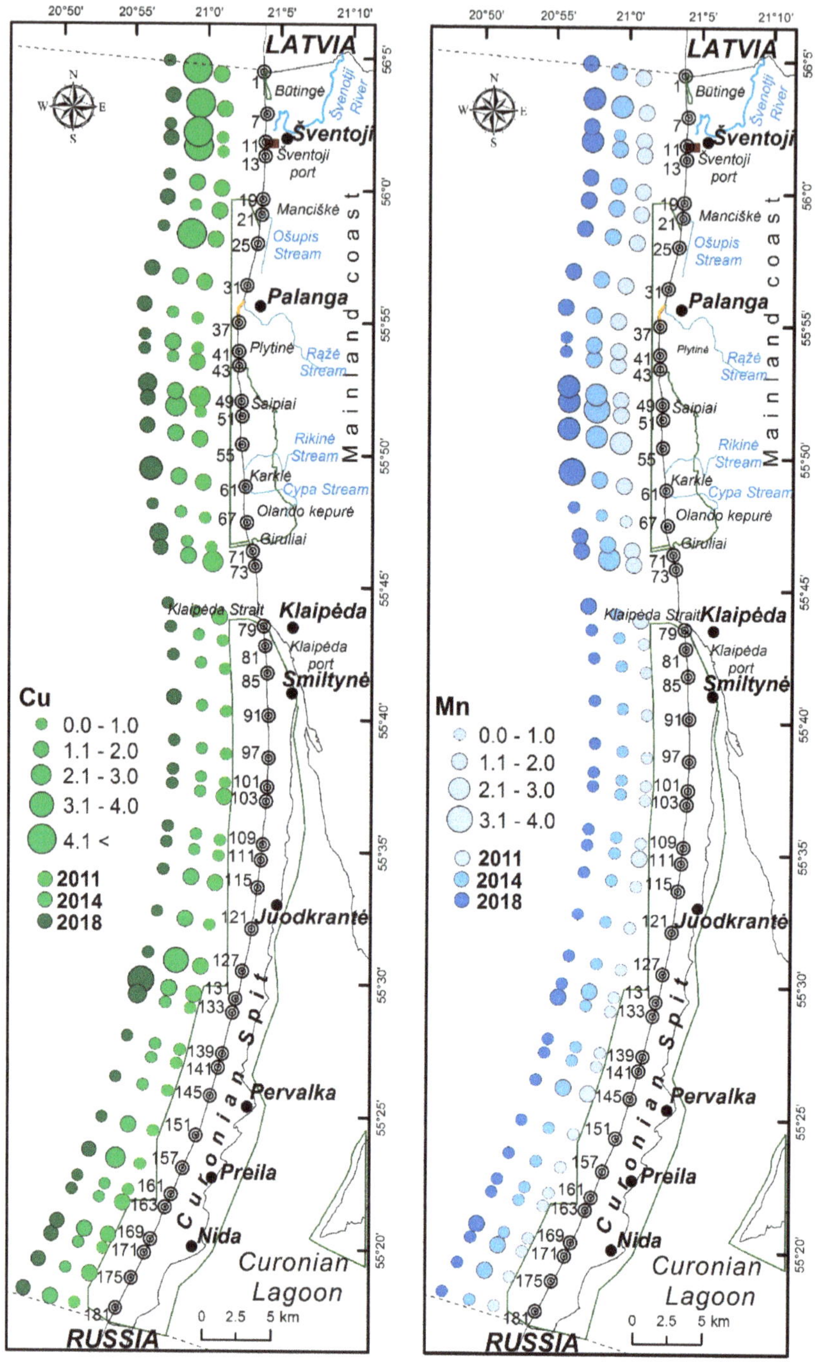

(a)

Figure 5. *Cont.*

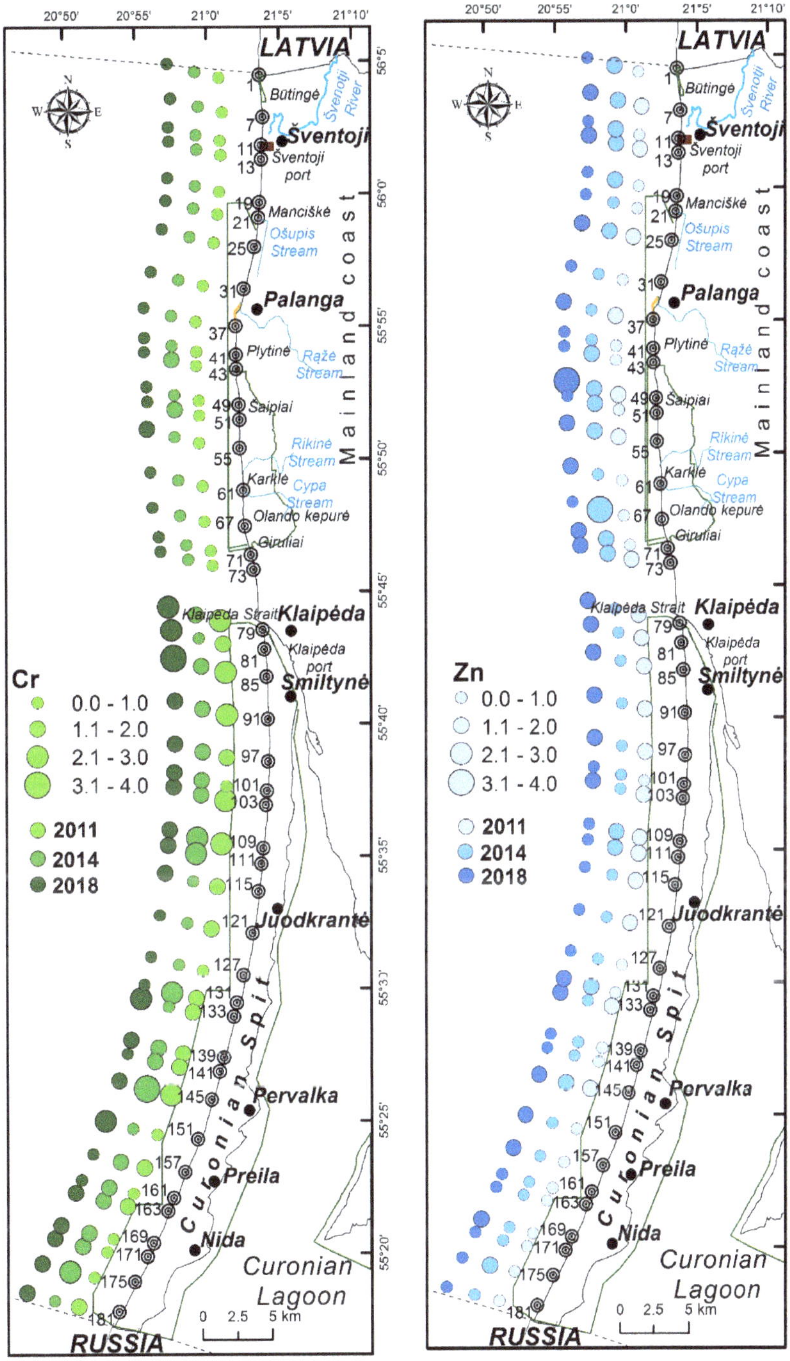

(b)

Figure 5. *Cont.*

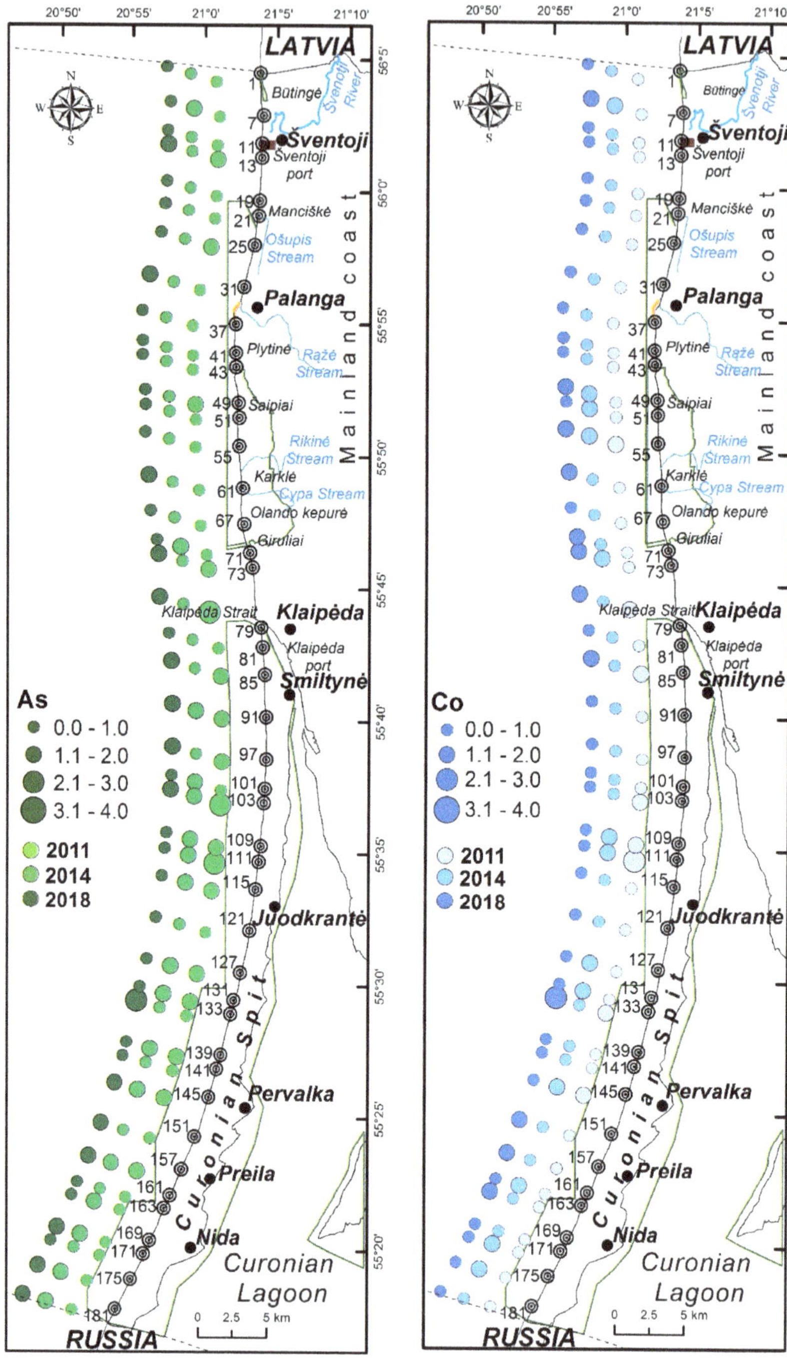

(c)

Figure 5. *Cont.*

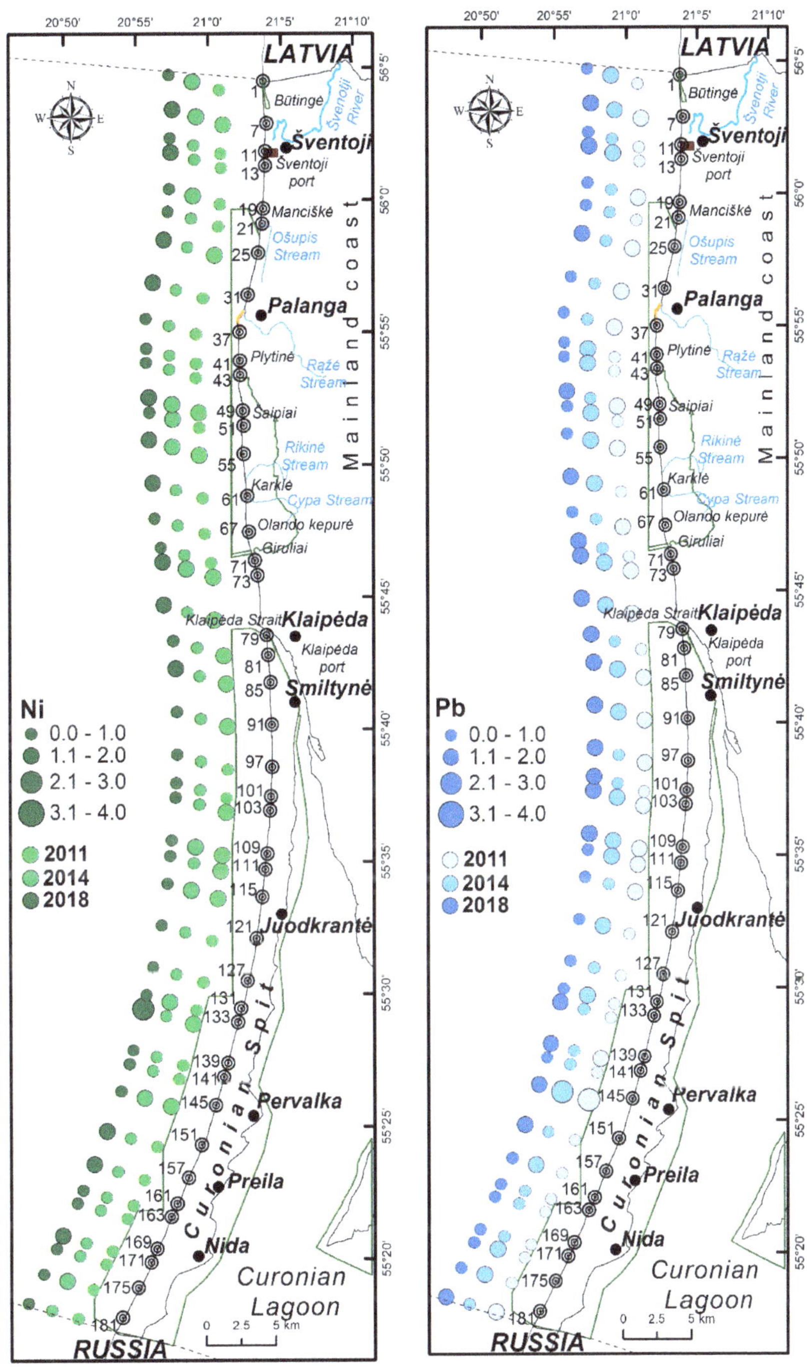

(**d**)

Figure 5. The distribution of concentration ratio (K_k) of (**a**) Cu and Mn, (**b**) Cr and Zn, (**c**) As and Co, and (**d**) Ni and Pb along the coast of the south-eastern Baltic Sea coast (Lithuania).

The Cu concentration ratio exceeding or equal to 1.5 was estimated in five sites (1st, 7th, 49th, 55th, and 73rd) on the mainland coast in 2011; in 2014, in three sites (51st, 127th and 157th), and in 2018, in four sites (49th, 61st, 71st and 133rd). The *Kk* of Cu in 2014 exceeded four in five sites located near Šventoji (1st, 7th, 11th, 13th, and 25th) in the north part of the mainland coast and two sites on the Curonian Spit between Juodkrantė and Pervalka (133rd and 141st sites) in 2018. The ratio *Kk* of Pb did not exceed 2.5 compared with Cu. In 2011 and 2014, these ratios were higher than 1.5 only in the 145th site in Pervalka and in none of the sites in 2018.

The higher concentration ratios of Mn were estimated on the mainland coast opposite to Cr. In all analysed years, higher Mn concentration ratios were estimated in the cliff area (49th–55th). Higher Mn ratio in 2014 and 2018 was also estimated in the site at 7th and also in the 13th and 73rd sites in 2018.

Zinc showed no clear pattern or difference between the mainland and Curonian Spit coasts (Figure 5). Overall, the highest Zn concentration ratios were determined on the mainland coast at the cliff area – in the 67th site in 2014 and the 49th in 2018. In 2011, the Zn concentration ratio was higher than 1.5 in eight sites (7th, 79th, 85th, 103rd–111th, and 145th), in 2014 in four sites (1st, 49th, 67th, and 71st) and in 2018, in six sites (7th, 49th, 61st, 71st, 85th, and 133rd).

Cobalt pattern differed from other elements because its concentration did not variate so much among the years. The concentration ratio exceeded 1.5 in 2011 only in two sites (111th and 145th); in 2014, in the sites 43rd, the 131st, and 145th; and in 2018, only in the 133rd site. The Ni as well as Co, Pb did not show a clear pattern, its ratio exceeded 1.5 in the 111th and 145th sites in 2011; 131st and 145th in 2014; and 55th, 61st, and 133rd sites in 2018.

To understand the overall distribution pattern of the analysed elements in 3 different years, we estimated and compared the mean of the concentration ratios (K_d) in 2011, 2014, and 2018 (Figure 6). In 2011, slightly higher mean concentration ratios were indicated in the area where coastal erosion is active; 1 km northwards from Šventoji port (7th site), in the cliff areas (49th and 55th) and near the Klaipėda strait (73rd site). On the Curonian Spit coast, the distribution pattern of the mean concertation ratio is more pronounced than on the mainland coast, and increases northwards (highest peak at the 111th site).

In 2014, the trend of K_d on the mainland coast distinguished from the pattern in 2011 and 2018. The mean concentration ratio was higher than 2 in 11 sites from 18 on the mainland coast and increased northwards. The mean concentration ratio exceeding 2 was determined in three sites located at the vicinity and northwards from the Šventoji port. On the Curonian Spit, the trend was opposite to the other years—the mean ratio slightly decreasing northwards.

In 2018, on the mainland coast, the estimated K_d ratio values tend to decrease northwards, the highest values were determined in the cliff area (49th and 61st sites), and only in three sites was higher compared to previous years (49th, 61st, and 71st sites). On the Curonian Spit coast, the trend of the mean ratio increases northwards; however, it is mostly expressed from the 121st to 79th sites and from the 181st to 127th site, and the highest rations were determined in the section between Juodkrantė and Pervalka (127th to 133rd sites and 141st site). In general, the trend is similar to 2011.

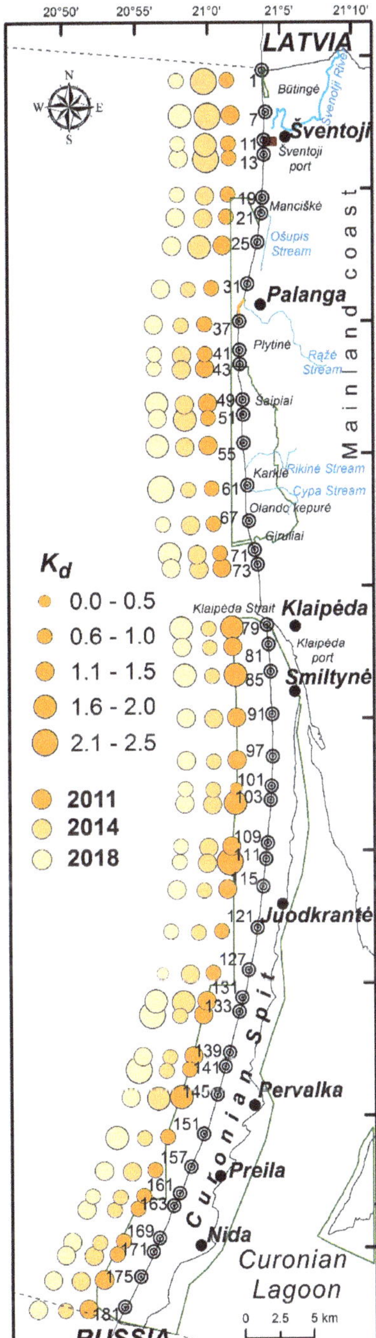

Figure 6. The distribution of the mean concentration ratio (K_d) of all analysed trace elements along the south-eastern Baltic Sea coast (Lithuania).

4.6. Correlation Analysis

The correlation analysis of trace elements and lithological and geomorphological factors showed that Cu content tends to accumulate in beach sediments where erosion processes are active as in 2011. In addition, it is depended on the sorting of the sediments ($r = 0.40$) (Table 2) and was lower in the well-sorted sediments and negatively correlated with beach volume. The copper concentration positively correlated with the MS values in 2011 and 2014 ($r = 0.59$ and $r = 0.46$).

Table 2. Pearson's correlation coefficient between trace metals and mean grain size (d, mm), sorting coefficient (S_o), beach sediment volume (Q, m3/m), and magnetic susceptibility (MS, µSI).

	2011				2014				2018			
Var	*d*	*So*	*Q*	**MS**	*d*	*So*	*Q*	**MS**	*d*	*So*	*Q*	**MS**
Cu	0.14	0.40 **	−0.31 *	0.59 **	0.00	0.05	−0.1	0.46 **	0.23	0.12	−0.18	0.16
Pb	−0.07	0.09	0.19	0.26	0.16	0.00	−0.1	0.24	0.38 *	0.29	0.06	0.33 *
Zn	−0.14	0.12	0.29	0.12	0.13	0.03	−0.22	0.45 **	0.28	0.26	−0.02	0.35*
Ni	−0.08	0.1	0.22	0.29	0.34 *	0.23	−0.27	0.23	0.59 **	0.48 **	−0.27	0.31 *
Co	−0.09	−0.06	0.27	0.04	0.39 **	0.32 *	−0.14	0.23	0.68 **	0.47 **	−0.16	0.33 *
Mn	−0.19	0.38 *	−0.46 **	0.79 **	0.27	0.24	−0.45 **	0.83 **	0.39 *	0.25	−0.45 **	0.36 *
As	−0.19	−0.2	0.49 **	−0.16	0.26	0.14	0.12	−0.28	0.45 **	0.34 *	0.04	0.22
Cr	−0.23	−0.3	0.47 **	−0.2	−0.11	−0.14	0.19	−0.3	−0.07	−0.11	0.39 *	0.04
d	1	0.41 **	0.04	0.07	1	0.79 **	−0.12	0.13	1	0.64 **	−0.43 **	0.35 *
So	0.41 **	1	−0.27	0.52 **	0.79 **	1	−0.05	0.1	0.64 **	1	−0.2	0.27
Q	0.04	−0.27	1	−0.41 **	−0.12	−0.05	1	−0.28	−0.43 **	−0.2	1	−0.39 *
MS	0.07	0.52 **	−0.41 **	1	0.13	0.1	−0.28	1	0.35 *	0.27	−0.39 *	1

** Correlation is significant at the 0.01 level (2-tailed). * Correlation is significant at the 0.05 level (2-tailed).

The manganese content in all analysed years negatively correlated with beach sediment volume ($r = -0.46$, $r = -0.45$, and $r = -0.45$) and positively correlated with MS values ($r = 0.79$, $r = 0.83$, and $r = 0.36$). In 2011, the Mn concentration positively correlated ($r = 0.38$) with the sorting coefficient and in 2018 with grain size ($r = 0.39$). The zinc concentration positively correlated with MS values in 2014 and 2018 ($r = 0.45$ and $r = 0.33$). In 2011, the concentration positively correlated with beach sand volume ($r = 0.49$). In 2018, the As content positively correlated with grain size and sorting coefficient ($r = 0.45$ and $r = 0.34$).

The grain size positively correlated with sorting coefficient in all investigated years; in 2018, grain size also positively correlated with MS values and negatively with beach volume (Table 2). In 2011, the sorting coefficient of beach sediments positively correlated with MS values. The measured beach volume negatively correlated with MS values in all years, but only in 2014 and 2018 it was statistically significant.

4.7. Multivariate Analysis

The PCA analysis was used to reduce the number of analysed variables and to identify the different factors that controlled the distribution of the elements in the beach sediments. For further analysis, we added macroelements as, Al, Ca, Fe, Mg, and Ti, which helped to explain the origin of the trace elements. The results present principal components whose eigenvalues were above 1.

The PCA analysis of 3 years data revealed that there are two dominant groups of elements that are defined by the eigenvalues and the correlation coefficients (Figure 7). In 2011, three components were extracted, the first group (PC1) Fe–Co–Al–Cr–Ni–Zn–As–Pb (explains 52.2% of the total variance with the highest eigenvalue 6.8), second (PC2) Mn–Ca–Mg–Ti (eigenvalue 3.4; explains 26.2% of the total variance), and third (PC3) Cu (eigenvalue 1.27; explains 9.7%). In 2014, only two components were extracted, the first group consisting of the following elements Fe–Co–Ni–Cr–As–Al–Pb (eigenvalue 5.3; explains 40.9% of the total variance) and second Mn–Ca–Mg–Ti (eigenvalue 4.2; explains 32.3% of the total variance). In 2018, the element distribution between groups was similar as in other years—PC1: Fe–Pb–As–Co–Cr–Ni–Al (eigenvalue 5.2; explains 40.3% of the total variance), PC2: Mn–Ca–Mg–Ti–Ni–Al (eigenvalue 4.7; explains 36.3% of the total variance), and PC3: Cu (eigenvalue 1.12; CV 8.6% of the total variance).

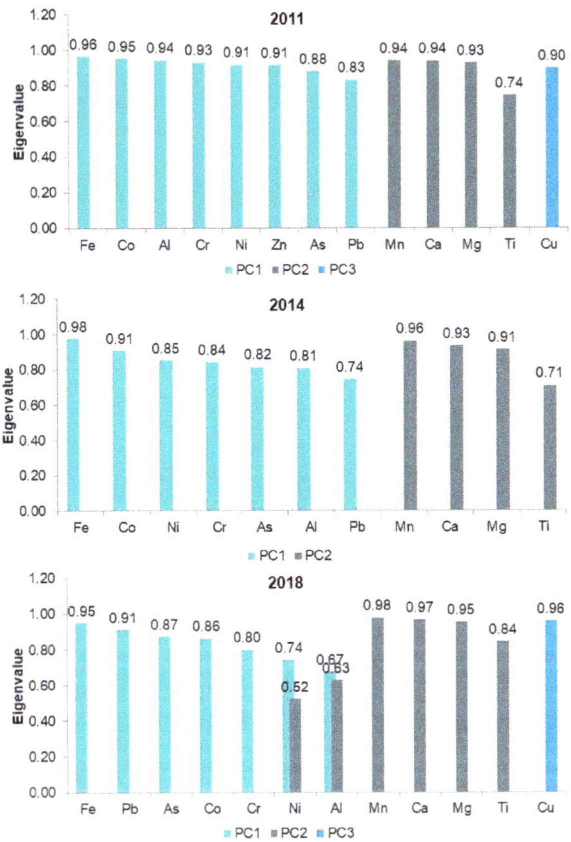

Figure 7. The loadings of analysed elements in different component groups after principal component analysis (Varimax rotation) in beach sediments of south-eastern Baltic Sea coast (Lithuania) in 2011, 2014, and 2018.

The PCA results showed that the elements from the first component group in 2011, 2014, and 2018 dominated in the Curonian Spit. The second group described sediments from the mainland coast (Figure 8).

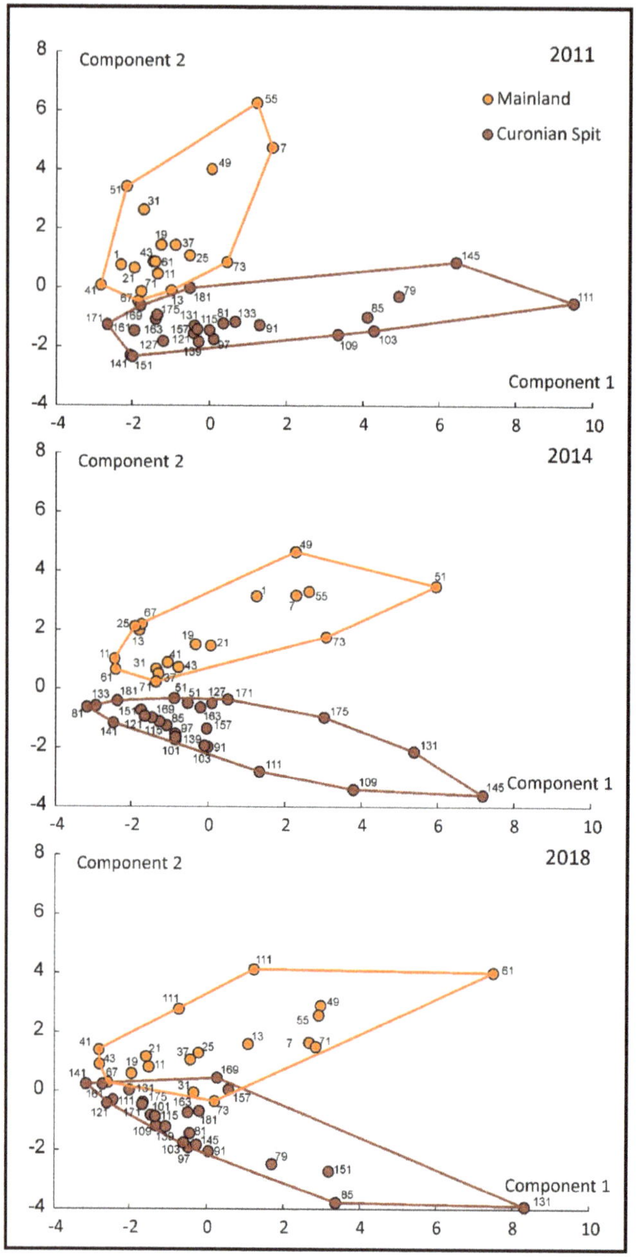

Figure 8. Principal component analysis after normalized varimax rotation, and factorial scores showed according to geographical location (Curonian Spit and mainland coast) in 2011, 2014, and 2018.

5. Discussion

It was found out that the trace element concentrations of beach sediments depend on the coastal processes and nonsignificantly varied among years, except for Cu and Zn in 2014. The Mn and Cr showed a clear distribution pattern along the coast during the years.

Manganese concentration ratio was higher on the mainland coast, and it tended slightly to decrease northwards. The Cr concentration ratio was higher in Curonian Spit and increased at the distal end of it where the unloading of the sediments is active. These tendencies indicate that the alongshore sediment transport play a significant role in the distribution of the elements along the coast. The decrease in grain size in the north direction shows that the accumulation processes are conditioned by alongshore sediment transport directed from south to north (Figure 2). The measured amounts of beach sediment volume agree with these statements. Poorly sorted sediments were found in anthropogenically affected sections of the coast (Būtingė, Šventoji, Palanga, and Klaipėda) and in places where the sand grain size depends on the geological framework (Juodkrantė–Pervalka stretch).

The higher Mn concentration was determined in the sites with higher heavy mineral content and with a reduced beach sediment volume. Finer light mineral fractions such as quartz, feldspar, and muscovite mica are usually washed away, and the fine but heavier particles related to heavy minerals co-occur with coarser quartz particles [35,61]. Manganese according to the PCA analysis is associated with the second component group—carbonates (Ca and Mg), and this association remained during the investigation years. The dominance of carbonates can be explained by the deposits of detrital dolomite and biogenic calcite found on the mainland coast [21,36]. The greatest source of calcite and dolomite was found at Klaipėda and gradually decreased towards the south of Palanga [21]. The area southward from Klaipėda strait is less abundant in calcite [21].

The Mn distribution pattern also showed that its main source is at the moraine cliff area (Šaipiai–Olando kepurė). Its concentration increased during the years probably due to more active erosion on the mainland coast. Carbonate rocks also tend to form compounds with cations of divalent metals such as Cu, Zn, Mn, Sr, Pb, etc. [54]. The Zn pattern shows that its concentration was also the highest close to the moraine cliff area. In addition, enlarged Zn concentration in 2014 and 2018 relates to higher heavy mineral content. The results indicated that its concentration is also linked to the erosion processes. The PCA analysis revealed that Zn concentration in 2011 is associated with sediments from the Curonian Spit, while in 2018 and 2014, there was no concrete association with any of groups. In 2014, the content of Zn was significantly higher than in other investigated years. This suggests that there might be additional sources besides natural ones. The cliff area (49th—61st sites) 30 years ago was an active military zone [75], which also could have affected sediments resulting in higher content of Zn. The trace elements emitted in the past could migrate into deeper layers and in erosive areas could re-enter the environment after storms [76]. The Zn, Mn, and Cu content is higher in the 71st and 73rd sites, which overlap with the dumping sites of the sediments dredged from the Klaipėda Strait (Figure 1). The dredged sediments are more enriched with trace elements than beach sediments [53].

Higher than average MS values are associated with an increase in heavy minerals, whose content depends on the geological framework, which could enhance after a storm [26]. On the distal end of Curonian Spit the sediment accumulation prevail due to sediments delivered by the alongshore sediment transport [27,57]. The Curonian Spit coast beaches are enriched with glauconite (greensand) originating from Neogene–Paleogene deposits from the Sambian Peninsula [27,48]. The trace metals tend to attach to clay particles such as glauconite, mica, and biotite, which are transported northwards and unload at the distal end of the spit [21,48,50,77–79]. This relation is confirmed by the PCA analysis results, it means Cr together with Pb, Ni, Co, and As in all analysed years form a group of metals associated with Al and Fe (Pb–Ni–Co–Fe–As–Cr–Al) which constitute the composition of glauconite [80]. Therefore, despite the sediments are enriched in trace elements in this part of the spit, they are characterised by decreased MS values as a result of coverage with quartz sand from aeolian processes.

As declared previously, the positive correlation with grain size and MS value indicated the coastal erosion. However, on the Curonian Spit, there is no active erosion [12,81]. The only anomaly of coarse sand in the Curonian Spit is located near Juodkrantė—Pervalka site. The long-term sediment grain-size study recorded the relict origin of the anomaly and general stability of coastal sections of the spit [12]. This is in line with our lithological analysis. The concentration of analysed elements was determined higher in this area especially in 2011 and 2014 at 145th site and in 2018 at 131rd—133th sites. Since these coast stretches of the Curonian Spit were also richer in heavy minerals, this could cause the higher concentration of elements [82]. The correlation analysis revealed relationship related to greater elements' concentration (Ni and Co in 2014 and Ni, Co, Pb, and As in 2018) in coarser beach sands.

In 2014, the analysed trace elements' content was higher than the concentrations determined in 2011 and 2018, specifically, the loading of Cu and Zn on the mainland coast. Similar to Zn, the increased Cu concentration was associated with a higher content of heavy minerals only in 2011 and 2014. However, in 2014, Cu concentration significantly increased to the north of Šventoji Port that could suggest an anthropogenic source. In this case, the sediments' geochemistry could be affected by the Šventoji Port reconstruction in 2011–2012, when the bottom sediments were dredged from the entrance channel and stored on the beach [83], and later washed away after the storm "Xaver" in 2013. The bottom surface sediments in the Šventoji Port basin were enriched with Pb, Ni, Cu, and Zn [84]. Additionally, the sediment in the north of Šventoji has greater absorption capacity due to the peat layer, which is exposed after stormy weather event [85].

The study has some limitations regarding ambiguous evidence of the anthropogenic impact. Our hypotheses are based on literature analysis. For example, our results show anomalies of trace elements southwards from Juodkrantė on the Curonian Spit coast. These anomalies could be also consequences of the local fisheries activities due to possible oil leakage or accidents at a D6 oil platform, coastal protection construction of tires, or cement block at the Sambian Peninsula in the provenance region of the spit sediments [27,67].

6. Conclusions

Distribution analysis of the trace elements on the south-eastern Baltic Sea coast indicated that the concentration mainly depended on the coastal processes (alongshore sediment transport, coastal erosion, and sediment accumulation). The differences in trace element concentrations and composition between the Curonian Spit and the mainland coast indicated various sources. The trace metal concentrations on the mainland coast mainly increased in the areas with active erosion processes when heavy minerals were exposed. On the Curonian Spit, trace element anomalies are associated with relict sands. The elements such as Cr and as tend to accumulate at the distal end of the spit where sediments most actively accumulate. Since the coast is a dynamic environment, the elements' concentration at the same sites may vary from year to year. For example, in 1 year, severe weather events may lead to more intense coast erosion. In another year, when calm weather is favourable for accretion, these sediments might be covered with fine quartz sand or reworked and redeposited. However, our results showed that beach sediment geochemical composition remains invariable in space and time. The anomalous concentration of Cu and Zn on the coast indicated the possible anthropogenic impact on sediment geochemical composition (such as former military activities and beach nourishment with dredged sediments from the port).

Author Contributions: Conceptualization, D.K. and D.P.; methodology, D.K., D.P. and D.J. and G.Ž.; formal analysis, D.J., D.P., and G.Ž.; investigation, D.K., D.P., and A.D.; data curation, D.K., A.D., D.J. and D.P.; writing—original draft preparation, D.K.; writing—review and editing, D.P., D.J., A.D. and G.Ž.; visualization, D.P. and D.K. All authors have read and agreed to the published version of the manuscript.

Funding: This research is a part of a project (duration 2017–2021) funded from the public budget of the Republic of Lithuania.

Institutional Review Board Statement: Not applicable.

Informed Consent Statement: Not applicable.

Acknowledgments: The authors would like to thank the reviewers for their constructive comments, which helped improve our manuscript. Authors are also very thankful for student Anna Cichon-Pupienis for fine tuning the manuscript.

Conflicts of Interest: The authors declare no conflict of interest.

References

1. Bełdowska, M.; Bolałek, J.; Bojakowska, I.; Burska, D.; Cyberski, J.; Dudzińska-Huczuk, B.; Ebbing, J.; Falkowska, L.; Graca, B.; Kramarska, R.; et al. *Geochemistry of Baltic Sea Surface Sediments*; Uścinowicz, S., Ed.; Polish Geological Institute-National Research Institute: Warsaw, Poland, 2011.
2. Harris, L.; Nel, R.; Holness, S.; Schoeman, D. Quantifying Cumulative Threats to Sandy Beach Ecosystems: A Tool to Guide Ecosystem-Based Management beyond Coastal Reserves. *Ocean Coast. Manag.* **2015**, *110*, 12–24. [CrossRef]
3. Trojanowski, J.; Bigus, K.; Trojanowska, C. Differences of Chemical Components in Beaches Sediments with Dissimilar Anthropopressure. *Balt. Coast. Zo. J. Ecol. Prot. Coastline* **2011**, *15*, 109–126.
4. Bird, E.C.F. *Coastal Geomorphology: An Introduction*; Wiley: Hoboken, NJ, USA, 2000.
5. Song, Y.; Choi, M.S. Assessment of Heavy Metal Contamination in Sediments along the Coast of South Korea Using Cs-Normalized Background Concentrations. *Mar. Pollut. Bull.* **2017**, *117*, 532–537. [CrossRef] [PubMed]
6. Ohta, T. Geochemistry of Jurassic to Earliest Cretaceous Deposits in the Nagato Basin, SW Japan: Implication of Factor Analysis to Sorting Effects and Provenance Signatures. *Sediment. Geol.* **2004**, *171*, 159–180. [CrossRef]
7. Weltje, G.J. Ternary Sandstone Composition and Provenance: An Evaluation of the Dickinson Model. *Geol. Soc. Lond. Spec. Publ.* **2006**, *264*, 79–99. [CrossRef]
8. Lee, Y.I. Geochemistry of Shales of the Upper Cretaceous Hayang Group, SE Korea: Implications for Provenance and Source Weathering at an Active Continental Margin. *Sediment. Geol.* **2009**, *215*, 1–12. [CrossRef]
9. Cox, R.; Lowe, D.R.; Cullers, R.L. The Influence of Sediment Recycling and Basement Composition on Evolution of Mudrock Chemistry in the Southwestern United States. *Geochim. Cosmochim. Acta* **1995**, *59*, 2919–2940. [CrossRef]
10. Abdel-Karim, A.A.M.; Zaid, S.M.; Moustafa, M.I.; Barakat, M.G. Mineralogy, Chemistry and Radioactivity of the Heavy Minerals in the Black Sands, along the Northern Coast of Egypt. *J. Afr. Earth Sci.* **2016**, *123*, 10–20. [CrossRef]
11. Armstrong-Altrin, J.S.; Lee, Y.I.; Kasper-Zubillaga, J.J.; Carranza-Edwards, A.; Garcia, D.; Eby, G.N.; Balaram, V.; Cruz-Ortiz, N.L. Geochemistry of Beach Sands along the Western Gulf of Mexico, Mexico: Implication for Provenance. *Geochemistry* **2012**, *72*, 345–362. [CrossRef]
12. Jarmalavičius, D.; Žilinskas, G.; Pupienis, D. Geologic Framework as a Factor Controlling Coastal Morphometry and Dynamics. Curonian Spit, Lithuania. *Int. J. Sediment Res.* **2017**, *32*, 597–603. [CrossRef]
13. Chavadi, V.C.; Hegde, V.S. A Note on the Textural Variation of Beach Sediments in the Vicinity of Gangavali River Mouth near Ankola, West Coast of India. *Mahasagar* **1989**, *22*, 89–98.
14. Carranza-Edwards, A.; Kasper-Zubillaga, J.J.; Rosales-Hoz, L.; Morales-dela Garza, E.A.; Lozano-Santa Cruz, R. Beach Sand Composition and Provenance in a Sector of the Southwestern Mexican Pacific. *Rev. Mex. Cienc. Geol.* **2009**, *26*, 433–447.
15. Barnard, P.L.; Foxgrover, A.C.; Elias, E.P.L.; Erikson, L.H.; Hein, J.R.; McGann, M.; Mizell, K.; Rosenbauer, R.J.; Swarzenski, P.W.; Takesue, R.K.; et al. Integration of Bed Characteristics, Geochemical Tracers, Current Measurements, and Numerical Modeling for Assessing the Provenance of Beach Sand in the San Francisco Bay Coastal System. *Mar. Geol.* **2013**, *345*, 181–206. [CrossRef]
16. Armstrong-Altrin, J.S.; Nagarajan, R.; Lee, Y.I.; Kasper-Zubillaga, J.J.; Córdoba-Saldaña, L.P. Geochemistry of Sands along the San Nicolás and San Carlos Beaches, Gulf of California, Mexico: Implications for Provenance and Tectonic Setting. *Turk. J. Earth Sci.* **2014**, *23*, 533–558. [CrossRef]
17. Armstrong-Altrin, J.S.; Nagarajan, R.; Balaram, V.; Natalhy-Pineda, O. Petrography and Geochemistry of Sands from the Chachalacas and Veracruz Beach Areas, Western Gulf of Mexico, Mexico: Constraints on Provenance and Tectonic Setting. *J. South Am. Earth Sci.* **2015**, *64*, 199–216. [CrossRef]
18. Dickinson, W.R.; Suczek, C.A. Plate Tectonics and Sandstone Compositions1. *Am. Assoc. Pet. Geol. Bull.* **1979**, *63*, 2164–2182. [CrossRef]
19. Basu, A.; Blanchard, D.P.; Brannon, J. Rare Earth Elements in the Sedimentary Cycle: A Pilot Study of the First Leg. *Sedimentology* **1986**, *29*, 737–742. [CrossRef]
20. Kasper-Zubillaga, J.J.; Dickinson, W.W. Discriminating Depositional Environments of Sands from Modern Source Terranes Using Modal Analysis. *Sediment. Geol.* **2001**, *143*, 149–167. [CrossRef]
21. Kairyte, M.; Stevens, R.L.; Trimonis, E. Provenance of Silt and Clay within Sandy Deposits of the Lithuanian Coastal Zone (Baltic Sea). *Mar. Geol.* **2005**, *218*, 97–112. [CrossRef]
22. Saha, S.; Banerjee, S.; Burley, S.D.; Ghosh, A.; Saraswati, P.K. The Influence of Flood Basaltic Source Terrains on the Efficiency of Tectonic Setting Discrimination Diagrams: An Example from the Gulf of Khambhat, Western India. *Sediment. Geol.* **2010**, *228*, 1–13. [CrossRef]
23. Kurian, N.P.; Prakash, T.N.; Jose, F.; Black, K.P. Hydrodynamic Processes and Heavy Mineral Deposits of the Southwest Coast, India. *J. Coast. Res.* **2001**, 154–163.
24. Kovaleva, O.; Chubarenko, B.; Pupienis, D. Grain Size Variability as an Indicator of Sediment Transport Alongshore the Curonian Spit (South-Eastern Baltic Sea). *Baltica* **2016**, *29*, 145–155. [CrossRef]

25. Morrone, C.; De Rosa, R.; Le Pera, E.; Marsaglia, K.M. Provenance of Volcaniclastic Beach Sand in a Magmatic-Arc Setting: An Example from Lipari Island (Aeolian Archipelago, Tyrrhenian Sea). *Geol. Mag.* **2017**, *154*, 804–828. [CrossRef]
26. Pupienis, D.; Buynevich, I.V.; Bitinas, A. Distribution and Significance of Heavy-Mineral Concentrations along the Southeast Baltic Sea Coast. *J. Coast. Res.* **2011**, *SI64*, 1984–1988.
27. Pupienis, D.; Buynevich, I.; Ryabchuk, D.; Jarmalavičius, D.; Žilinskas, G.; Fedorovič, J.; Kovaleva, O.; Sergeev, A.; Cichoń-Pupienis, A. Spatial Patterns in Heavy-Mineral Concentrations along the Curonian Spit Coast, Southeastern Baltic Sea. *Estuar. Coast. Shelf Sci.* **2017**, *195*, 41–50. [CrossRef]
28. Caredda, A.M.; Cristini, A.; Ferrara, C.; Lobina, M.F.; Baroli, M. Distribution of Heavy Metals in the Piscinas Beach Sediments (SW Sardinia, Italy). *Environ. Geol.* **1999**, *38*, 91–100. [CrossRef]
29. Vidinha, J.M.; Rocha, F.; Silva, E.; Patinha, C.; Andrade, C. Geochemical Beach Sediments Studies-A Contribution to a Standard Definition Useful for Public Health. *J. Coast. Res.* **2009**, *2009*, 905–908.
30. Jarmalavičius, D.; Pupienis, D.; Buynevich, I.V.; Žilinskas, G.; Fedorovič, J. Aeolian Sand Differentiation along the Curonian Spit Coast, Baltic Sea, Lithuania. *Coast. Sediments* **2015**, *2015*, 1–10. [CrossRef]
31. Yalcin, F.; Nyamsari, D.G.; Paksu, E.; Yalcin, M.G. Statistical Assessment of Heavy Metal Distribution and Contamination of Beach Sands of Antalya-Turkey: An Approach to the Multivariate Analysis Techniques. *Filomat* **2016**, *30*, 945–952. [CrossRef]
32. Nordstrom, K.F.; Lampe, R.; Vandemark, L.M. Reestablishing Naturally Functioning Dunes on Developed Coasts. *An Int. J. Decis. Makers Sci. Environ. Audit.* **2000**, *25*, 37–51. [CrossRef]
33. Ariza, E.; Jiménez, J.A.; Sardá, R. An Interdisciplinary Analysis of Beach Management in the Catalan Coast (North-Western Mediterranean). *Coast. Manag.* **2012**, *40*, 442–459. [CrossRef]
34. Defeo, O.; McLachlan, A.; Schoeman, D.S.; Schlacher, T.A.; Dugan, J.; Jones, A.; Lastra, M.; Scapini, F. Threats to Sandy Beach Ecosystems: A Review. *Estuar. Coast. Shelf Sci.* **2009**, *81*, 1–12. [CrossRef]
35. Hanamgond, P.T.; Gawali, P.B.; Lakshmi, B.V.; Babu, J.L.V.M.; Deendayalan, K. Sediment Texture and Geochemistry of Beaches between Redi-Vengurla, Sindhudurg, West Coast of India. *J. Coast. Res.* **2017**, *33*, 1135–1147. [CrossRef]
36. Yalcin, M.G. Heavy Mineral Distribution as Related to Environmental Conditions for Modern Beach Sediments from the Susanoglu (Atakent, Mersin, Turkey). *Environ. Geol.* **2009**, *58*, 119–129. [CrossRef]
37. Sun, W.; Sang, L.; Jiang, B. Trace Metals in Sediments and Aquatic Plants from the Xiangjiang River, China. *J. Soils Sediments* **2012**, *12*, 1649–1657. [CrossRef]
38. Ra, K.; Kim, E.-S.; Kim, K.-T.; Kim, J.-K.; Lee, J.-M.; Choi, J.-Y. Assessment of Heavy Metal Contamination and Its Ecological Risk in the Surface Sediments along the Coast of Korea. *J. Coast. Res.* **2013**, *65*, 105–110. [CrossRef]
39. Jonathan, M.P.; Roy, P.D.; Thangadurai, N.; Srinivasalu, S.; Rodríguez-Espinosa, P.F.; Sarkar, S.K.; Lakshumanan, C.; Navarrete-López, M.; Muñoz-Sevilla, N.P. Metal Concentrations in Water and Sediments from Tourist Beaches of Acapulco, Mexico. *Mar. Pollut. Bull.* **2011**, *62*, 845–850. [CrossRef]
40. Nagarajan, R.; Jonathan, M.P.; Roy, P.D.; Wai-Hwa, L.; Prasanna, M.V.; Sarkar, S.K.; Navarrete-López, M. Metal Concentrations in Sediments from Tourist Beaches of Miri City, Sarawak, Malaysia (Borneo Island). *Mar. Pollut. Bull.* **2013**, *73*, 369–373. [CrossRef]
41. Vetrimurugan, E.; Shruti, V.C.; Jonathan, M.P.; Roy, P.D.; Kunene, N.W.; Villegas, L.E.C. Metal Concentration in the Tourist Beaches of South Durban: An Industrial Hub of South Africa. *Mar. Pollut. Bull.* **2017**, *117*, 538–546. [CrossRef]
42. Díaz Rizo, O.; Buzón González, F.; Arado López, J.O. Assessment of Ni, Cu, Zn and Pb Levels in Beach and Dune Sands from Havana Resorts, Cuba. *Mar. Pollut. Bull.* **2015**, *100*, 571–576. [CrossRef]
43. Bramha, S.N.; Mohanty, A.K.; Satpathy, K.K.; Kanagasabapathy, K.V.; Panigrahi, S.; Samantara, M.K.; Prasad, M.V.R. Heavy Metal Content in the Beach Sediment with Respect to Contamination Levels and Sediment Quality Guidelines: A Study at Kalpakkam Coast, Southeast Coast of India. *Environ. Earth Sci.* **2014**, *72*, 4463–4472. [CrossRef]
44. Dobaradaran, S.; Schmidt, T.C.; Nabipour, I.; Khajeahmadi, N.; Tajbakhsh, S.; Saeedi, R.; Javad Mohammadi, M.; Keshtkar, M.; Khorsand, M.; Faraji Ghasemi, F. Characterization of Plastic Debris and Association of Metals with Microplastics in Coastline Sediment along the Persian Gulf. *Waste Manag.* **2018**, *78*, 649–658. [CrossRef] [PubMed]
45. Retama, I.; Jonathan, M.P.; Roy, P.D.; Rodríguez-Espinosa, P.F.; Nagarajan, R.; Sarkar, S.K.; Morales-García, S.S.; Muñoz-Sevilla, N.P. Metal Concentrations in Sediments from Tourist Beaches of Huatulco, Oaxaca, Mexico: An Evaluation of Post-Easter Week Vacation. *Environ. Earth Sci.* **2016**, *75*, 1–10. [CrossRef]
46. Nour, H.E.; El-Sorogy, A.S. Distribution and Enrichment of Heavy Metals in Sabratha Coastal Sediments, Mediterranean Sea, Libya. *J. Afr. Earth Sci.* **2017**, *134*, 222–229. [CrossRef]
47. Vetrimurugan, E.; Shruti, V.C.; Jonathan, M.P.; Roy, P.D.; Rawlins, B.K.; Rivera-Rivera, D.M. Metals and Their Ecological Impact on Beach Sediments near the Marine Protected Sites of Sodwana Bay and St. Lucia, South Africa. *Mar. Pollut. Bull.* **2018**, *127*. [CrossRef]
48. Krek, A.; Krechik, V.; Danchenkov, A.; Krek, E. Pollution of the Sediments of the Coastal Zone of the Sambia Peninsula and the Curonian Spit (Southeastern Baltic Sea). *PeerJ* **2018**, *2018*, 1–17. [CrossRef]
49. Bigus, K.; Astel, A.; Niedzielski, P. Seasonal Distribution of Metals in Vertical and Horizontal Profiles of Sheltered and Exposed Beaches on Polish Coast. *Mar. Pollut. Bull.* **2016**, *106*, 347–359. [CrossRef]
50. Stauskaitė, R. Mineralogical Composition of Sand from the Baltic Sea Coastal Stretch between Šventoji- Jantarnoe (Palvininkai). *Proc. Sci. Acad. Lith. Ser. B* **1962**, *4*, 83–105.

51. Jarmalavičius, D.; Satkūnas, J.; Žilinskas, G.; Pupienis, D. The Influence of Coastal Morphology on Wind Dynamics. *Est. J. Earth Sci.* **2012**, *61*, 120–130. [CrossRef]
52. Radzevicius, R. Main Trends in Accumulation of Trace Elements from Surface Sediments of the Baltic Sea (Lithuanian Waters). *Baltica* **2002**, *15*, 63–73.
53. Remeikaitė-Nikienė, N.; Garnaga-Budrė, G.; Lujanienė, G.; Jokšas, K.; Stankevičius, A.; Malejevas, V.; Barisevičiūtė, R. Distribution of Metals and Extent of Contamination in Sediments from the South-Eastern Baltic Sea (Lithuanian Zone). *Oceanologia* **2018**, *60*, 193–206. [CrossRef]
54. Galkus, A.; Jokšas, K. *Nuosėdinė Medžiaga Tranzitinėje Akvasistemoje=: Sedimentary Material in the Transitional Aquasystem;* Geografijos Inst: Vilnius, Lithuania, 1997.
55. Pustelnikovas, O. *Geochemistry of Sediments of the Curonian Lagoon (Baltic Sea)*; Inst. of Geography: Vilnius, Lithuania, 1998.
56. Pupienis, D.; Buynevich, I.V.; Jarmalavičius, D.; Žilinskas, G.; Fedorovič, J. Regional Distribution of Heavy-Mineral Concentrations along the Curonian Spit Coast of Lithuania. *J. Coast. Res.* **2013**, *65*, 1844–1849. [CrossRef]
57. Žilinskas, G.; Jarmalavičius, D.; Pupienis, D. The Influence of Natural and Anthropogenic Factors on Grain Size Distribution along the Southeastern Baltic Spits. *Geol. Q.* **2018**, *62*, 375–384. [CrossRef]
58. Žilinskas, G.; Janušaitė, R.; Jarmalavičius, D.; Pupienis, D. The Impact of Klaipėda Port Entrance Channel Dredging on the Dynamics of Coastal Zone, Lithuania. *Oceanologia* **2020**, *62*, 489–500. [CrossRef]
59. Blott, S.J.; Pye, K. Gradistat: A Grain Size Distribution and Statistics Package for the Analysis of Unconsolidated Sediments. *Earth Surf. Process. Landforms* **2001**, *26*, 1237–1248. [CrossRef]
60. Sandgren, P.; Snowball, I. *Application of Mineral Magnetic Techniques to Paleolimnology BT-Tracking Environmental Change Using Lake Sediments: Physical and Geochemical Methods*; Last, W.M., Smol, J.P., Eds.; Springer Netherlands: Dordrecht, The Netherlands, 2001; pp. 217–237. [CrossRef]
61. Hunt, C.P.; Moskowitz, B.M.; Banerjee, S.K. Magnetic Properties of Rocks and Minerals. In *Rock Physics & Phase Relations*; Wiley: Hoboken, NJ, USA, 1995; pp. 189–204. [CrossRef]
62. Dearing, J. Environmental Magnetic Susceptibility. In *Using the Bartington MS2 System*; Chi Publ: Kenilworth, UK, 1999; Volume 43.
63. Hatfield, R. Particle Size-Specific Magnetic Measurements as a Tool for Enhancing Our Understanding of the Bulk Magnetic Properties of Sediments. *Minerals* **2014**, *4*, 758–787. [CrossRef]
64. Eberhards, G.; Lapinskis, J.; Saltupe, B. Hurricane Erwin 2005 Coastal Erosion in Latvia. *Baltica* **2006**, *19*, 10–19.
65. Garnaga, G. Integrated Assessment of Pollution in the Baltic Sea. *Ekologija* **2012**, *58*, 331–355. [CrossRef]
66. Liaghati, T.; Preda, M.; Cox, M. Heavy Metal Distribution and Controlling Factors within Coastal Plain Sediments, Bells Creek Catchment, Southeast Queensland, Australia. *Environ. Int.* **2004**, *29*, 935–948. [CrossRef]
67. Krek, A.; Danchenkov, A.; Ulyanova, M.; Ryabchuk, D. Heavy Metals Contamination of the Sediments of the South-Eastern Baltic Sea: The Impact of Economic Development. *Baltica* **2019**, *32*, 51–62. [CrossRef]
68. Kadūnas, V. *Technogeninė Geochemija=Technogenic Geochemistry*; Geology Institute: Vilnius, Lithuania, 1998.
69. Krek, A.; Ulyanova, M.; Koschavets, S. Influence of Land-Based Kaliningrad (Primorsky) Amber Mining on Coastal Zone. *Mar. Pollut. Bull.* **2018**, *131*, 1–9. [CrossRef] [PubMed]
70. Hakanson, L. An Ecological Risk Index for Aquatic Pollution Control.a Sedimentological Approach. *Water Res.* **1980**, *14*, 975–1001. [CrossRef]
71. Abrahim, G.; Parker, R. Assessment of Heavy Metal Enrichment Factors and the Degree of Contamination in Marine Sediments from Tamaki Estuary, Auckland, New Zealand. *Int. J. Devoted to Prog. Use Monit. Data Assess. Environ. Risks to Man Environ.* **2008**, *136*, 227–238. [CrossRef] [PubMed]
72. Turekian, K.K.; Wedephol, K.H. Distribution of the Elements in Some Major Units of the Earth's Crust. *GSA Bull.* **1961**, *72*, 175–192. [CrossRef]
73. Taylor, S.R. Abundance of Chemical Elements in the Continental Crust: A New Table. *Geochim. Cosmochim. Acta* **1964**, *28*, 1273–1285. [CrossRef]
74. Putys, P. *Geocheminiai Tyrimai Šilutės Plote. = Geochemical Analysis in Šilutė Area*; Lithuanian Geological Survey: Vilnius, Lithuania, 1999.
75. Baubinas, R.; Taminskas, J. *Karinė Gamtonauda Lietuvoje Sovietmečiu: Ekologinės Pasekmės =: Military Nature Use in Lithuania in the Soviet Years: Ecological Consequences*; Geografijos Institutas: Vilnius, Lithuania, 1998.
76. Bełdowska, M.; Jędruch, A.; Łęczyński, L.; Saniewska, D.; Kwasigroch, U. Coastal Erosion as a Source of Mercury into the Marine Environment along the Polish Baltic Shore. *Environ. Sci. Pollut. Res.* **2016**, *23*, 16372–16382. [CrossRef]
77. Gudelis, V.; Kirlys, V.; Stauskaitė, R.; Jankevičiūtė-Močiekienė, S. Dynamics of Swash (Wave Run-up) and the Lithodynamics of Sand Beaches on the Lithuanian Coast of the Baltic Sea. In О динамике отмелого песчаного берега (по наблюдениям на побережьеВосточной Балтики)//Развитие морских берегов в условиях колебательных движений земной коры.= *On the Dynamics of a Shallow Sandy Coast (According to Observations on the Coast of the Eastern Baltic)//Development of Sea Coast in a Vibrational Motion of the Earth's Crust*; Aibulatov, N.A., Dolotov, Y.S., Orlova, G.A., Yurkevich, M.G., Orvik, K.K., Orlova, G.A., Eds.; Valgus: Tallinn, Estonia, 1966; p. 240. Available online: https://erb.nlib.ee/?kid=13831161&fbclid=IwAR0Pd6ZUD77efQgwik5WFKezhWUxmVrHafO_gzNLtbjCh2dubRkkiYkJOY4 (accessed on 25 January 2021).
78. Apanaviciute, J.; Simkevicius, P. Distribution of Heavy Minerals in Surficial Bottom Sediments of the Nida—Klaipėda Mapping Area in the Baltic Sea. *Geologija* **2001**, *33*, 29–39.

79. Sergeev, A. *The History of Geological Development of the Curonian Spit in the Holocene and Modern Lithodynamic Processes in the Coastal Zone*; A.P. Karpinsky Russian Geological Research Institute: St. Petersberg, Russia, 2015. (In Russian)
80. Thompson, G.R.; Hower, J. The Mineralogy of Glauconite. *Clays Clay Miner.* **1975**, *23*, 289–300. [CrossRef]
81. Jarmalavicius, D.; Pupienis, D.; Zilinskas, G.; Janusaite, R.; Karaliunas, V. Beach-Foredune Sediment Budget Response to Sea Level Fluctuation. Curonian Spit, Lithuania. *Water* **2020**, *12*, 583. [CrossRef]
82. Salomons, W.; Förstner, U. *Metals in the Hydrocycle*; Springer Science & Business Media: Berlin/Heidelberg, Germany, 1984.
83. Jokšas, K.; Galkus, A.; Stakėnaitė, R.; Lagunavičienė, L. Heavy Metals in Bottom Sediments of Water System Transformed by Human Activity: Šventoji Port, Lithuania. In Proceedings of the 3rd International Conference in Lithuania "Metals in the Environment", Vilnius, Lithua, 22–23 May 2006; pp. 56–58. [CrossRef]
84. Galkus, A.; Joksas, K.; Stakeniene, R.; Lagunaviciene, L. Heavy Metal Contamination of Harbor Bottom Sediments. *Pol. J. Environ. Stud.* **2012**, *21*, 1583–1594.
85. Bitinas, A.; Damušytė, A.; Žaromskis, R.; Gulbinskas, S.; Žilinskas, G.; Jarmalavičius, D. Baltijos jūros Lietuvos krantų geologinis atlasas. *Geol. Akiračiai* **2005**, *2*, 14–22.

applied sciences

MDPI

Article

Sources and Metal Pollution of Sediments from a Coastal Area of the Central Western Adriatic Sea (Southern Marche Region, Italy)

Federico Spagnoli [1,2,3], Rocco De Marco [1], Enrico Dinelli [4], Emanuela Frapiccini [1,*], Fabrizio Frontalini [5] and Patrizia Giordano [6]

1 National Research Council-Institute of Marine Biological Resources and Biotechnologies (CNR-IRBIM), Largo Fiera della Pesca, 60125 Ancona, Italy; federico.spagnoli@cnr.it (F.S.); rocco.demarco@cnr.it (R.D.M.)
2 Istituto per le Scienze Marine (ISMAR), Consiglio Nazionale delle Ricerche, Largo Fiera della Pesca 2, 60125 Ancona, Italy
3 Geology Division, School of Science and Technology, University of Camerino, Via Gentile III da Varano, 62032 Camerino, Italy
4 Dipartimento di Scienze Biologiche, Geologiche e Ambientali, Alma Mater Studiorum Università di Bologna, Piazza di Porta S. Donato 1, 40126 Bologna, Italy; enrico.dinelli@unibo.it
5 Dipartimento di Scienze Pure e Applicate (DiSPeA), Campus Scientifico Enrico Mattei, Università degli Studi di Urbino "Carlo Bo", Località Crocicchia, 61029 Urbino, Italy; fabrizio.frontalini@uniurb.it
6 Istituto di Scienze Polari, Consiglio Nazionale delle Ricerche, Via Gobetti 101, 40129 Bologna, Italy; patrizia.giordano@cnr.it
* Correspondence: emanuela.frapiccini@cnr.it

Abstract: Sediments represent a critical compartment of marine coastal ecosystems due to the toxic and long-lasting effects of the contaminants buried therein. Here, we investigated the properties of surficial sediments in front of the Southern Marche Region coast (Central Adriatic Sea, Italy). The grain size of the surficial sediments was determined by X-ray sedigraphy. TN and OC contents were determined by elemental analysis. The concentrations of Al, Fe, Mg, K, S, Ca, Ti, P, Na, Mn, Mg, Li, As, Ba, Ga, Pb, Sr, and Zn were determined by ICP-OES to evaluate their spatial patterns and temporal trends. A Q-mode Factor Analyses was applied and resulted in the identification of three compositional facies (Padanic, Coastal, and Residual) characterized by common biogeochemical, mineralogical, sedimentological properties, transport pathway, and source. Some pollution indicators, such as the enrichment factor, the geoaccumulation index, and the pollution load index were calculated to assess the deviation from the natural background levels. The results showed a pollution by As and Ba due to the human activities in the 20th century. Furthermore, a general decreasing of Al, Ti, P, Co, Cr, Cu, Ga, Ni, Pb, Sc, V, and Y concentrations from the background levels suggested a change in the sedimentation processes during the last decades.

Keywords: pollution indicators; surficial sediments; central western Adriatic Sea

Citation: Spagnoli, F.; De Marco, R.; Dinelli, E.; Frapiccini, E.; Frontalini, F.; Giordano, P. Sources and Metal Pollution of Sediments from a Coastal Area of the Central Western Adriatic Sea (Southern Marche Region, Italy). *Appl. Sci.* **2021**, *11*, 1118. https://doi.org/10.3390/app11031118

Academic Editor: Mauro Marini
Received: 22 December 2020
Accepted: 22 January 2021
Published: 26 January 2021

1. Introduction

The biogeochemical and sedimentological properties of bottom marine sediments are the results of complex interactions between their composition, sourcing areas, transport, depositional, and early diagenesis processes [1]. Additionally, anthropic activities by discharging contaminants can further influence the natural composition of the sediments. Considering all the factors affecting the biogeochemical and the sedimentological composition of the bottom marine sediments, information about the source area, the sedimentological processes, and the anthropic inputs can be inferred [2–7]. Wastewater discharges such as domestic wastes, drugs, fertilizers, and zootechnical byproducts represent the major anthropogenic contributions in marine coastal area [8]. Anthropogenic chemicals can be introduced into the marine environments by different sources (i.e., effluent treatment plants, accidental discharges, dumping, riverine inputs, surface runoff, atmospheric deposition, etc.) and then accumulate in sediments [9–15]. In coastal marine sediment, however, heavy metals might be sourced by anthropic activity or by natural sources [8,16] and could represent a serious hazard due to their toxicity, bioavailability and persistence [17,18]. Because of the adsorption, hydrolysis, and co-precipitation of metal ions, most of the water column metals are deposited in the marine sediment and only a small portion of these ions remains dissolved in the water column [19]. Then, marine sediments enriched in metals can release them by various remobilization processes to the water column [1].

Marine sediments are preferred over seawater for monitoring environmental quality because pollutants are much higher and less variable in time and space than in seawater, and they reflect in an integrated manner the pollution level in the area [7,20–22]. Marine sediments act not only as a reservoir of contaminants, but also serve as a source of toxicants to marine fauna. This can be due to the ingestion of bottom or resuspended sediment particles [23,24], or to the adsorption of solutes present in the pore waters and on the bottom. The solutes present near the bottom can be the consequence of diffusive or bio-irrigation fluxes or resuspension [1], resuspension that can be due to natural (by storms, waves, tidal currents) or anthropogenic (dredging, trawling) processes [25,26]. For these reasons, sediments are commonly targeted in marine pollution surveillance programs [27–29]. Generally, natural metal concentrations in sediments are not detrimental to inhabiting organisms [30]. For metabolism, organisms essentially require some key micronutrients (i.e., iron, copper, zinc, manganese, cobalt) [31] but at high metals concentration can pose a toxicological threat to marine organisms [8,32,33]. Sediment features such as mineralogy, sediment texture, metal properties, pH, organic matter, oxidation–reduction potential, and adsorption and desorption processes are important controlling factors for the accumulation and availability of heavy metals in the sediment [34–42]. Thus, sediments are considered as sources of heavy metals in marine environments and play a key role in their deposition and transmission along the trophic chain and must be therefore monitored. The current European Union (EU) legislation for the protection and conservation of the marine environment emphasizes the need to evaluate and limit the concentrations of contaminants, and to undertake preventive measures to re-establish a Good Environmental Status in impacted marine areas, as requested by Marine Strategy Framework Directive (MSFD, 2008/56/EC, Commission Decision 2010/477/EU).

The present study aims to recognize the sedimentological processes that drove the formation of the present sediment bodies on the base of the geochemical and sedimentological properties of the superficial marine sediments in a central Italian Adriatic Sea coastal area. In addition, the trace element pollution in the surface sediment of this area is investigated. The level of pollution has been evaluated by the elaboration of contamination index and by comparison with the limits of the Italian legislation [43]. The comparison with these limits needs to be considered as an indication due to the different methods used to measure the element concentrations. The research has been carried out by using a statistical elaboration of the biogeochemical and sedimentological variables. The statistical approach allowed the determination of compositional facies, that is, areas with common geochemical, mineralogical, and sedimentological properties. Previous studies have investigated the sedimentological and geochemical processes and recognized the main source area of sediments as well as anthropic heavy metal inputs in different offshore areas of the Adriatic Sea [4,13,44–46] but little is known about the present study area that instead shows a specific role as consequence of the coastal morphology on the Adriatic Sea circulation [47].

2. Study Area

The Adriatic Sea is located between the Italian peninsula and the Balkan Region; it is an elongated basin (about 800 km long and 200 km wide) oriented NW/SE and is connected to the Mediterranean Sea through the Otranto Strait. The Adriatic sea is historically divided into three distinct sectors [48]: the northern sector is characterized by a shallow continental shelf with a very gentle slope of about 0.02° until the isobath of 100 m; the middle sector, between the 100 m isobath and the Gargano Promontory, where a basin, called Mid-Adriatic Depression (MAD), or Jabuka Pit or Pomo Pit, occurs; this basin exhibits three sub-trenches, with maximum depths of 255 m (Western Pit), 270 m (Central Pit), and 240 m (Eastern Pit) [49]; the southern sector, between the Gargano Promontory and the Otranto Strait, which shows a maximum depth of 1260 m in the South Adriatic Depression (SAD). The hydrodynamic of the Adriatic Sea is characterized by a general anticlockwise circulation, aligned to the coasts that reaches its maximum in winter and spring, and by weak currents resulting in a series of clockwise and anti-clockwise gyres, in summer (Figure 1). In the eastern side of the basin, the current is directed northward (The East Adriatic Current: EAC). The EAC is characterized by warmer and high saline waters coming from the Levantine Basin (Levantine intermediate water, LIW), and from Ionian Basin (Ionian surface water: ISW) and in some cases from the Modified Atlantic Waters (MAW) following the pattern of the Bimodal Oscillating System (BiOS) [50–53]. The Adriatic Sea circulation is complicated by the formation of dense waters in the Northern Adriatic Sea (North Adriatic Dense Water: AdDW). These are cold and high salinity waters forming in winter in the Northern Adriatic Sea when the Bora wind blows [54,55]. The AdDW annually or biennially, generally in spring, flows southward close to the coast and reaches partly the MAD and flowing further south [56] (Figure 1).

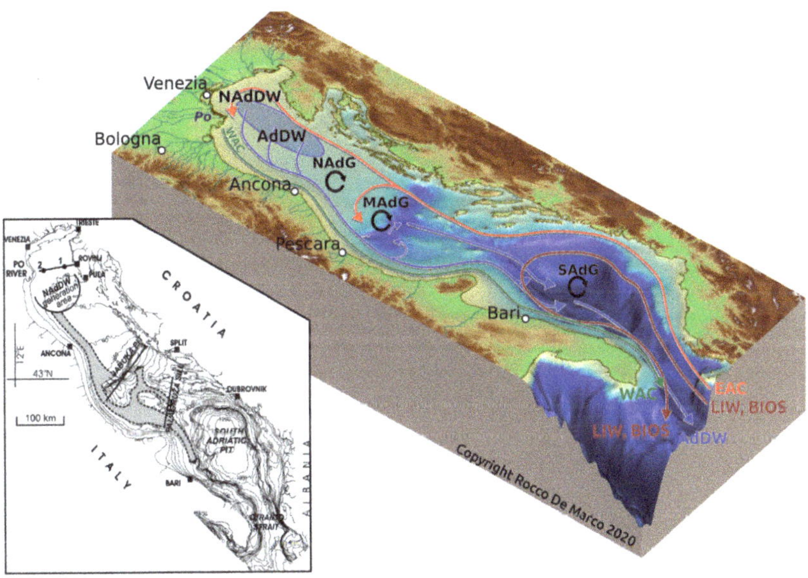

Figure 1. Bathymetric map of the Adriatic Sea with the three morphological sectors (lower left corner), the general Adriatic Sea circulation (in the center) and the study area (on the upper right corner). WAC: Western Adriatic Current; EAC: Eastern Adriatic Current; NAdG: North Adriatic Gyre; MAdG: Mid Adriatic Gyre; SAdG = South Adriatic Gyre. AdDW: Adriatic Dense Water (in blue North Adriatic formation area of AdW-NAdDW). LIW: Levantine intermediate Water. BIOS: Adriatic-Ionian bimodal oscillating system [50–53].

In this context, waters and the fine sedimentary load of the Po and Apennine rivers as well as of the local rivers flowing along the South Marche coast, tend to branch out offshore in summer and to be confined near the coast in winter [4,57]. Overall, the combined action of currents and waves induces a redistribution of bottom sediments towards the south and, to a certain extent, towards the open sea, in the Northern and Central Adriatic Sea [2,4]. The main sedimentary inputs of the northern and central sectors of the Adriatic Sea are located along the western coasts since the contributions from the Balkan rivers are limited, due to the prevalent carbonatic composition of the rocks, and the confinement within the inshore basins parallel to the coast [58]. The sediment contribution of the Northern and Western Adriatic Sea rivers is represented by the northern Alpine rivers that contribute for 8 MT/y, the Po basin, contributing for 13 MT/y, and the Apennine rivers that contribute for 22 MT/y, for a total of 43 MT/y [59]. The catchment area of the Po River includes igneous, metamorphic and sedimentary rocks of the western and central Alps, carbonatic rocks mainly from the northern Alpine rivers, and a mixture of carbonatic and silicoclastic sediments from the Po alluvial System [42]. South of the Po River mouths the solid inputs of these rivers are composed mainly of terrigenous sediments with varying contents of carbonates [60]. The solid load of the minor rivers Southern of the Po River is estimated to be 16.9×10^6 MT/y, between the Po River and Ancona, and 7.8×10^6 MT/y, between Ancona and Punta Penne [61].

The current sea bottom sedimentological features of the northern and central sectors of the Adriatic Sea are the result of the last sea level rise and the following and present general Adriatic Sea hydrodynamics and riverine inputs. Indeed, the northern and central Adriatic Sea bottom is characterized by a Holocene accretionary wedge parallel to the Italian shoreline. This wedge is made up of coarse sediments near the coast, followed by finer sediments eastward. Further offshore, the Holocene accretionary wedge gives way to coarser sediments again consisting mainly of reworked sands. These sandy substrates were left in place by the last fast transgression and are due to the outcropping of partial reworked beach sands, paralic sediments and old filled valleys [58,62,63].

The study area is located in the central part of the western Adriatic Sea, in front of the southern part of the Marche Region (from the Chienti River to the Tronto River), from the shoreline to a depth of about 50 m (Figure 2). The main tributaries in the area are the Chienti, Tenna, Aso and Tronto Rivers. These rivers are located in an area with neighboring industrial activities, are contaminated with domestic and industrial wastewater and are exposed to pesticides used in surrounding agricultural areas [64]. From west to east, these rivers drain the carbonatic rocks of the Apennines, followed by the arenitic and clay Miocene rocks, the Plio-Pleistocene clays, and the arenitic and conglomeratic Pleistocene formations (Figure 2a). From a morphological point of view, the study area is characterized by a flat bottom gently sloping eastward, with weak dipping near the coast, slightly higher dipping in the center and again a less sloping to the East (Figure 2b). Only two minor river catchments, namely Ete and Tesino, do not include the Apennine carbonatic rocks (Figure 2). In the study area, the Holocene wedge is formed by an accretionary prism that tends to become thinner towards the coast and offshore [47]. Moreover, this coastal area undergoes intensive bottom sediment resuspension due to strong autumn and winter storms [65–67].

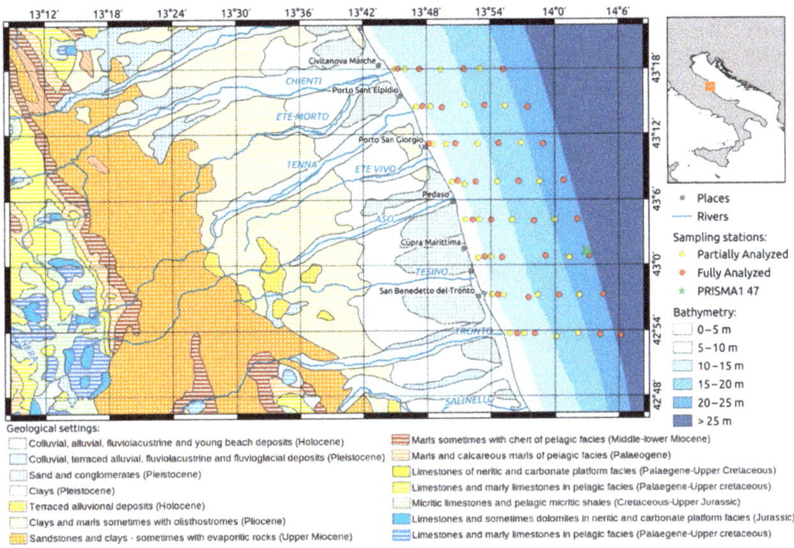

(**a**)

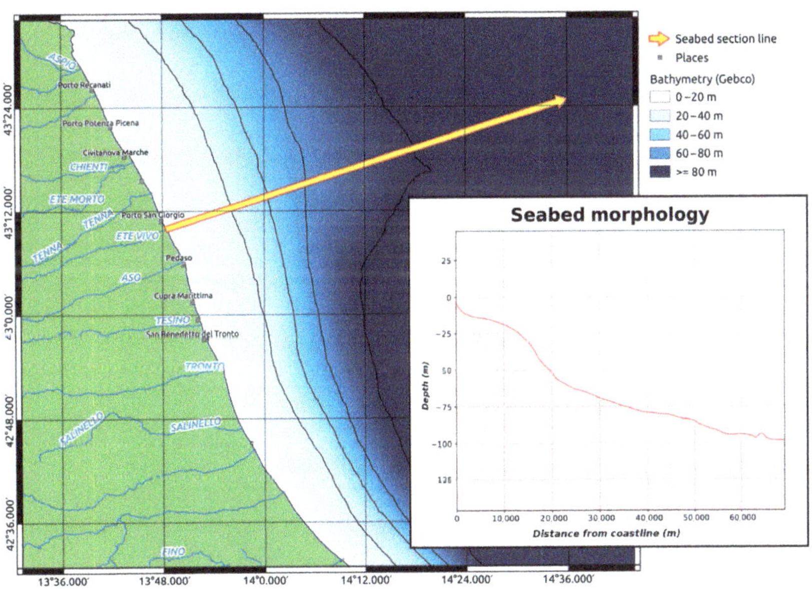

(**b**)

Figure 2. Study area with the mainland geological setting and sampling stations, red and yellow dots, respectively (**a**); topographic section along the study area (**b**). The map has been generated using QGIS 3.10.4-A Coruña. Rivers courtesy of ISPRA (http://www.sinanet.isprambiente.it/it/sia-ispra/download-mais/reticolo-idrografico/view), and the geological setting courtesy of Geoportale Nazionale (http://www.pcn.minambiente.it/viewer/).

3. Materials and Methods

3.1. Sampling

Sediment samples were collected by a box-corer (sizes 101 × 17 w × 25 d cm) following a grid of 64 stations located in front of the Southern Marche Region coast, from shoreline to offshore (Figure 2a). The sediment samples were collected during the CASE1 cruise, in April 2010, carried out by the CNR-ISMAR of Ancona on the M/N G. Dallaporta. Each box-core was described for the macroscopic characteristics (color by Munsell tables, presence of organism and bioturbation, hydration and texture, sedimentary structures, grain-size) of the lateral and superficial surfaces. pH and Eh were measured in the sediment by punching in electrodes in the first 2 cm. The sediment samples were then collected at different depth on the base of the texture of the sediment core. On the basis of present research, only the superficial samples (0–2 cm) have been considered. Each sediment sample was split in aliquots immediately after the sub-sampling for the various analyses. The aliquot for the grain-size and water-content analyses was stored at 4 °C, while samples for geochemical analyses were stored at −20 °C and then freeze-dried.

Some samples (red dot in Figure 2a) were analyzed for grain-size and biogeochemical variables (total carbon (TC), total nitrogen (TN), organic carbon (OC), $\delta^{13}C$, and chemical elements); while other samples (yellow dot in Figure 2a) were analyzed only for grain-size, TC, TN, and OC.

3.2. Grain-Size and Organic Geochemistry

Grain-size analyses were carried out on wet sediment samples pre-treated with H_2O_2-16 vol. solution to remove organic matter. The coarser fractions were dry-sieved over the range 2 mm down to 0.063 mm by a stack of sieves ISO 3310 in accordance with UKAS Traceability Test Sieve LAB22-1. Bioclast component (>2 mm fraction) was separated by sands (2 mm < sands > 0.063 mm), while the finer fractions were analyzed by X-ray sedigraph (Micrometrics 5000D) [45].

The TC and TN were measured with a FisonsNA2000 elemental analyzer on the freeze-dried bulk sediment, while the OC was analyzed after acidification in hydrochloric acid 1.5 M [68] by the same instrument. The inorganic carbon (IC) was determined from the difference between TC and OC.

Stable isotopic analyses of OC ($\delta^{13}C$) were carried out on samples by using a FINNIGAN Delta Plus mass spectrometer directly coupled to the elemental analyzer. Stable-isotope data were expressed in ‰ relative to the variation (δ) from the international PDB standard.

3.3. Inorganic Geochemistry

The analyses of chemical elements (Ag, Al, As, Ba, Be, Bi, Ca, Cd, Co, Cr, Cu, Fe, Ga, Hg, K, Mg, Li, Mn, Mo, Na, Ni, P, Pb, Sb, S, Sc, Sr, Te, Ti, Tl, U, V, W, Y, Zn, and Zr) were carried out on freeze-dried sediment sample after acid digestion employing HF, $HClO_4$, HNO_3, and HCl and the following determination of the concentration of each element in the eluate solution by Inductively Coupled Plasma-Optical Emission Spectroscopy (Activation Laboratory LTD, Ancaster, ON, Canada).

To identify possible anthropic contributions, some pollution indicators, such as the contamination factor (C_f), the degree of contamination index (C_d), the enrichment factor (EF), the geoaccumulation index (I_{geo}), and the pollution load index (PLI) of heavy metals and other trace elements were calculated [69] (Table 1). Furthermore, element concentrations were compared to the threshold values of the Italian legislation [43], to the general abundance data reported for the marine shale [70] (Table 2) and the ecotoxicological risks with the Sediment Quality Guidelines (SQGs) indicated in Rachel and Wasserman, 2015 [71].

Table 1. Summary of the classification categories of sediment contamination derived from the different applied indicators.

C_f	Classification	EF	Classification	I_{geo}	Classification	mC_d	Classification	PLI	Classification
<8	Low	≤2	Low enrichment	<0	Unpolluted	<1.5	Very low	<1	Unpolluted
8–16	Moderate	2–5	Moderate enrichment	0–1	Unpolluted to moderately polluted	$1.5 < mC_d < 2$	Low	1–2	Unpolluted to moderately polluted
16–32	Considerable	5–20	High enrichment	1–2	Moderately polluted	$2 < mC_d < 4$	Moderate	2–3	Moderately polluted
>32	Very high	20–40	Very high enrichment	2–3	Moderately to strongly polluted	$4 < mC_d < 8$	High	3–4	Moderately to highly polluted
		>40	Extreme enrichment	3–4	Strongly polluted	$8 < mC_d < 16$	Very high	4–5	Highly polluted
				4–5	Strongly to extremely strongly polluted	$16 < mC_d < 32$	Extremely high	>5	Very highly polluted
				>5	Extremely polluted	$32 < mC_d$	Ultra high		

Table 2. Reference values, local background, normative limits and ecotoxicological reference concentrations for the elements investigated in the present study, when available 1 [70].

	Unit	Marine Shales (1)	Local Background	L1 DM 173/2016	L2 DM 173/2016	DM 173/2016	TEL	ERL	PEL	ERM
Ag	ppm	0.07								
Al	%	8	6.9			1				
As	ppm	13	10	12	20	1	5.9	33	17	85
Ba	ppm	580	295							
Be	ppm	3								
Ca	%	1.6	11							
Cd	ppm	0.3		0.3	0.8	0.03	0.6	5	3.53	9
Co	ppm	19	13							
Cr	ppm	90	128	50	150	1	37	70	90	145
Cu	ppm	45	40	40	52	1	36	70	197	390
Fe	%	4.72	3.5			1				
Ga	ppm	19	16.2							
Hg	ppm	0.04		0.3	0.8	0.03	0.2	0.2	0.49	1.3
K	%	2.66	1.6							
Li	ppm	66	2.2							
Mg	%	1.5								
Mn	ppm	850	774.5							
Mo	ppm	2.6								
Na	%	0.96								
Nb	ppm	11								
Ni	ppm	68	73	30	75	1	18	30	36	50
P	ppm	700	0.06							
Pb	ppm	20	18	30	70	1	35	35	91.3	110
S	ppm	2400	n.a							
Sc	ppm	13	29							
Sr	ppm	300	348							
Te	ppm	0.08	n.a							
Ti	ppm	4600	0.3							
V	ppm	130	116			1				
Y	ppm	26	25							
Zn	ppm	95	48	100	150	1	123	120	315	270
Zr	ppm	160	97							

Contamination factor (C_f). The C_f [5] is an index developed to evaluate the contamination of a given toxic substance in a basin, and it is calculated as:

$$C_{fi} = \frac{C_{ei}}{C_{bi}} \quad (1)$$

where $C_{ei}$is the concentration of the substance i in sediment samples, and C_{bi} is the background values of the same substance indicated in Table 2. In this work, the C_{fi} was calculated for all elements even if only trace elements that may be affected by human activities (As, Ba, Co, Cr, Cu, Ni, Pb, V, and Zn) or that can be indicative of peculiar processes (Ga, Sc, Sr, Y, and Zr) have been discussed.

The background values (C_{bi}) of the element i and of the other elements not considered for the calculation of C_{fi} (Table 2) were deduced by using the concentrations in the sample collected at a depth of 105 cm of the sediment core 47 recovered in the study area (Figure 2a, green star) during the PRISMA1 project [4]. The element concentration in the core of the PRISMA1 project were determined by XRF following the method specified in [5] so that the concentrations can be compared to those determined by strong acid dissolution, both referring to the total concentrations. The core 47 is located inside the study area and the level 105 is referred to a time previous the industrial age as deduced by the radiometric and metal anthropogenic data of the PRISMA1 and by the accumulation rates determined in this area by previous studies [72] so that is not affected by anthropic inputs. Furthermore, by the normalization respect to the Al or Ti content determined in the same sample, it can be compared to samples with different grain size composition. Other methods such as those used in an area in front of the Abruzzo coast [73] can be useful applied when available only superficial samples.

Enrichment factor (EF). The EF has been developed to evaluate the metal contamination. The EF normalizes the trace element content with respect to a sample reference metal, in this case the Al:

$$EF = \frac{\frac{M}{Al}Ss}{\frac{M}{Al}Bs} \quad (2)$$

where (M/Al)Ss is the ratio of each metal and Al concentrations of the sediment sample and (M/Al)Bs is the same ratio in the background sample (Table 2). The ecological risks based on EF values are categorized according to Table 1.

Geoaccumulation index (I_{geo}). The I_{geo} is an indicator of heavy metal contamination of sediments with respect to the background natural levels (C_{bi}). I_{geo} is expressed for each metal as:

$$I_{geo} = \frac{C_{ei}}{1.5 * C_{bi}} \quad (3)$$

Based on the I_{geo} values sediments are classified according to the classes shown in Table 1.

Modified degree of contamination index (mC_d). The mC_d, is the sum of all contamination factors of various heavy metals. It is obtained as follow:

$$mC_d = \frac{\sum_{i=1}^{i=n} C_f}{n} \quad (4)$$

in which n is the number of analyzed elements and i = ith element. The classification of mC_d is shown in Table 1. In this work, the mC_d has been calculated considering only metals or trace elements that show almost one value of $C_f > 1$ (As, Ba, Ga, Pb, Sr, and Zn).

Pollution load index (PLI). The Pollution Load Index (PLI) evaluates the heavy metal contamination of sediment samples as:

$$PLI = (C_{f1} \times C_{f2} \times C_{f3} \times \ldots \times C_{fn})^{1/n}, \quad (5)$$

where C_f is the contamination factor and n is the number of metals. The PLI was calculated retaining only metals or trace elements that show at least one value of $C_f > 1$ (As, Ba, Ga, Pb, Sr and Zn). According to the contamination degree, the PLI data were classified as reported in Table 1 [74].

3.4. Statistical Analyses

Selected variables (i.e., major and trace element, grain-size, OC, IC, $\delta^{13}C$, TN, OC/TN, pH, and Eh) were processed using a multivariate statistical analysis (i.e., Factor Analysis, FA). This analysis allowed us to recognize the primary relationships between a series of samples, greatly reducing the size of the multidimensional system without losing a significant amount of information. The analysis consisted of a Q-mode FA applied to the sediment samples and their compositional variables [75]. The FA was carried out using the following steps [76]: (1) standardization of the initial data between the minimum and the maximum value for each variable; (2) correlation of the variables using the similarity coefficient $\cos\theta$ [77]; (3) choice of a number of factors for which the sum of the individual variances ranges between 90–95% of the total variance of the system so that not much significant information is lost; and (4) final rotation of the axes-factors of the new system keeping them orthogonal. This last step was carried out to obtain better guidance on the original samples and to preserve the independence of the variables (Varimax criterion).

The FA was carried out to avoid an excessive number of variables respect to the number of samples. The selected variables of each sample were processed to obtain statistical factors clustering similar geochemical and sedimentological compositions. These factors represent biogeochemical-sedimentary facies (BSF), that is, the component of each sample that has been subjected to the same transport and depositional processes and whose sediments are from the same source areas. The areal mapping of the score of each factor extracted for each sample allows us to obtain the areal distribution of the different BSF and then to infer the main sedimentological and biogeochemical processed that took place in the area. The areal distribution maps of the BSF as well as of the single variables were drafted by using the tension continuous curvature spline method [78]. All the biogeochemical and sedimentological variables were plotted according to their areal distribution following a standardized method that creates a grid file with GMT v. 6.0.0 [79] and plots the map by QGIS v. 3.10.4-A Coruña. The areal distribution of each variable is shown in Figure S1 (Supplementary Materials).

4. Results and Discussion

A summary of statistics of the investigated variables is reported in Table 3. Selected chemical elements (i.e., Cd, Sb, Tl, U, W) have not been considered as their concentrations were either below the detection limits or above detection limits in a very limited number of samples (in parenthesis after the element): Bi (1), Hg (2).

Table 3 also includes information about Ag (9), Mo (16), Te (19), Be (29), and As (41) although they were not observed in some of the samples. All these elements were excluded from the multivariate statistical analysis (the complete dataset is reported in Supplemental Table S1). Table 2 presents some reference values used in the discussion, the background concentrations considered for some elaborations, regulatory values and the ecotoxicological sediment values. Meanwhile, Table 4 includes basic statistics relative to the various indexes calculated for describing the sediment status.

Table 3. Summary statistics for the investigated variables.

	Unit	Mean	Median	Min	Max	Dev. St	*n*
Water content	%	35.8	36.4	20.2	59.1	11.3	64
Porosity	%	56.3	58.2	38.1	77.9	11.9	64
Bioclasts	%	1.1	0.1	0.0	19.6	3.1	64
Sand	%	33.5	17.3	0.2	96.7	34.2	64
Silt	%	34.9	38.1	2.4	61.8	18.5	64
Clay	%	30.5	31.7	0.8	66.2	19.9	64
TN	%	0.1	0.1	0.0	0.1	0.0	64
IC	%	4.3	4.4	3.1	5.5	0.6	64
OC	%	0.4	0.4	0.1	1.0	0.2	64
$\delta^{13}C_{PDB}$	‰	−24.0	−23.7	−26.4	−21.2	0.8	64
OC/TN		8.8	8.5	4.9	11.5	1.2	64
carbonates	%	3.7	3.6	2.4	5.9	0.7	63
pH		7.3	7.3	6.8	8.1	0.3	64
Eh	mV	−98.1	−125.9	−250.0	143.7	88.3	64
Ag	ppm	0.2	0.2	<0.3	1.6	0.2	9
Al	%	4.0	3.7	2.5	6.0	1.2	48
As	ppm	9.1	9	<3	31	5.8	41
Ba	ppm	272.8	271	234	329	21.6	48
Be	ppm	1.1	1	<1	2	0.6	29
Ca	%	12.8	13.55	8.9	15.4	2.1	48
Cd	ppm			<0.3			0
Co	ppm	7.8	7	3	13	3.2	48
Cr	ppm	61.5	61	27	101	15.9	48
Cu	ppm	14.4	12	3	32	8.0	48
Fe	%	1.9	1.7	0.79	3.3	0.8	48
Ga	ppm	12.3	12.5	5	22	5.2	48
Hg	ppm	0.6	0.5	0.5	2.0	0.3	2
K	%	1.4	1.24	1	2.4	0.4	48
Mg	%	1.2	1.095	0.58	2.0	0.5	48
Li	ppm	26.8	23	8	53	14.8	48
Mn	ppm	703.9	714.5	506	873	67.6	48
Mo	ppm	1.1	0.5	<1	4.0	1.0	16
Na	%	1.6	1.52	1.02	2.4	0.3	48
Ni	ppm	32.7	28.5	10	63	17.3	48
P	%	0.0	0.049	0.026	0.1	0.0	48
Pb	ppm	12.1	11	6	24	4.7	48
S	ppm	0.2	0.2	0.14	0.4	0.0	48
Sc	ppm	6.8	6.5	<4	12	3.0	42
Sr	ppm	352.9	354	280	436	34.0	48
Te	ppm	3.0	1	1	12	3.0	19
Ti	%	0.2	0.185	0.11	0.3	0.0	48
V	ppm	49.4	44.5	14	105	26.8	48
Y	ppm	13.7	14	9	16	1.7	48
Zn	ppm	49.0	45	15	93	25.2	48
Zr	ppm	35.3	35.5	11	60	13.8	48

Table 4. Statistical values (mean, median, min, max and dev.st.) of the contamination indexes data and pollution indicators.

		Mean	Median	Min	Max	Dev. Std.	*n*
As	C_f	1.0	1.0	0.3	3.1	0.5	41
	EF	1.8	1.6	0.7	4.8	0.8	41
	I_{geo}	−0.7	−0.6	−2.3	1.0	0.6	41
Ba	C_f	0.9	0.9	0.8	1.1	0.1	48
	EF	1.7	1.8	1.0	2.7	0.5	48
	I_{geo}	−0.7	−0.7	−0.9	−0.4	0.1	41
Co	C_f	0.6	0.5	0.2	1.0	0.2	48
	EF	1.0	1.0	0.6	1.2	0.2	48
	I_{geo}	−1.5	−1.5	−2.7	−0.6	0.6	48
Cr	C_f	0.5	0.5	0.2	0.8	0.1	48
	EF	0.9	0.8	0.5	1.6	0.2	48
	I_{geo}	−1.7	−1.7	−2.8	−0.9	0.4	41
Cu	C_f	0.4	0.3	0.1	0.8	0.2	48
	EF	0.6	0.6	0.2	1.2	0.2	48
	I_{geo}	−2.3	−2.3	−4.3	−0.9	0.9	48
Ga	C_f	0.8	0.8	0.3	1.4	0.3	48
	EF	1.2	1.3	0.8	1.6	0.2	48
	I_{geo}	−1.0	−0.9	−2.0	−0.1	0.6	41
Ni	C_f	0.4	0.4	0.1	0.9	0.2	48
	EF	0.7	0.7	0.3	1.0	0.2	48
	I_{geo}	−1.9	−1.7	−3.5	−0.8	0.8	41
Pb	C_f	0.7	0.6	0.3	1.3	0.3	48
	EF	1.1	1.1	0.7	2.4	0.2	48
	I_{geo}	−1.2	−1.3	−2.2	−0.2	0.5	41
Sc	C_f	0.3	0.3	0.1	0.4	0.1	42
	EF	0.4	0.4	0.3	0.5	0.0	42
	I_{geo}	−2.6	−2.4	−3.4	−1.9	0.5	37
Sr	C_f	1.0	1.0	0.8	1.3	0.1	48
	EF	1.9	1.9	1.0	2.9	0.6	48
	I_{geo}	−0.6	−0.6	−0.9	−0.3	0.1	41
V	C_f	0.4	0.4	0.1	0.9	0.2	48
	EF	0.7	0.7	0.3	1.1	0.2	48
	I_{geo}	−1.9	−1.7	−3.6	−0.7	0.9	41
Y	C_f	0.5	0.6	0.4	0.6	0.1	48
	EF	1.0	1.0	0.7	1.4	0.2	48
	Igeo	−1.5	−1.4	−2.1	−1.2	0.2	41
Zn	C_f	0.7	0.6	0.2	1.3	0.3	48
	EF	1.1	1.0	0.5	1.5	0.3	48
	I_{geo}	−1.4	−1.3	−2.9	−0.3	0.8	48
Zr	C_f	0.4	0.4	0.1	0.6	0.1	48
	EF	0.6	0.6	0.3	0.9	0.2	48
	I_{geo}	−2.2	−2.0	−3.7	−1.3	0.7	48
mC_d	0	0.8	0.8	0.0	1.3	0.3	50
PLI	0	0.7	0.7	0.0	1.1	0.4	50

4.1. Sedimentological Processes and Facies Recognition

The statistical elaboration of the biogeochemical and sedimentological variables highlights the presence of three main factors explaining, on overall, 94.6% of the total variance. These factors mainly represent the three key BSFs in the coastal marine area of the Southern Marche.

The most significant *facies* (BSF1) explains 79.2% of the total variance and is characterized by sediments with (Figure 3a) high contents of clay and fine silt, some elements (i.e., Al, Ti, Mg, K, Na, P, Fe, Co, Cu, Zn, Ni, Pb, V, Ga, Li, Y, and Zr mainly linked to clay minerals), OC that is directly linked to the Organic Matter (OM), and TN that is contained both in the OM and in the clay minerals. The SBF1 is therefore sedimentological facies consisting primarily of fine pelitic sediments with abundant clay minerals and OM. The extracted scores of these facies increase seaward, starting from values close to zero, that imply absence of pelitic sediment, clay minerals and OM, near coast, to values close to 1 (i.e., sediments made up of almost all fine sediments) (Figure 3b). By considering the areal distribution of the fine pelitic, organic and clay-rich sediments of BSF1 and the general cyclonic circulation of the Adriatic Sea as well as the reworking and removing action of the wave near the coast, the sediments of the BSF1 are inferred to be mainly sourced from the fine load of the Po River as well as from that of the Apennine rivers in the northern part of the study area. For this reason, the BSF1 facies can be assimilated and named as the *Padanic Facies* of Spagnoli et al. (2014) [4]. The fine particles introduced in the Adriatic Sea in front of the river mouths and are then resuspended by the wave action and transported southwards by the WAC [2]. As a result, the fine particles settle at greater depths in a belt area aligned to the coastline, where the action of the wave motion on the seabed is weaker [4]. Furthermore, the WAC confines the suspended sediments near the coast, preventing their spreading eastward limiting their sedimentation towards the center of the basin. The fine sediments of the BSF1 are subject to a continuous reworking process acting on the sea bottom. This consists in repeated resuspension and redeposition processes, due to strong storms that result in the transportation and dilution processes of the BSF1 sediments southward. As a consequence of this sedimentological and hydrographic set up, the present offshore sediment accumulation is very feeble, and the relict sediments connected to the last sea level rise tend to surface in the deeper side of the north-central Adriatic Sea. The BSF1 facies approaches towards the coast between the Chienti and the Ete Vivo rivers and in front of the Tronto river where it also shows a maximum (Figure 3b). This distribution near the coast suggests that the Chienti, Tenna and, to a greater extent, the Tronto rivers discharge more fine sediment than the other rivers in the southern Marche.

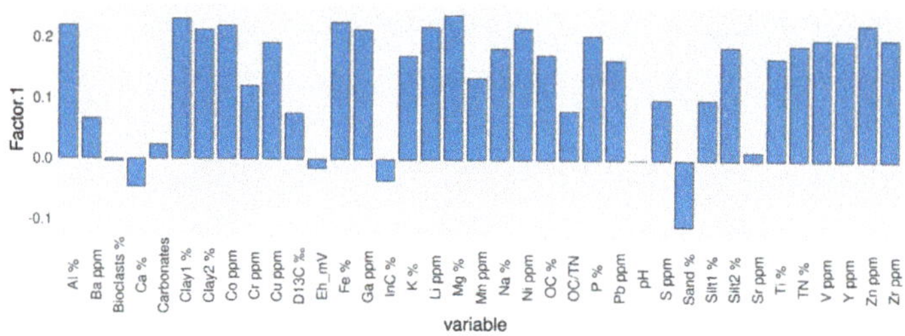

(a)

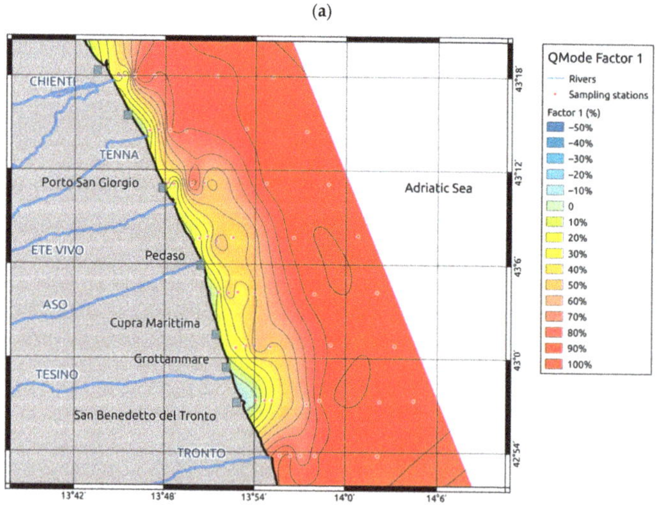

(b)

Figure 3. Histogram of the factor scores (**a**) and map of the areal distribution of the of the BSF1 (**b**).

The second facies (BSF2) represents 14.5% of the total variance and is marked by high contents of sand, IC, Ca, Sr, OC/TN, pH, and Eh (Figure 4a). Calcium and IC are typical of sediments with abundant carbonate, while high values of pH and Eh are associated to coarse grain-sizes. Based on these characteristics, it can be inferred that the BSF2 is mainly made up of a coarse sandy sediment enriched in carbonates. The areal distribution pattern of the BSF2 (Figure 4b) is complementary to that of the BSF1, with values that suddenly increase the near the coast, where they show values close to 1, meaning sediments consisting almost exclusively of carbonate sandy materials and decrease offshore where they reach value lower than 0.1. The origin of the BSF2 sediments is ascribed to the coarse sandy contributions of the local rivers of the Marche Region. They drain the carbonate rocks of the central Apennine and discharge their coarse sediments near the coast, while finer sediments are removed by wave action. The BSF2 facies can be named as *Coastal Facies* being mainly affected by coastal processes and inputs. The BSF2 presents a smaller extension towards the offshore in front of the Chienti and Tenna rivers and, to a greater extent, in front of the Tronto River where, there is also an area marked by the lowest values. This trend in the areal distribution of BSF2 confirms the greater inputs of fine material from the Chienti, Tenna and, to a greater extent, from the Tronto rivers.

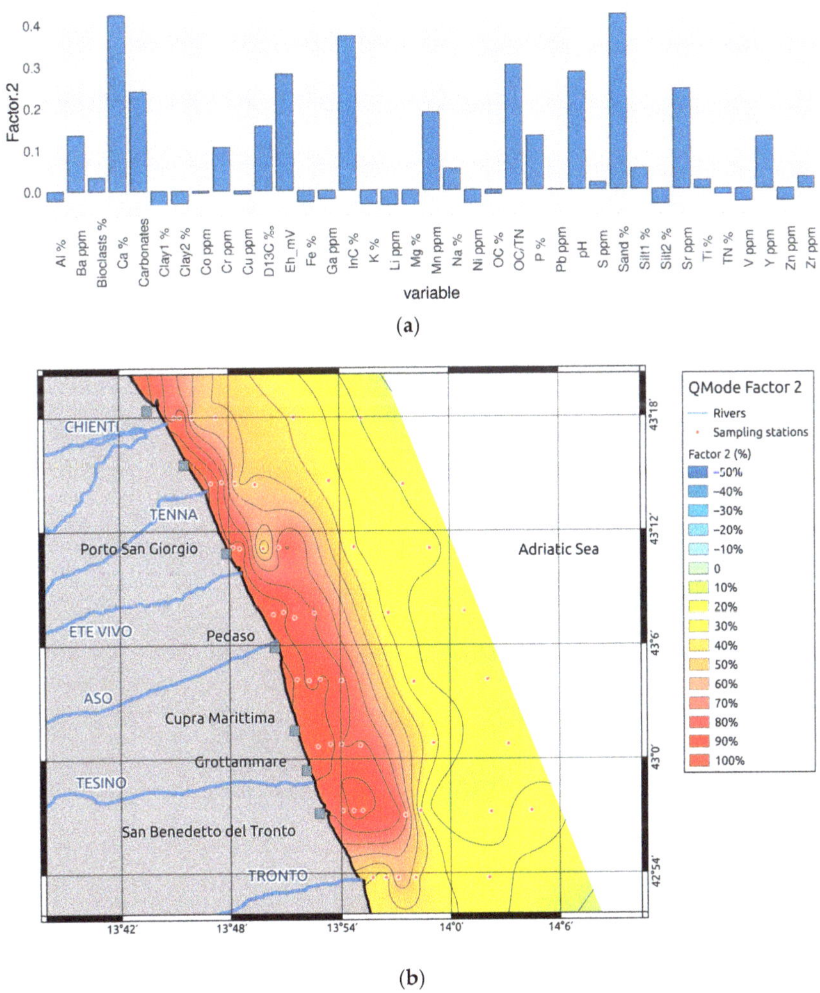

Figure 4. Histogram of the factor scores (**a**) and map of the areal distribution of the of the BSF2 (**b**).

The third facies (BSF3) explains 1.9% of the total variance and is defined by abundant silts, P, Mn, Ba, Sr, and partially Zr (Figure 5a). The highest values of the BSF3 are mostly found along a belt parallel to the coast in an intermediate position between the BSF1 and BSF2 maxima (Figure 5b) with three maximum areas. By considering the distribution of the BSF3 sediments in relation with the hydrography of the central Adriatic Sea, the silty sediments of the BSF3 are the results of the accumulation at depth where the wave action can remove the finer granulometric fractions, but where the coarse sandy coastal material supplied by the local rivers and reworked by the wave motion, is not deposited. Furthermore, the presence Zr and P also suggest the presence of heavy minerals like zircon and apatite, while the occurrence of Sr and Mn as well as IC suggest the presence of carbonates. These minerals support the residual nature of these sediments in which heavy minerals and carbonate concentrations increase for the removing of lighter minerals and finer grains. The characteristics of the BSF3 are similar to the *Residual Facies* of Spagnoli et al. (2014) [4] and can then be partially considered as the same *Residual Facies*. The two minima near the coast are in front of the Ete Vivo and Tesino rivers: the two rivers that do not drain the limestone rocks of the Sibillini mountains. This means that the residual sediments are partially associated with the carbonate rocks.

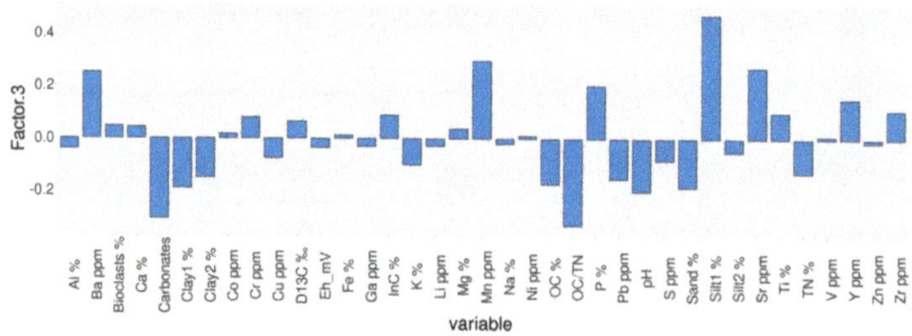

(a)

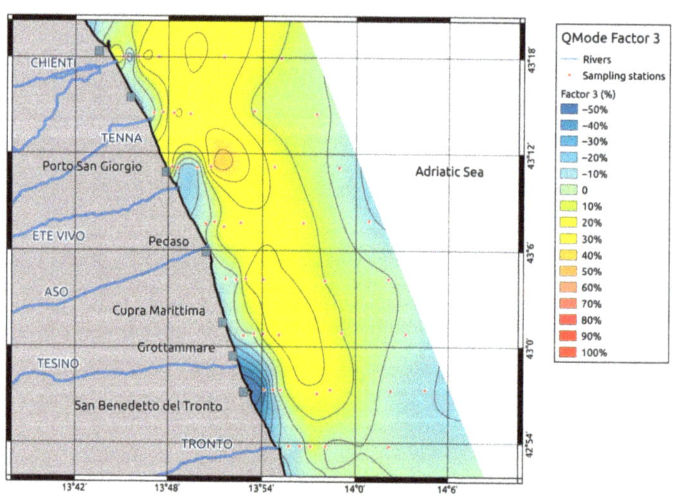

(b)

Figure 5. Histogram of the factor scores (**a**) and map of the areal distribution of the of the BSF3 (**b**).

By considering single elements, it is worthy to note that the elements that present high affinity with the clay minerals (i.e., Al, Co, Cr, Cu, Fe, Ga, K, Li, Mg, Mn, Na, Ni, P, Pb, Sc, Ti, V, Y, Zn, Zr) show a distribution pattern very similar to that of the BSF1 *facies* (*Padanic Facies*) (Figures 6–8) because they naturally tend to accumulate with finer fractions. On the other hand, elements having high affinity with the carbonate minerals (i.e., Ca and Sr) show distribution patterns (Figures 6f and 7e) similar to that of the *Coastal Facies* BSF2 (Ca) and BSF3 (Sr). This is in full agreement with the findings of the statistical elaboration that connects the same elements with the clay minerals and with the carbonate minerals. For this reason, the anomalies regarding these patterns have been considered to infer anthropic inputs or local natural processes. In this context, an anomaly is recorded in front of the Tronto River where a siliciclastic fine sediment enriched in S is recorded (Figure 6). This is also supported by the sediment description of the box-core upon collection that consisted of compact gray mud with blackish veins and burrows. Following the bathymetric reliefs that delineate a depression, the geochemical composition of this area is related to a strong erosion of the bottom that allows the outcropping of older, weak diagenized siliciclastic sediments.

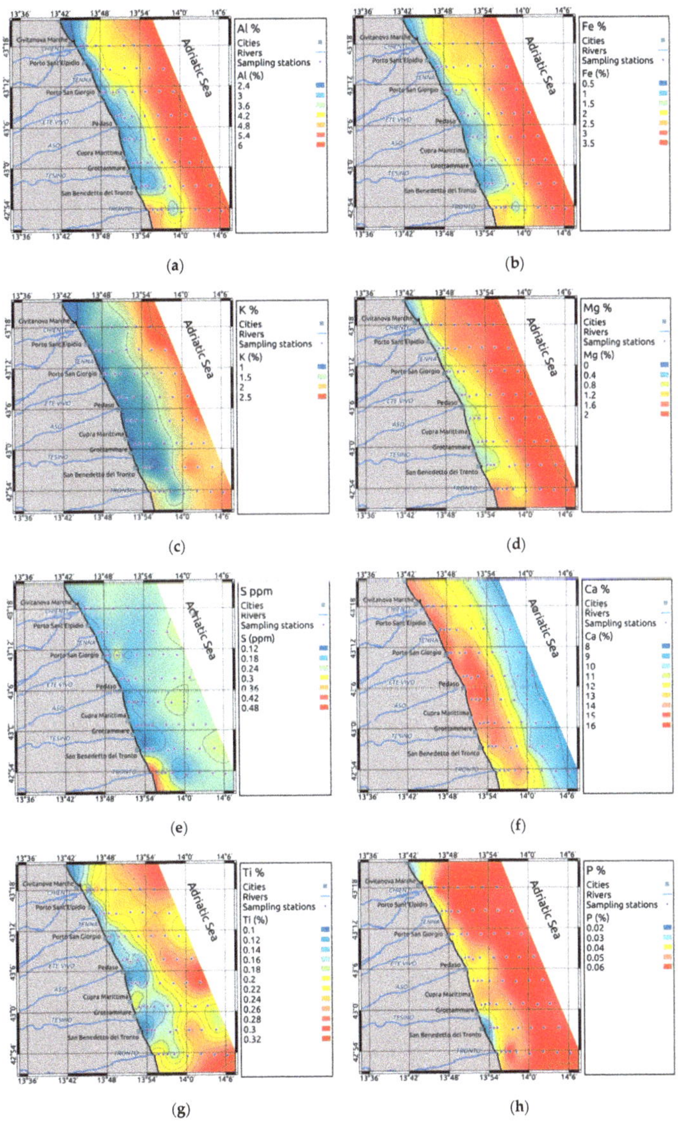

Figure 6. *Cont.*

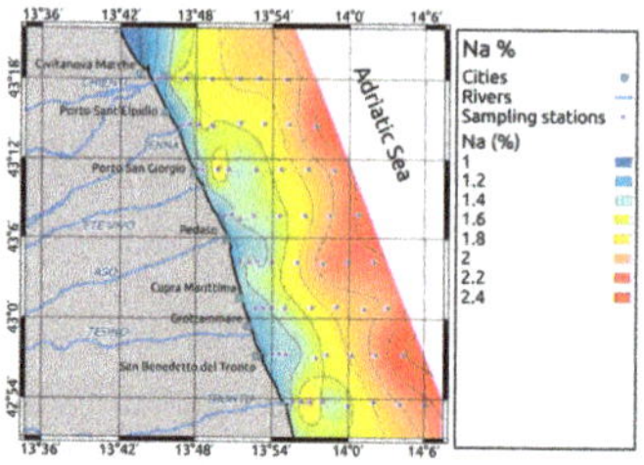

(i)

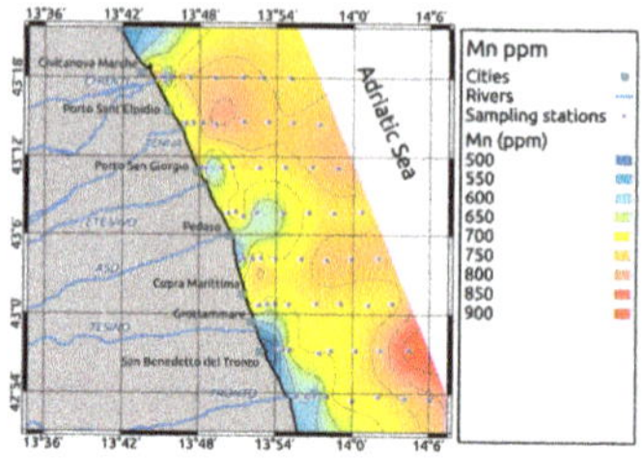

(j)

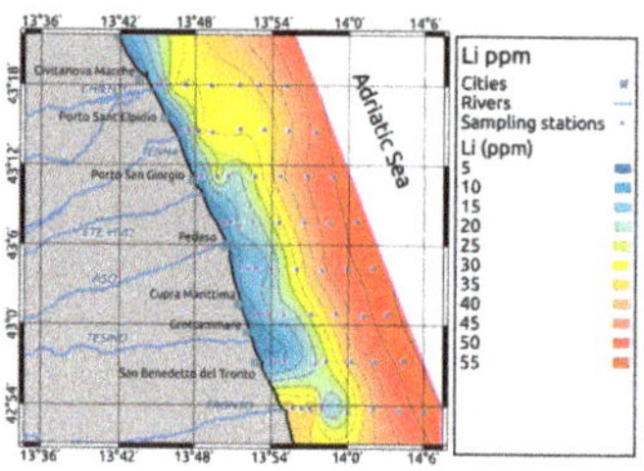

(k)

Figure 6. Areal distribution of the elements in surface sediments. (**a**) Al, (**b**) Fe, (**c**) K, (**d**) Mg, (**e**) S, (**f**) Ca, (**g**) Ti, (**h**) P, (**i**) Na, (**j**) Mn, and (**k**) Li.

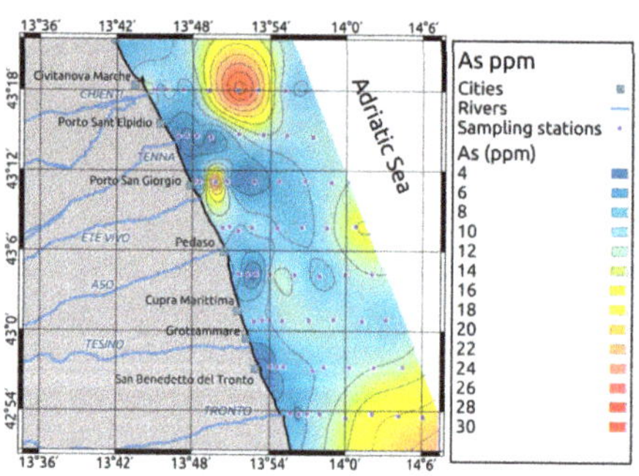

(a)

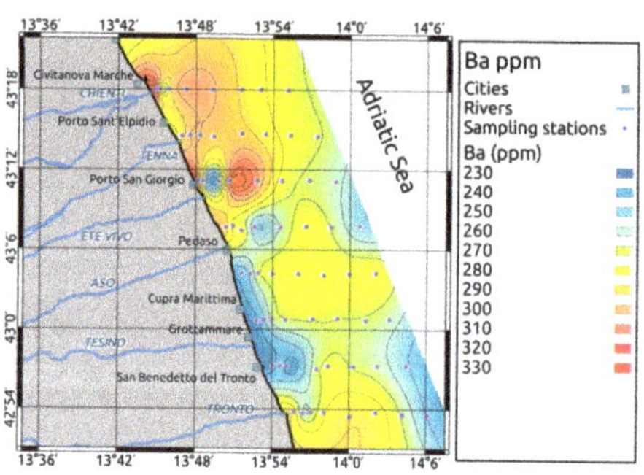

(b)

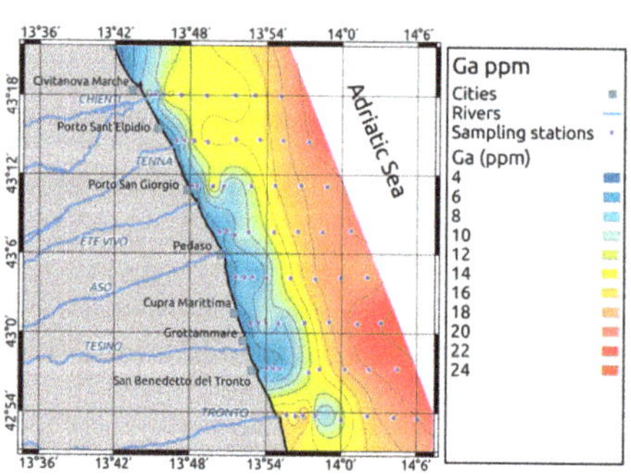

(c)

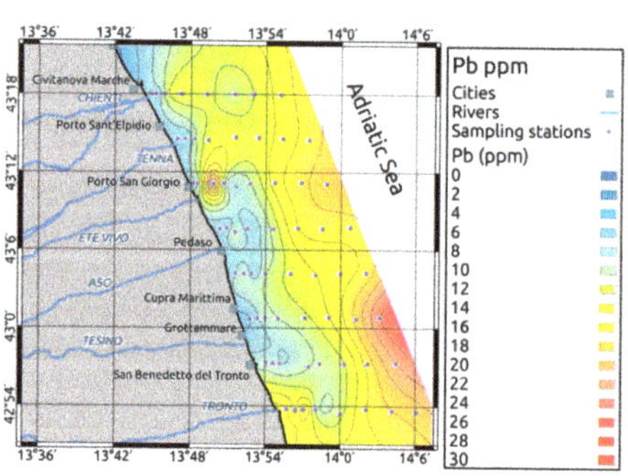

(d)

Figure 7. *Cont.*

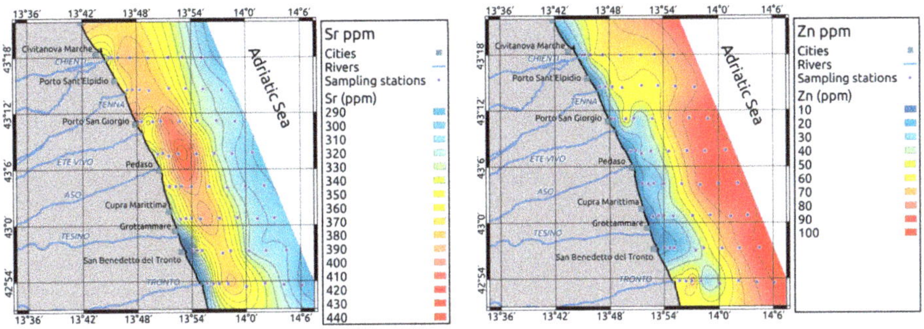

(**e**) (**f**)

Figure 7. Areal distribution of the elements in surface sediments with $C_f > 1$ in some samples. (**a**) As, (**b**) Ba, (**c**) Ga, (**d**) Pb, (**e**) Sr, and (**f**) Zn.

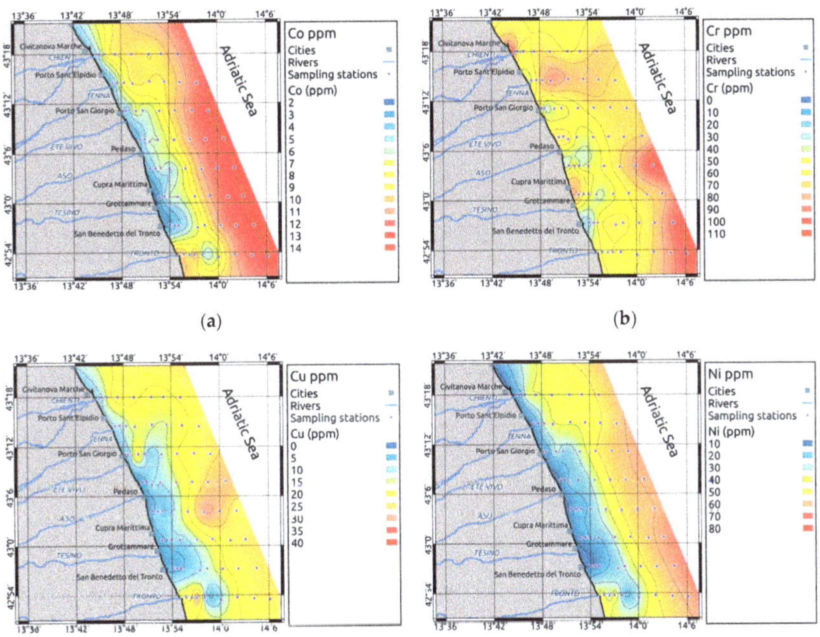

(**a**) (**b**)

(**c**) (**d**)

Figure 8. *Cont.*

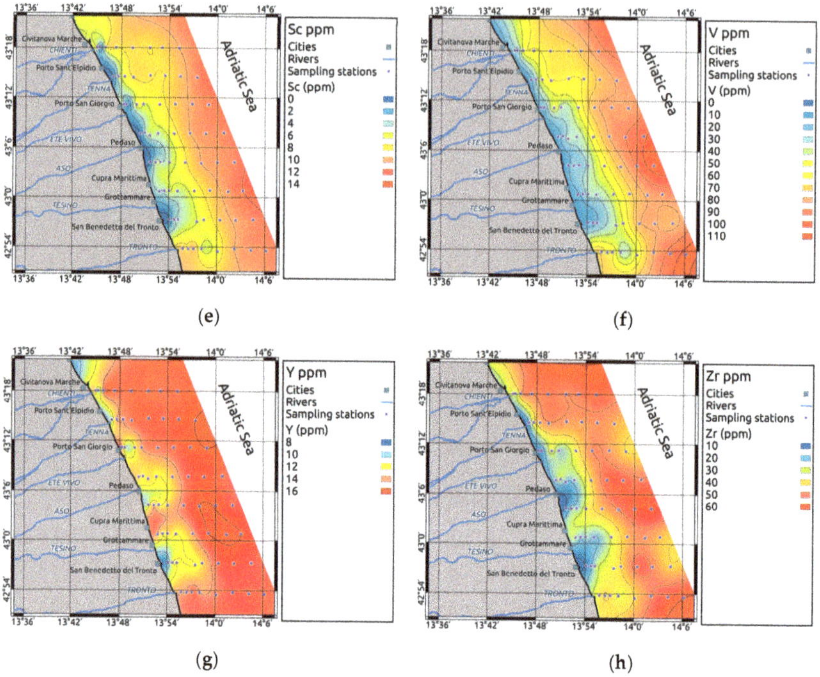

Figure 8. Areal distribution of the elements in surface sediments with $C_f < 1$ in all samples. (**a**) Co, (**b**) Cr, (**c**) Cu, (**d**) Ni, (**e**) Sc, (**f**) V, (**g**) Y, and (**h**) Zr.

4.2. Evaluation of the Contamination and Ecotoxicological Risks

4.2.1. Contamination Assessment

Only elements with Cf higher than one in some samples are herein considered.

As. Arsenic mean values are in the order of 9 ppm (Table 3) that are slightly lower than the average marine shale composition (Table 2) and are in line with the local background value (Table 2) of the southern Marche Holocene pelitic wedge (10 ppm at −105 cm). The spatial distribution indicates the occurrence of relatively high concentrations (up to 31 ppm) in the northern area (Figure 7a).These data suggest that this area has, on average, high As values likely related to the composition of the source rocks coming from North [4]. Arsenic was included in the multivariate analysis but shows a good affinity with elements of the BSF1. In two sandy areas located inshore in front of the Chienti and Ete Vivo river mouths, as well as in a pelitic belt directed north-south, in front of the Chienti River (Figure 7a), the As values are equal to or higher than the Italian thresholds (Table 2), its C_f is higher than 1 and the EF and I_{geo} fall in the second class (Table 4).The mainly sandy sites, located inshore, in front of the Chienti River (station 101) and of the Ete Vivo River (station 303) fall in the moderate enrichment class (2–5) for EF index and unpolluted to moderately polluted (0–1) for I_{geo}, respectively. Furthermore, the concentration of As at station 303 is over the second threshold of the Italian legislation (20 ppm, Table 2). In addition, the station 106, in the pelitic belt in front of the Chienti River, is moderately polluted (1–2) for I_{geo} and shows a concentration over the second threshold of the Italian legislation (20 ppm, Table 2). Unpolluted to moderately polluted (0–1) I_{geo} values are also recorded offshore and to the Southeast of the study area, in correspondence with mainly clayey sediment, but in this case the high values are due to the inputs of fine materials coming from North, that however influence also other proximal areas, further south [73]. The As pollution in the north-western area could be ascribed to local discharges of waste of fertilizer productions along the southern Marche coast in the past decades.

Ba. The average concentrations of barium are around 270 ppm (Table 3) reaching a maximum of 329 ppm, far below the average marine shale composition and in line with local background (Table 2). The highest values of Ba are found near the coast between the Chienti and Aso Rivers and in a belt along the central area that mainly corresponds to the BSF3 (Figure 7b). In some stations of these areas, the EF of Ba slightly exceeds 2, resulting in the moderate enrichment class, while the absolute concentrations (average 272.8 ppm; Table 2) are higher than the background value of 295 ppm in the high Al content stations (Table 2). The high Ba concentrations near the coast and in the BSF3 *facies*, particularly in the northern side of the study area, suggest a provenance of Ba from North. It also suggests a scavenging process, due to high specific weight of the barium sulphate, a heavy mineral that tends to remain in situ after reworking processes. The barium sulphate is a component of the drilling muds so that the high Ba concentrations could be due to discharges into the sea in the past of these drilling muds from hydrocarbon platform perforation as reported for other areas [2,46,80].

Ga. The values of Ga in the whole area (average 12.3 ppm, Table 3) with a maximum of 22 ppm are in line with average marine shale and with local background concentrations (16.2 ppm, Table 2) even if some stations show values of C_f slightly above 1. The EF and I_{geo} indexes are constantly lower than the minimum threshold values. Its distribution describes an almost regular increase seaward (Figure 7c).

Pb. The average values of Pb in the whole area (12.1 ppm, Table 3), with a maximum of 24 ppm, match well with the background data (18 ppm, Table 2) and the average marine shale. Some samples, however, show a C_f slightly over 1 (Table 4). The ER and I_{geo} indexes are permanently less than the minimum threshold values (Table 4). Both the present average Pb contents and the background values are strongly lower than the L1 threshold level of 30 ppm (Table 2) suggesting limited anthropogenic Pb inputs.

Sr. Strontium has an average concentration of 353 ppm (maximum of 436 ppm, Table 3) that are comparable to the local background and slightly higher than average marine shale (Table 2). This element has a strong affinity for carbonates and its concentrations show higher values in correspondence of the residual facies BSF3. Among the sedimentary quality index, the I_{geo} is constantly lower than minimum threshold value, but the EF index is higher than 2 (low enrichment, Table 4) in many samples, likely due for the way the index is calculated on a normalized base.

Zn. Zinc concentration records an average of 49 ppm and a maximum of 95 ppm (Table 3) that are, in general, slightly lower than local background and average marine shale values (Table 2). The sediment quality index displays few observations of the C_f value of 1, whereas the EF and I_{geo} indexes are constantly lower than minimum threshold values (Table 4). All the observations are below the L1 threshold levels (100 ppm, Table 2) pointing therefore to a limited anthropogenic contribution for this element.

As regard the cumulative impact of the elements it is important to recall that the mC_d and the PLI were calculated considering only metals or trace elements that show almost one value of $C_f > 1$ (As, Ba, Ga, Pb, Sr, and Zn). Despite the cumulative impact of more anthropogenic substances, the values of mC_d are always below the threshold of 1.5, so all sample sites fall in the very low polluted class (Table 4). On the other hand, the PLI shows values slightly higher than 1 in front of the Civitanova Marche, at the higher depths towards the offshore, particularly on the south-eastern part of the study area, and in front of the Tronto River. While the high values of PLI present in the offshore are attributed to the fine grain size sediments coming from the north and belonging to the BSF1 Facies (Table 3, Figure 9), those occurring in the area between Chienti and Ete Vivo Rivers could be the results of local pollution inputs, in particular the pollution in this area could have been higher if it were related to finer grain-size. In this coastal area, the pollution is due to high values of As, followed by Pb and Zn. The high concentrations of heavy metals, mainly As, could be caused by the discharge of processing residues of fertilizer production industries that were operative along the southern Marche coast in the last century.

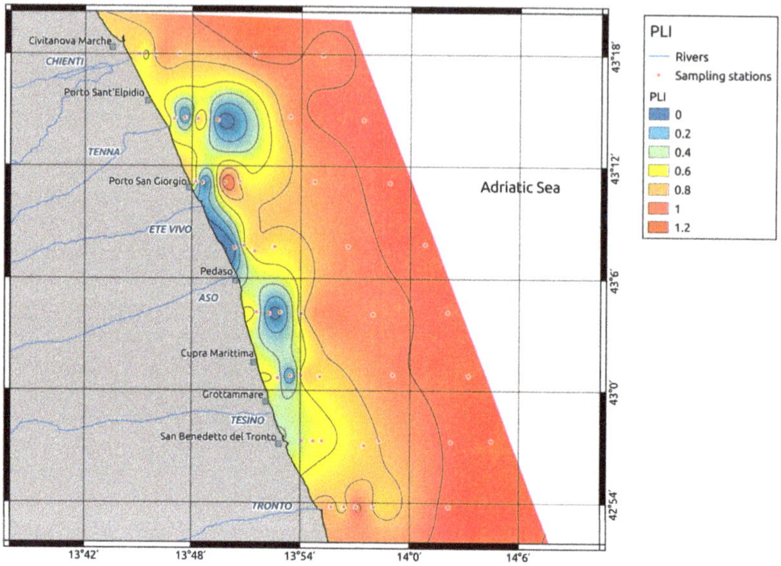

Figure 9. Areal distribution of the PLI values.

The following elements exhibit a Cf lower than one.

Co. The concentrations of Co in the whole area display an average value of 7.8 ppm and reach a maximum of 13 ppm (Table 3), in general below the local background and the average marine shale (Table 2), that suggest a regional depletion compared to global references. Its distribution displays an almost regular increase seaward (Figure 8a), common to all the elements included in the BSF1. The C_f values are constantly less than 1.

Cr. Chromium average concentration is 61.5 ppm with a maximum of 101 ppm (Table 3). The average marine shale (90 ppm Cr, Table 2) and the local background (128 ppm Cr, Table 2) values are comparable or higher indicating that there is no enrichment respect to the pre-industrial period. In some stations close to the coast, mainly sandy, some higher values are recorded and could be either related to possible anthropic inputs or to selective enrichment in heavy minerals, given the high values of other elements such as Zr and Ti. Many of the observations are above the L1 threshold level of the Italian legislation (50 ppm, Table 2). It should however be considered that, both for Cr and all the other considered elements, the values of the national legislation refer to an aqua regia digestion and not to a multi-acid digestion including HF, as the one applied in the present study.

Cu. Copper concentrations have a mean value of 14.4 ppm and a maximum value of 32 ppm Cu (Table 3), the latter is lower than the average marine shale (45 ppm) and the local background (40 ppm) ones. (Table 2). This element does not, therefore, represent an important issue for this area. This is further confirmed by the C_f values of Cu (Table 4) constantly lower than 1 as well as by the values of the EF and I_{geo} indexes, permanently lower than the minimum threshold values (Table 4). Considering this, there has not been any enrichment with respect to the pre-industrial periods. Indeed, the weak decrease in the last decades could be related to the reduction of northern riverine inputs.

Ni. Nickel has an average concentration of 32.7 ppm and a maximum of 63 ppm, which are both lower than the average marine shale and the local background. Despite of it, many stations show values above the L1 threshold level of the Italian legislation (30 ppm, Table 2). The comparison between the average values and the background data in the study area suggests that L1 threshold level of the Ni could be underestimated for this area, likely due to the different analytical method. A support is given by the C_f values always below 1 and by the EF and I_{geo} indexes constantly lower than the minimum threshold values (Table 4).

Sc. Scandium has an average concentration of 6.8 ppm and a maximum of 12 ppm (Table 3) against an average marine shale concentration of 13 ppm and a local background of 29 ppm (Table 2). It is not an element considered by the legislative side, and its C_f values are constantly less than 1 as well as the EF ER and I_{geo} indexes are permanently lower than minimum threshold values (Table 4). These data indicate that there is not Sc enrichments respect to the pre-industrial period and its decrease in the last decades could reflect the decrease of northern riverine inputs.

V. Vanadium has an average concentration of 49 ppm and a maximum value of 105 ppm. Its spatial distribution displays a clear seaward increase (Figure 8f). The values are below the average marine shale and the local background values (Table 2), likely for the diluting effect of the carbonate fraction and the relative importance of sandy sediments, both causing a general depletion of V. The C_f value are always lower and the EF and I_{geo} indexes and lower than the minimum threshold values (Table 4). This indicates that there are no enrichments of V respect to the pre-industrial period but an overall decrease. These decreases could be due to the decrease of northern riverine inputs in the last decades.

Y. Yttrium concentrations have an average concentration of 13.7 ppm and a maximum of 16 ppm, all values are well below the global reference and the local background (Table 2). The dilution effect of carbonate material is the likely explanation for this difference. The C_f values are always lower than 1; similarly, EF and I_{geo} indexes are constantly lower than the minimum threshold values (Table 4). All these evidence points to a lack of t enrichments of this element with respect to the pre-industrial period.

Zr. Zirconium has average value of 35 ppm Zr and a maximum value of 65 ppm that are well below the average marine shale (160 ppm, Table 2) and the local background (97 ppm, Table 2). The values of C_f value are always <1, and EF and I_{geo} indexes are permanently lower than minimum threshold values (Table 4). This decrease respect to the pre-industrial period is attributed to the decrease of northern riverine inputs in the last decades.

The analysis of the core 47 also allows us to establish the background values of other elements other than those considered for the contamination assessment (Table 2). An important finding arises for the comparison between the present concentrations and the background values. The concentrations of most of the elements with a high affinity to clay and silicate minerals show a decrease respect to the background values (Table 5a), while other elements with an affinity with carbonate minerals exhibit an increasing respect to the past (Table 5b). These trends suggest a change in the sedimentation pattern in the area in the last decades. Specifically, this change consists in a percentage decrease of the clay contributions coming from the North and a percentage increase in the carbonate fraction due to the local river inputs. This change could be ascribed to the lower fine solid inputs of Po River as a result of the regimentation works and the lower quantity of water flowing of the Po River due to both the pumping water for zootechnical, agricultural and industrial uses and for the climate changes in the last decades [81].

4.2.2. Ecotoxicology of Central Western Adriatic Sea Bottom Sediments

The Numerical Sediment Quality Guidelines (SQGs) developed for marine and estuarine ecosystems as suggested by Rachel and Wasserman (2015) [71], are used here to evaluate the ecotoxicological state of the bottom sediments. According to Rachel and Wasserman SQGs (2015) [71], the following thresholds can be considered (Table 2): the TEL (Threshold Effect Level) that is the concentrations below which sediment-dwelling organism do not exhibit toxic effects; the ERL (effect range low) that is the concentrations below which sediment-dwelling organism rarely exhibit toxic effects; the PEL (Probable effect level) and the ERM (Effect Range Median) that are the concentrations above which toxic effects are detected frequently [82].

Table 5. Elements that decrease respect to the past (a); elements that increase respect to the past (b); elements for which only the background value is available (c).

Element	Present Average Values (ppm)	Background Values (ppm)
	(a)	
Al	4.03	6.9
Ti	0.2	0.3
P	0.05	0.06
Co	7.8	13.0
Cr	61.5	128.0
Cu	14.4	40.0
Fe	1.93	3.49
Ga	12.3	16.2
K	1.43	1.64
Mg	1.23	2.22
Mn	703.9	774.5
Ni	32.7	73.0
P	0.05	0.06
Pb	12.1	18
Sc	7.4	29.0
Ti	0.19	0.32
V	49.4	116.0
Y	13.7	25.0
Zn	49.0	74.0
	(b)	
Ca	12.0	10.0
Sr	353.0	348.0
	(c)	
Ce		59
Fe		3.49
Nb		1.65
Rb		115

Most of the samples show As values exceeding the TEL; this is the consequence of the background values of As that is rather high in this area and due to the inputs of sediments coming from the North that present high As source catchment areas. Furthermore, two samples (303 and 106 sites) show As values exceeding the PEL. In this case, as previously noted for the contamination indexes, the exceeding of the SQLs thresholds is probably due to anthropic causes due to the discharges into the sea of waste of the fertilizer production of firms that have been operative along the Southern Marche coast during the 20th century. As regards the Cr, almost all samples exceed the TEL, 11 samples exceed the ERL and 3 samples overcame the PEL. The high contents of Cr that generally exceed the lower levels of some SQGs are ascribed to the composition of the sediments coming from the North that exhibits high Cr levels [14] and are a well-known geochemical signature sediment associated to the Po river in northern areas [83–85]. As regard the Ni, 34 stations exceed the TEL, 25 the ERL, and 16 the PEL. In this case, as for the Cr, these excesses could be ascribed to the composition of the sediments coming from the North, which naturally show high Ni levels, or due to the sediments affected by anthropogenic input in the North Adriatic. Finally, in the cases of Cu, Pb e Zn no station presents contents higher than the TEL.

5. Conclusions

The present work analyzes the bottom sediments in front of the Marche Region by using a multi-proxies and integrated approach. This allows us to recognize and to map the distribution of three main sedimentological and geochemical facies: the *Padanic Facies* (BSF1), the *Coastal Facies* (BSF2) and the *Residual Facies* (BSF3). The Padanic BSF1 *Facies* is characterized by fine siliciclastic sediments coming from Northern areas, reflecting the regional sediment dispersal pattern, mostly originating from the Po River, and transported southwards by the WAC and towards offshore by wave motion of the strongest storms. The BSF2 *Coastal Facies* is mainly defined by coarser sediments enriched in carbonatic minerals, coming from local rivers and partly mixed, by the wave motion of the stronger storms, with the sediments of the BSF1 *Facies*. The BSF3 *Residual Facies* is mainly composed by silty sediments coming from local rivers that are partially sorted from the clayey from the wave motion. Towards the offshore, this facies mixes progressively with sediments of Padanic sources coming from the North. The distribution of these three facies is the result of two principal processes. The first is the predominance of sedimentary inputs coming from the North and from local rivers, hence, the prevalence of sediments coming from source areas with different mineral-petrographic composition. The second regards the general hydrodynamic of the Adriatic Sea such as the cyclonic circulation that mainly flows southward near the Italian coast and the wave motion of the stronger storms directed perpendicularly to the coast.

The data processing of trace elements and heavy metals concentrations also allow us to determine the background reference values for the study area. It is noteworthy to stress that the Cr and Ni values recorded in the past and in the surface sediments of this study area are above the L1 threshold levels of the Italian legislation (DM 173/2016). On the other hand, the present and past concentrations of Pb are generally lower than it. This suggests a possible underestimation of Cr and Ni thresholds and an overestimation of the Pb threshold. Therefore, a further study in order to validate or reject these discrepancies on Cr, Ni and Pb reference values should be carried out.

The present study also identifies a small coastal area, between the Chieti and Ete Vivo rivers, that is probably affected by an As and Ba local pollution. The high contents of As could be ascribed to local discharges of waste of fertilizer productions during the 20th century. Moreover, the pollution of Ba is probably related to past discharges into the sea of drilling muds coming from oil platform perforation as reported for other area on the Central Adriatic Sea. Finally, the comparison between the background concentrations and the current average concentrations point to a general decrease in Al, Ti, P, Co, Cr, Cu, Ga, Ni, Pb, Sc, V, Y, and Zn and an increase in Ca and Sr. These trends suggest a change in the sedimentation in the last decades characterized by a decrease in the clayey contributions coming from the north and an increase in the carbonate fraction coming from the local rivers. This could be ascribed to the lower solid inputs of the Po River as a result of the regimentation works and the lower water flow prompted by both climate changes and pumping for irrigation, farm, civil and industrial uses.

Supplementary Materials: The following are available online at https://www.mdpi.com/2076-3417/11/3/1118/s1. Figure S1: Areal distribution of the elements in surface sediments. Ag (a), Mo (b), Be (c) and Te (d); Table S1: Values of the biogeochemical and sedimentological data and pollution indicators.

Author Contributions: Conceptualization, F.S.; Data curation, F.S., R.D.M., E.D., E.F., F.F. and P.G.; Formal analysis, F.S., E.D., E.F., F.F. and P.G.; Funding acquisition, F.S.; Investigation, F.S., F.F. and P.G.; Methodology, F.S., R.D.M., E.D., F.F. and P.G.; Project administration, F.S.; Resources, F.S.; Software, F.S., R.D.M. and P.G.; Supervision, F.S.; Validation, F.S., E.D., F.F. and P.G.; Visualization, F.S.; Writing—original draft, F.S., R.D.M., E.D., E.F., F.F. and P.G.; Writing—review & editing, F.S., R.D.M., E.D., E.F., F.F. and P.G. All authors will be informed about each step of manuscript processing including submission, revision, revision reminder, etc. via emails from our system or assigned Assistant Editor. All authors have read and agreed to the published version of the manuscript.

Funding: This study has been conducted with the financial support of PERSEUS Project, grant number 287600

Institutional Review Board Statement: Not applicable.

Informed Consent Statement: Not applicable.

Data Availability Statement: Data is contained within the article and the Supplementary Materials.

Acknowledgments: We want to give our thanks to the crew of the M/N Dallaporta, to the Giuseppe Corti and the Stefania Cocco of the Università Politecnica delle Marche, to Fabio Zaffagnini, Laura Borgognoni, Eva Turicchia and Piero Ferracuti, to Giuseppe Caccamo and Elisa Ghetti of the IRBIM Ancona, for their valuable help given during the sampling, analysis and data processing phases.

Conflicts of Interest: The authors declare no conflict of interest.

References

1. Spagnoli, F.; Bergamini, M.C. Water-solid exchanges of nutrients and trace elements during early diagenesis resuspension of anoxic shelf sediments. *Water Air Soil Pollut.* **1997**, *99*, 541–556. [CrossRef]
2. Marcaccio, M.; Spagnoli, F.; Frascari, F. Drilling discharges as tracer of sedimentation and geochemical processes in Adriatic, Sea. *J. Coast. Res.* **2003**, *19*, 89–100.
3. Spagnoli, F.; Bartholini, G.; Dinelli, E.; Giordano, P. Geochemistry particles size of surface sediments of Gulf of Manfredonia (Southern Adriatic, Sea). *Estuar. Coast. Shelf Sci.* **2008**, *80*, 21–30. [CrossRef]
4. Spagnoli, F.; Dinelli, E.; Giordano, P.; Marcaccio, M.; Zaffagnini, F.; Frascari, F. Sedimentological biogeochemical mineralogical facies of Northern and Central Western Adriatic, Sea. *J. Mar. Syst.* **2014**, *139*, 183–203. [CrossRef]
5. Spagnoli, F.; Andresini, A. Biogeochemistry sedimentology of Lago di Lesina, (Italy). *Sci. Total Environ.* **2018**, *643*, 868–883. [CrossRef]
6. Lopes-Rocha, M.; Langone, L.; Miserocchi, S.; Giordano, P.; Guerra, R. Spatial patterns and temporal trends of trace metal mass budgets in the western Adriatic sediments (Mediterranean Sea). *Sci. Total Environ.* **2017**, *599*, 1022–1033. [CrossRef] [PubMed]
7. Lopes-Rocha, M.; Langone, L.; Miserocchi, S.; Giordano, P.; Guerra, R. Detecting long-term temporal trends in sediment-bound metals in the western Adriatic (Mediterranean, Sea). *Mar. Pollut. Bull.* **2017**, *124*, 270–285. [CrossRef]
8. Bastami, K.D.; Neyestani, M.R.; Shermirani, F.; Soltani, F.; Haghparast, S.; Akbari, A. Heavy metal pollution assessment in relation to sediment properties in the coastal sediments of the southern Caspian, Sea. *Mar. Pollut. Bull.* **2015**, *92*, 237–243. [CrossRef]
9. Burton, G.A.; Denton, D.L.; Ho, K.; Ireland, D.S. Sediment toxicity testing: Issues and methods. In *Handbook of Ecotoxicology*, 2nd ed.; Hoffman, D.J., Rattner, B.A., Burton, G.A., Jr., Cairns, J., Jr., Eds.; Lewis Publishers: Boca Raton, FL, USA, 2003; pp. 111–150.
10. Focardi, S.; Specchiulli, A.; Spagnoli, F.; Fiesoletti, F.; Rossi, C. A combinated approach to investigate the biochemistry hydrology of a shallow bay in the South Adriatic Sea: The Gulf of Manfredonia, (Italy). *Environ. Monit. Assess.* **2009**, *153*, 209–220. [CrossRef]
11. Pérez-Albaladejo, E.; Rizzi, J.; Fernandes, D.; Lille-Langøy, R.; Karlsen, O.A.; Goksøyr, A.; Oros, A.; Spagnoli, F.; Porte, C. Assessment of the environmental quality of coastal sediments by using a combination of in vitro bioassays. *Mar. Pollut. Bull.* **2016**, *108*, 53–61. [CrossRef]
12. Roussiez, V.; Ludwig, W.; Radakovitch, O.; Probst, J.L.; Monaco, A.; Charrière, B.; Buscail, R. Fate of metals in coastal sediments of a Mediterranean flood-dominated system: An approach based on total and labile fractions. *Estuar. Coast. Shelf Sci.* **2011**, *92*, 486–495. [CrossRef]
13. Spagnoli, F.; Dell' Anno, A.; De Marco, A.; Dinelli, E.; Fabiano, M.; Gadaleta, M.V.; Ianni, C.; Loiacono, F.; Manini, E.; Marini, M.; et al. Biogeochemistry, grain size and mineralogy of the central and southern Adriatic Sea sediments: A review. *Chem. Ecol.* **2010**, *26*, 1944. [CrossRef]
14. Spagnoli, F.; Kaberi, H.; Giordano, P.; Zeri, C.; Borgognoni, L.; Bortoluzzi, G.; Campanelli, A.; Ferrante, V.; Giuliani, G.; Martinotti, V.; et al. Benthic fluxes of dissolved heavy metals in polluted sediments of the Adriatic Sea. In Proceedings of the Integrated Marine Research in the Mediterranean and the Black Sea, Bruxelles, Belgium, 7–9 December 2015; pp. 301–302, 399.
15. Neal, C.; Robson, A.J.; Jeffery, H.A.; Harrow, M.L.; Neal, M.; Smith, C.J.; Jarvie, H.P. Trace element inter-relationships for the Humber rivers: Inferences for hydrological and chemical controls. *Sci. Total Environ.* **1997**, *194/195*, 321–343. [CrossRef]
16. Callender, E. Heavy Metals in the Environment-Historical, Trends. *Treatise Geochem.* **2003**, *9*, 67–105.
17. Gao, X.; Chen, C.A. Heavy metal pollution status in surface sediments of the coastal Bohai, Bay. *Water Res.* **2012**, *46*, 1901–1911. [CrossRef]
18. Illuminati, S.; Annibaldi, A.; Truzzi, C.; Tercier-Waeber, M.L.; Nöel, S.; Braungardt, C.B.; Achterberg, E.P.; Howell, K.A.; Turner, D.; Marini, M.; et al. In-situ trace metal (Cd, Pb, Cu) speciation along the Po River plume (Northern Adriatic Sea) using submersible systems. *Mar. Chem.* **2019**, *212*, 47–63. [CrossRef]
19. Olsen, C.R.; Cutshall, N.H.; Larsen, I.L. Pollutant-particle association and dynamics in coastal marine environments: A review. *Mar. Chem.* **1982**, *11*, 501–533. [CrossRef]
20. Bellas, J.; Nieto, O.; Beiras, R. Integrative assessment of coastal pollution: Development and evaluation of sediment quality criteria from chemical contamination and ecotoxicological data. *Cont. Shelf Res.* **2011**, *31*, 448–456. [CrossRef]
21. Martins, M.A.; Mane, M.A.; Frontalini, F.; Santos, J.F.; Silva, F.S.; Terroso, D.; Miranda, P.; Figueira, R.; Laut, L.L.M.; Bernardes, C.; et al. Early diagenesis and adsorption by clay minerals important factors driving metal pollution in sediments. *Environ. Sci. Pollut. Res.* **2015**, *22*, 10019–10033. [CrossRef]
22. Rial, D.; León, V.M.; Bellas, J. Integrative assessment of coastal marine pollution in the Bay of Santander and the Upper Galician Rias. *J. Sea Res.* **2017**, *130*, 239–247. [CrossRef]

23. Frapiccini, E.; Panfili, M.; Guicciardi, S.; Santojanni, A.; Marini, M.; Truzzi, C.; Annibaldi, A. Effects of biological factors and seasonality on the level of polycyclic aromatic hydrocarbons in red mullet (*Mullus barbatus*). *Environ. Pollut.* **2020**, *258*, 113742. [CrossRef] [PubMed]
24. Baldrighi, E.; Semprucci, F.; Franzo, A.; Cvitkovic, I.; Bogner, D.; Despalatovic, M.; Berto, D.; Formalewicz, M.M.; Scarpato, A.; Frapiccini, E.; et al. Meiofaunal communities in four Adriatic ports: Baseline data for risk assessment in ballast water management. *Mar. Poll. Bull.* **2019**, *147*, 171–184. [CrossRef] [PubMed]
25. Fichet, D.; Radenac, G.; Miramand, P. Experimental studies of impacts of harbour sediments resuspension to marine invertebrates larvae: Bioavailability of Cd, Cu, Pb and Zn and toxicity. *Mar. Pollut. Bull.* **1998**, *36*, 509–518. [CrossRef]
26. Rial, D.; Beiras, R. Prospective ecological risk assessment of sediment resuspension in an estuary. *J. Environ. Monit.* **2012**, *14*, 2137–2144. [CrossRef] [PubMed]
27. Directive 2010/75/UE of the European Parliament and of the Council of the European Union, 2010, November 24 2010. Industrial emissions (integrated pollution prevention and control) (Recast). *Off. J. Eur. Union* **2010**, *L334*, 17–119.
28. OSPAR Commission. *Protecting and Conserving the North-East Atlantic and Its Resources*; Annual Report 2012/2013 OSPAR Secretariat; Publication Number: 620/2013; OSPAR Commission: London, UK, 2012.
29. Camp Dresser & McKee, Inc. *Guidelines for Water Reuse*; EPA/625/R-04/108 (NTIS PB2005 106542); U.S. Environmental Protection Agency: Washington, DC, USA, 2004.
30. Stankovic, S.; Kalaba, P.; Stankovic, A.R. Biota as toxic metal indicators. *Environ. Chem. Lett.* **2014**, *12*, 63–84. [CrossRef]
31. Stankovic, S.; Stankovic, A.R. Bioindicators of toxic metals. In *Green Materials for Energy, Products and Depollution*; Springer: Dordrecht, The Netherlands, 2013; pp. 151–228. [CrossRef]
32. Frontalini, F.; Greco, M.; Di Bella, L.; Lejzerowicz, F.; Reo, E.; Caruso, A.; Cosentino, C.; Maccotta, A.; Scopelliti, G.; Nardelli, M.P.; et al. Assessing the effect of mercury pollution on cultured benthic foraminifera community using morphological and eDNA metabarcoding approaches. *Mar. Pollut. Bull.* **2018**, *129*, 512–524. [CrossRef]
33. Frontalini, F.; Semprucci, F.; Di Bella, L.; Caruso, A.; Cosentino, C.; Maccotta, A.; Scopelliti, G.; Sbrocca, C.; Bucci, C.; Balsamo, M.; et al. The response of cultured meiofaunal and benthic foraminiferal communities to lead contamination: Results from mesocosm experiments. *Environ. Toxicol. Chem.* **2018**, *37*, 2439–2447. [CrossRef]
34. Bastami, K.D.; Bagheri, H.; Haghparast, S.; Soltani, F.; Hamzehpoor, A.; Bastami, M.D. Geochemical and geo-statistical assessment of selected heavy metals in the surface sediments of the Gorgan Bay, Iran. *Mar. Pollut. Bull.* **2012**, *64*, 2877–2884. [CrossRef]
35. Buccolieri, A.; Buccolieri, G.; Cardellicchio, N.; Dell' Atti, A.; Di Leo, A.; Maci, A. Heavy metals in marine sediments of Taranto Gulf (Ionian Sea, southern Italy). *Mar. Chem.* **2006**, *99*, 227–235. [CrossRef]
36. de Souza Machado, A.A.; Spencer, K.; Kloas, W.; Toffolon, M.; Zarfl, C. Metal fate and effects in estuaries: A review and conceptual model for better understanding of toxicity. *Sci. Total Environ.* **2016**, *541*, 268–281. [CrossRef] [PubMed]
37. El Nemr, A.M.; El Sikaily, A.; Khaled, A. Total leachable heavy metals in muddy sandy sediments of Egyptian coast along Mediterranean, Sea. *Environ. Monit. Assess.* **2007**, *129*, 151–168. [CrossRef] [PubMed]
38. Hakanson, L. An ecological risk index for aquatic pollution control. A sedimentological approach. *Water Res.* **1980**, *14*, 975–1001. [CrossRef]
39. Manahan, S. *Chimica Dell'ambiente*, 6th ed.; Piccin: Padova, Italy, 2000.
40. Wright, P.; Mason, C.F. Spatial and seasonal variation in heavy metals in the sediments and biota of two adjacent estuaries, the Orwell and the Stour, in eastern England. *Sci. Total Environ.* **1999**, *226*, 139–156. [CrossRef]
41. Tam, N.F.Y.; Wong, Y.S. Spatial variation of heavy metals in surface sediments of Hong Kong mangrove swamps. *Environ. Pollut.* **2000**, *110*, 195–205. [CrossRef]
42. Natali, C.; Bianchini, G. Natural vs anthropogenic components in sediments from the Po River delta coastal lagoons (NE Italy). *Environ. Sci. Pollut. Res.* **2018**, *25*, 2981–2991. [CrossRef]
43. DM 173/2016. Ministero Dell'ambiente e della Tutela del Territorio e del Mare. Decreto 15 luglio 2016, n. 173. Regolamento Recante Modalità e Criteri Tecnici Per L'autorizzazione All'immersione in Mare dei Materiali di Escavo di Fondali Marini. (16G00184) (GU Serie Generale n.208 del 06-09-2016-Suppl. Ordinario n. 40). Available online: https://www.gazzettaufficiale.it/eli/gu/2016/09/06/208/so/40/sg/pdf (accessed on 26 January 2021).
44. Annibaldi, A.; Illuminati, S.; Truzzi, C.; Scarponi, G. Heavy Metals in Spring and Bottled Drinking Waters of Sibylline Mountains National Park (Central Italy). *J. Food Prot.* **2018**, *81*, 295–301. [CrossRef]
45. Droghini, E.; Annibaldi, A.; Prezioso, E.; Tramontana, M.; Frapiccini, E.; De Marco, R.; Illuminati, S.; Truzzi, C.; Spagnoli, F. Mercury content in Central and Southern Adriatic Sea sediments in relation to seafloor geochemistry and sedimentology. *Molecules* **2019**, *24*, 4467. [CrossRef]
46. Frascari, F.; Marcaccio, M.; Spagnoli, F.; Modica, A. Effects of offshore drilling activities on the geochemical and sedimentological processes in the Northern Adriatic coastal area. *Period. Biol.* **2000**, *102*, 225–241.
47. Curzi, P.V.; Tomadin, L. Dinamica della sedimentazione pelitica attuale ed olocenica nell'Adriatico centrale. *G. Di Geol.* **1987**, *49*, 101–111.
48. Artegiani, A.; Bregant, D.; Paschini, E.; Pinardi, N.; Raicich, F.; Russo, A. The Adriatic Sea General Circulation. Part I: Air–Sea Interactions and Water Mass Structure. *J. Phys. Oceanogr.* **1997**, *27*, 1492–1514. [CrossRef]
49. Marini, M.; Russo, A.; Paschini, E.; Grilli, F.; Campanelli, A. Short-term physical and chemical variations in the bottom water of middle Adriatic depressions. *Clim. Res.* **2006**, *31*, 227–237. [CrossRef]

50. Artegiani, A.; Paschini, E.; Russo, A.; Bregant, D.; Raicich, F.; Pinardi, N. The Adriatic Sea General Circulation. Part II: Baroclinic Circulation Structure. *J. Phys. Oceanogr.* **1997**, *27*, 1515–1532. [CrossRef]
51. Civitarese, G.; Gačvić, M.; Lipizer, M.; Borzelli, G. On the impact of the Bimodal Oscillating System (BiOS) on the biogeochemistry and biology of the Adriatic and Ionian Seas (Eastern Mediterranean). *Biogeosci. Discuss.* **2010**, *7*. [CrossRef]
52. Lipizer, M.; Partescano, E.; Rabitti, A.; Giorgetti, A.; Crise, A. Qualified temperature, salinity and dissolved oxygen climatologies in a changing Adriatic Sea. *Ocean Sci.* **2014**, *10*, 771–797. [CrossRef]
53. Vilibić, I.; Supić, N. Dense water generation on a shelf: The case of the Adriatic Sea. *Ocean Dyn.* **2005**, *55*, 403–415. [CrossRef]
54. Boldrin, A.; Carniel, S.; Giani, M.; Marini, M.; Bernardi Aubry, F.; Campanelli, A.; Grilli, F.; Russo, A. Effects of bora wind on physical and biogeo chemical properties of stratified waters in the northern Adriatic. *J. Geophys. Res.* **2009**, *114*, C08S92. [CrossRef]
55. Vilibić, I.; Mihanović, H.; Janeković, I.; Šepić, J. Modelling the formation of dense water in the northern Adriatic: Sensitivity studies. *Ocean Model.* **2016**, *101*, 17–29. [CrossRef]
56. Marini, M.; Maselli, V.; Campanelli, A.; Foglini, F.; Grilli, F. Role of the Mid-Adriatic deep in dense water interception and modification. *Mar. Geol.* **2016**, *375*, 5–14. [CrossRef]
57. Frascari, F.; Spagnoli, F.; Marcaccio, M.; Giordano, P. Anomalous Po river flood event effects on sediments and water column of the Northwestern Adriatic Sea. *Clim. Res.* **2006**, *31*, 151–165. [CrossRef]
58. Van Straaten, L.M.J.U. Holocene and late-Pleistocene sedimentation in the Adriatic Sea. *Geol. Rundsch.* **1970**, *60*, 106–131. [CrossRef]
59. Syvitski, J.P.; Kettner, A.J. On the flux of water and sediment into the Northern Adriatic Sea. *Cont. Shelf Res.* **2007**, *27*, 296–308. [CrossRef]
60. Dinelli, E.; Lucchini, F. Sediment supply to the Adriatic Sea basin from the Italian rivers; geochemical features and environmental constraints. *G. Di Geol.* **1999**, *61*, 121–132.
61. Frignani, M.; Langone, L.; Ravaioli, M.; Sorgente, D.; Alvisi, F.; Albertazzi, S. Fine-sediment mass balance in the western Adriatic continental shelf over a century time scale. *Mar. Geol.* **2005**, *222–223*, 113–133. [CrossRef]
62. Colantoni, P.; Tramontana, M.; Tedeschi, R. Contributo alla conoscenza dell'Avampaese Apulo: Struttura del Golfo di Manfredonia (Adriatico Meridionale). *G. Di Geol.* **1990**, *52*, 19–32.
63. Ronchi, L.; Fontana, A.; Correggiari, A.; Asioli, A. Late Quaternary incised and infilled landforms in the shelf of the northern Adriatic Sea (Italy). *Mar. Geol.* **2018**, *405*, 47–67. [CrossRef]
64. Palermo, F.; Mosconi, G.; Angeletti, M.; Polzonetti-Magni, A.M. Assessment of water pollution in the Tronto River (Italy) by applying useful biomarkers in the fish model *Carassius auratus*. *Archives Environ. Contam. Toxicol.* **2008**, *55*, 295–304. [CrossRef]
65. Acciarri, A.; Bisci, C.; Cantalamessa, G.; Di Pancrazio, G.; Spagnoli, F. Gli effetti antropici nell'evoluzione storica della costa "Picena". *Studi Costieri* **2017**, *24*, 3–10.
66. Acciarri, A.; Bisci, C.; Cantalamessa, G.; Di Pancrazio, G.; Spagnoli, F. Tendenza evolutiva della spiaggia della Riserva Naturale della Sentina (San Benedetto del Tronto, AP). *Studi Costieri* **2017**, *24*, 11–16.
67. Acciarri, A.; Bisci, C.; Cantalamessa, G.; Cappucci, S.; Conti, M.; Di Pancrazio, G.; Spagnoli, F.; Valentini, E. Metrics for short-term coastal characterization, protection and planning decisions of Sentina Natural Reserve, Italy. *Ocean Coast. Manag.* **2021**, *201*, 105472. [CrossRef]
68. Froelich, P.N. Analysis of organic carbon in marine sediments. *Limnol. Oceanogr.* **1980**, *25*, 564–572.
69. Shirani, M.; Afzali, K.N.; Jahan, S.; Soleimani-Sardo, M. Pollution and contamination assessment of heavy metals in the sediments of Jazmurian playa in southeast Iran. *Sci. Rep.* **2020**, *10*, 4775. [CrossRef] [PubMed]
70. Li, Y.H. Distribution Patterns of the Elements in the Ocean: A Synthesis. *Geochim. Cosmochim. Acta* **1991**, *55*, 3223–3240. [CrossRef]
71. Rachel, T.; Wasserman, H. Sediment Quality Guidelines (SQGs): A Review and Their Use in Practice. *Geoenviron. Eng.* **2015**. Available online: https://www.geoengineer.org/education/web-class-projects/cee-549-geoenvironmental-engineering-fall-2015/assignments/sediment-quality-guidelines-sqgs-a-review-and-their-use-in-practice (accessed on 26 January 2021).
72. Palinkas, C.; Nittrouer, C. Clinoform sedimentation along the Apennine shelf, Adriatic Sea. *Mar. Geol.* **2006**, *234*, 245–260. [CrossRef]
73. Surricchio, G.; Pompilio, L.; Novelli, A.A.; Scamosci, E.; Marinangeli, L.; Tonucci, L.; D'Alessandro, N.; Tangari, A.C. Evaluation of heavy metals background in the Adriatic Sea sediments of Abruzzo region, Italy. *Sci. Total Environ.* **2019**, *684*, 445–457. [CrossRef]
74. Wu, W.; Wu, P.; Yang, F.; Sun, D.; Zhang, D.; Zhou, Y. Assessment of heavy metal pollution and human health risks in urban soils around an electronics manufacturing facility. *Sci. Total Environ.* **2018**, *630*, 53–61. [CrossRef]
75. Davis, J.C. *Statistics and Data Analysis in Geology*, 3rd ed.; John Wiley & Sons: New York, NY, USA, 2002; 638p.
76. Klovan, J.E.; Imbrie, J. An algorithm and FORTRAN-IV program for large-scale Q-mode factor analysis and calculation of factor scores. *J. Int. Assoc. Math. Geol.* **1971**, *3*, 61–77. [CrossRef]
77. Imbrie, J.F. Purdy. Classification of modern Bahamian carbonate sediments. In classification of carbonate rocks. *Am. Assoc. Rel. Geol. Mem.* **1982**, *7*, 257–272.
78. Wessel, P.; Bercovici, D. Interpolation with splines in tension: A Green's function approach. *Math. Geol.* **1998**, *30*, 77–93. [CrossRef]
79. Wessel, P.; Luis, J.F.; Uieda, L.; Scharroo, R.; Wobbe, F.; Smith, W.H.F.; Tian, D. The Generic Mapping Tools version 6. *Geochem. Geophys. Geosyst.* **2019**, *20*, 5556–5564. [CrossRef]

80. Frontalini, F.; Cordier, T.; Balassi, E.; Armynot du Châtelet, E.; Cermakova, K.; Apothéloz-Perret-Gentil, L.; Alves Martins, M.V.; Bucci, C.; Scantamburlo, E.; Treglia, M.; et al. Benthic foraminiferal metabarcoding and morphology-based assessment around three offshore gas platforms: Congruence and complementarity. *Environ. Int.* **2020**, *144*, 106049. [CrossRef] [PubMed]
81. Price, N.B.; Mowbray, S.; Giordani, P. Sedimentation and heavy metal input changes on the Northwest Adriatic Sea shelf: A consequence of anthropogenic activity. Proceeding of the 10th Meeting of the Italian Association of Oceanology and Limnology, Alassio, Italy, 4–6 November 1999; pp. 23–33.
82. Burton, G.A., Jr. Sediment quality criteria in use around the world. *Limnology* **2002**, *3*, 65–76. [CrossRef]
83. Amorosi, A. Chromium and nickel as indicators of source-to-sink sediment transfer in a Holocene alluvial and coastal system (Po Plain, Italy). *Sediment. Geol.* **2012**, *280*, 260–269. [CrossRef]
84. Amorosi, A.; Centineo, M.C.; Dinelli, E.; Lucchini, F.; Tateo, F. Geochemical and mineralogical variations as indicators of provenance changes in Late Quaternary deposits of SE Po Plain. *Sediment. Geol.* **2002**, *151*, 273–292. [CrossRef]
85. Greggio, N.; Giambastiani, B.M.S.; Campo, B.; Dinelli, E.; Amorosi, A. Sediment composition, provenance, and Holocene paleoenvironmental evolution of the Southern Po River coastal plain (Italy). *Geol. J.* **2018**, *53*, 914–928. [CrossRef]

MDPI
St. Alban-Anlage 66
4052 Basel
Switzerland
www.mdpi.com

Applied Sciences Editorial Office
E-mail: applsci@mdpi.com
www.mdpi.com/journal/applsci

www.ingramcontent.com/pod-product-compliance
Lightning Source LLC
Chambersburg PA
CBHW061502210926
51850CB00008B/1794
* 9 7 8 3 0 3 6 5 8 8 1 6 2 *